6th International Conference on Applications of Optics and Photonics (AOP2024)

EPJ Web of Conferences Volume 305 (2024)

Aveiro, Portugal
16-19 July 2024

Editors:

Manuel F. M. Costa
Carlos Marques

ISBN: 979-8-3313-0690-8

Printed from e-media with permission by:

Curran Associates, Inc.
57 Morehouse Lane
Red Hook, NY 12571

Some format issues inherent in the e-media version may also appear in this print version.

No changes have been made to the content of these proceedings. There may be changes to pagination and minor adjustments for aesthetics.

Printed with permission by Curran Associates, Inc. (2025)

For additional information, please contact EDP Sciences – Web of Conferences
at the address below.

EDP Sciences – Web of Conferences
17, Avenue du Hoggar
Parc d'Activité de Courtabœuf
BP 112
F-91944 Les Ulis Cedex A
France

Phone: +33 (0) 1 69 18 75 75
Fax: +33 (0) 1 69 28 84 91

contact-edps@webofconferences.org

Additional copies of this publication are available from:

Curran Associates, Inc.
57 Morehouse Lane
Red Hook, NY 12571 USA
Phone: 845-758-0400
Fax: 845-758-2633
Email: curran@proceedings.com
Web: www.proceedings.com

TABLE OF CONTENTS

ORGANIZATION

In cooperation with

SUPPORT & SPONSORS

Statement of Peer review

In submitting conference proceedings to *Web of Conferences*, the editors of the proceedings certify to the Publisher that

1. They adhere to its **Policy on Publishing Integrity** in order to safeguard good scientific practice in publishing.
2. All articles have been subjected to peer review administered by the proceedings editors.
3. Reviews have been conducted by expert referees, who have been requested to provide unbiased and constructive comments aimed, whenever possible, at improving the work.
4. Proceedings editors have taken all reasonable steps to ensure the quality of the materials they publish and their decision to accept or reject a paper for publication has been based only on the merits of the work and the relevance to the journal.

Title, date and place of the conference

6th International Conference on Applications of Optics and Photonics, AOP2024, July 16-19, 2024, Aveiro, Portugal

Proceedings editor(s):

Manuel F. M. Costa, University of Minho, Portugal
Carlos Marques, University of Aveiro, Portugal

Date and editor's signature

Preface

The Portuguese Society for Optics and Photonics, SPOF (*Sociedade Portuguesa para a Investigação e Desenvolvimento em Óptica e Fotónica*), successfully organized, July 16 to 19, 2024, in Aveiro, Portugal, its 6th International Conference on Applications of Optics and Photonics, AOP2024.

Meeting with colleagues from all over the world, the Portuguese Optics and Photonics community celebrated the remarkable contribution of Optics and Photonics to the development of our societies and humankind, reviewing the state-of-the-art and foreseeing the future evolution and developments in the field.

In a friendly relaxed and nurturing environment, we foster the establishment of a large range of cooperation projects and relationships with colleagues and institutions from all around the world while increasing the external visibility of Portugal' Optics and Photonics research.

174 effective participants from 31 countries attended the conference that counted 185 presentations including 5 plenaries, 13 keynote and 32 invited lectures by world renowned researchers and scholars as well as top level young researchers in diverse fields of Optics and Photonics.

From these, 27 full papers in different fields of optics and photonics were submitted and accepted, after careful peer review by independent reviewers and revision, for inclusion at the EDP Sciences proceedings book of the AOP2024 conference as follows:

- D-Shaped Photonic Crystal Fiber SPR Sensor for Humidity Monitoring in Oils
  Amanda F. Romeiro, Hudson J. B. Rodrigues, Cauã C. Miranda, Markos P. Cardoso, Anderson O. Silva, João C. W. A. Costa, M. Thereza R. Giraldi, Jose L. Santos, and Ariel Guerreiro

- Coil-shaped Optical Fiber Sensor for Compression Measurements
  Amanda F. Romeiro, Victor H. R. Cardoso, Marcos F. C. de Souza, Paulo Caldas, M. Thereza R. Giraldi, Orlando Frazão, José L. Santos and João C. W. A. Costa

- Enzymatic Plasmonic Optical Fiber Biosensor for Uric Acid Detection
  Ana Sofia Assunção, Mariana Sebastião, António J. S. Fernandes, António J. S. Teixeira, Florinda M. Costa, Cátia Leitão, and Sónia O. Pereira

- Vegetation health evaluation using cost-effective aerial reflectance measurements
  Anaísa Lucena, Rodrigo Rendeiro, and Carlos Marques

- Manipulation of Microparticles in Optofluidic Devices Fabricated by Femtosecond Laser Micromachining
  Carolina Cameira, João M. Maia, and P. V. S. Marques

- Design and Development of a Color Picker System to Integrate in POC Device Systems
  Caterina Serafinelli, Alessandro Fantoni, Miguel Tavares Fernandes, Elisabete C. B. A. Alegria and Manuela Vieira

- Numerical evaluation of strain transfer model for steel-reinforced optical fiber cable embedded in a cylindrical concrete beam with two void inclusions
Mira Kabbara, Xavier Chapeleau, Qinghua Zhang and Frédéric Bourquin

- Development of a High-Precision, Portable and Automated Mobile Laser Scanner for the Recording and Digitation of Texture and Micro-marks in Archaeological and Heritage Stone
Pablo Pardiñas, David S. Canzobre, Javier Lamas, Micaela Gómez-Coronel, Alberto Ramil, David M. Freire Lista and Ana J. López

- Refractive index and thickness analysis of planar interfaces by prism coupling technique
Paulo Lourenço, Yuri Vygranenko, Miguel Fernandes, Alessandro Fantoni and Manuela Vieira

- Monitoring Reinforced Concrete Structures Using Iron Thin Film Coated Optical Fibre Sensors
Pedro M. da Silva, João P. M. Carvalho, João P. Mendes, José M. M. M. de Almeida and Luís C. C. Coelho

- An affordable optical detection scheme for LSPR sensors
Rima Mansour, Catarina Serafinelli, Alessandro Fantoni and Rui Jesus

- Lighting preferences of Iberian sausages for Spanish observers
Shino Okuda, Masahito Nagata, Katsunori Okajima, Francisco José Rodríguez-Pulido, Francisco José Heredia, Rocío Ortega-Velázquez, María Dolores Gaviño-Nadal, Jesús Casado-González, Pedro Antonio García, María José Rivas, María Tejada-Casado, Ana Yebra and Manuel Melgosa

- A LiDAR based obstacle detection framework for railway
Susana Dias, Pedro Sousa, Francisco Afonso, Nuno Viriato, Paulo Tavares and Pedro Moreira

- Design and fabrication of asymmetric Mach-Zehnder interferometers based on EpoClad and EpoCore strip waveguides
Tiago E.C. Magalhães, Jérôme Borme, Temple Douglas, Christian Maibohm and Jana B. Nieder

The high-quality standard of a varied and exciting scientific program was complemented with an enjoyable social program creating the right atmosphere for a most fruitful exchange of ideas and experiences.

In a novel special session "Optics and Photonics Research Labs and Centers in Portugal and International Cooperation", participated by eleven major Portuguese and one foreign research institute center or lab, the current research achievements were presented and the prospects of future cooperation and development were explored and discussed with the conference participants.

The AOP20204 was organized by SPOF, the University of Aveiro, I3N and the ICO' Territorial Committee of Portugal in cooperation with the International Commission for Optics, ICO, the IberoAmerican Optics Network, RIAO, the European Optical Society, EOS, and the Portuguese Physics Society, SPF. IberOptics, MTB, ADMedida were the main sponsors of the conference that was also supported and or endorsed by the Portuguese universities of Minho, Porto, Lisbon, Algarve, Madeira, ISEL, Nova of Lisbon and Beira Interior as well as CICECO – University of Aveiro, IT the Institute of Telecommunications, INESCtec, IPFN, Instituto de Astrofísica e Ciências do Espaço, CENIMAT, the OPTICA student' chapter of the University of Aveiro and the SPIE' student' chapter of the University of Porto, FISUA, FÁBRICA Centro Ciência Viva, SPM, Ordem dos Engenheiros, and at international level by SEDOPTICA Sociedad

Española de Óptica (Spain), CTOM Mexico' Territorial Committee of ICO (Mexico), RCO-Red Colombiana de Óptica (Colombia), OPSS Optics and Photonics Society of Singapore (Singapore), CVO Comité Venezolano de Optica (Venezuela), Societé Tunisiene d' Optique (France), Societé Française d' Optique (France), Photonics Finland (Finland), PHOTONICS21, Hands-on Science Network, Laser World of Photonics (Germany), MDPI' journals Sensors, Applied Sciences, Photonics and Micromachine (Switzerland) as well as the Journal of the European Optical Society-Rapid Publications (France) and EDP Sciences (France).

The committees that served the conference were,

Conference Chair

Manuel Filipe P. C. M. Costa (University of Minho, Portugal)

Technical Chairs

Optical Communications

Rogerio Nogueira (University of Aveiro), Ana Rocha (IT-Aveiro), Gil Fernandes (IT)

Optical Fibers, Sensors and Applications

Orlando Frazão (INESC-Tec), Susana Silva (INESC-Tec), Marta Ferreira (i3N/UA), Nélia Alberto (IT)

Lasers and Nonlinear Optics

Gonçalo Figueira (University of Lisbon), Helder Crespo (University of Porto), Margarida Facão (i3N/UA)

Theoretical and Computational Optics

Manuel F. M. Costa (University of Minho), Ariel Guerreiro (University of Porto)

Quantum Optics

Nelson Muga (IT), Nuno Silva (IT)

Optical Metrology and Image Processing

Paulo Tavares (University of Porto), Manuel Filipe Costa (University of Minho)

Optical Design and Illumination

João Pinto Coelho (University of Lisbon)

Industrial Applications

Paulo Antunes (i3N/UA), Luis Rino (i3N/UA)

Biomedical and Medical Applications of Optics & Photonics

Susana Novais (INESC-Tec) Cátia Leitão (i3N/UA), Pedro Jorge (University of Porto), Fátima Domingues (IT/Khalifa University)

Optometry, Ophthalmic Optics, Color and visual sciences

Sandra Franco (University of Minho), António Baptista (University of Minho)

Nano-Photonics and Optoelectronics

José Manuel Baptista (University of Madeira) [* in memoriam], José Figueiredo (University of Lisbon), Paulo Marques (University of Porto)

Optical Instrumentation for Space and Astronomy

Carlos Marques (CICECO/UA), Manuel Abreu (University of Lisbon), Alexandre Cabral (University of Lisbon)

Microwave Photonics

Miguel Drummond (IT), Bruno Romeira (INL)

Materials Processing and Characterization

Sónia Pereira (i3N/UA), Teresa Monteiro (i3N/UA)

Energy, Sustainability and Smart Cities

José António Rodrigues (University of Algarve), Micael Nascimento (i3N/UA), João Leça (i3N/UA)

Optics and Photonics Education and Outreach

João Lemos Pinto (i3N/UA) Alessandra Kalinowski (i3N/UA)

Other topics

Manuel Filipe Costa (University of Minho)

Scientific Committee

Amália Martinez (CIO, Mexico)
Anna Consortini (University of Florence, Italy)
Anand Krishna Asundi (dOptron Pte Ltd, Singapore)
Andrea Cusano (University of Sannio, Italy)
Andrew Moore (Herriot-Watt University, Scotland)
Andrés Márquez Ruiz (Universidade de Alicante, Spain)
Angel Augier Calderin (INSTEC, Cuba)
Angel I. Negueruela (University of Zaragosa, Spain)
Angela M. Guzman (CREOL, USA)
Armando Nolasco Pinto (Universidade de Aveiro/IT, Portugal)
Asticio Vargas (Universidad de La Frontera, Chile)
Benoît Boulanger (Néel Institute / Société Française d'Optique, France)
Carlos Ferreira(Universidad de Valencia, Spain)
Carlos Saavedra Rubilar (Universidad de Concepción, Chile)
Cesar Augusto Costa Vera (Escuela Politécnica Nacional, Ecuador)
Clementina Timus (INFLR, Romania)
Cristiano M. B. Cordeiro (University of Campinas, Brazil)
Cristina Margarita Gomez-Sarabia (Universidad de Guanajuato, Mexico)
Daniel Malacara Hernández (CIO - Center for Optics Research, Mexico)
Efraín Solarte Rodriguez (RCO, Colombia)
Eric Rosas (CIO, Mexico)
Gonçalo Figueira (University Nova de Lisboa, Portugal)
Guillermo Baldwin (University Pontificia Catholic University of Peru)
Hai-Ning Cui (University of Nanjin, China)
Hector Rabal (CIOP - Center for Optics Research, Argentina)
Humberto Michinel (University of Vigo, Spain)
Hypolito Kalinowski (UTFPR, Brazil)
Ignacio Moreno Soriano (Universidad Miguel Hernandez, Spain)
Imrana Zahid (ICO-TC, Pakistan)
James Wyant (University of Arizona, USA)
Jana Nieder (INL, Portugal)
Jesús Lancis (Universitat Jaume I, Spain)
Joaquín Campos Acosta (Instituto de Óptica, CSIC, Spain)
João Lemos Pinto (I3N, Aveiro, Portugal)
João Manuel Tavares (University of Porto, Portugal)

Program Committee

Alessandro Fantoni (Portugal), Alexandre Cabral (Portugal), Amit Garg (India), Anand Krishna Asundi (Singapore), Angel Augier Calderin (Cuba), Angela M. Guzman (USA), Ana Maria Rocha (Portugal), Antonio Batista (Portugal), António Lobo (Portugal), Carla Carmelo Rosa (Portugal), Carlos Saavedra Rubilar (Chile), Cesar Augusto Costa Vera (Ecuador), Clementina Timus (Romania), Efraín Solarte Rodriguez (Colombia), Eric Rosas (Mexico), Gerardo Ávila (ESO, Germany), Gonçalo Figueira (Portugal), Hai-Ning Cui (China), Humberto Michinel (Spain,) Iran Rocha Segundo (Brasil), Ireneu Dias (Portugal), João M. P. Coelho (Portugal), João Manuel Tavares (Portugal), Joaquim Carneiro (Portugal), Jose Benito Vazquez-Dorrio (Spain), José Figueiredo (Portugal), José Luis Paz (Venezuela) José Luís Santos (Portugal), José Manuel Baptista (Portugal), José R. Salcedo (Portugal), Lúcia Bilro (Portugal), Luis Miguel Bernardo (Portugal), Kim Chew Ng (Australia), Manuel Filipe P. C. M. Costa (Portugal) Manuel Joaquim Marques (Portugal), Michael Scott Belsley (Portugal), Orlando Frazão (Portugal), Paulo António Ribeiro (Portugal), Pedro Serra (Portugal), Karola Panke (Latvia), Robert Liberman (USA), Rogério Nunes Nogueira (Portugal), Sandra Franco (Portugal), Susana Novais (Portugal), Susana Silva (Portugal).

Local Organizing Committee

Paulo Antunes (i3N/UA) * Chair of the Local Organizing Committee

Carlos Marques (CICECO/UA), Alessandra Kalinowski (i3N/UA), Cátia Leitão (i3N/UA), João Leça (i3N/UA), João de Lemos Pinto (i3N/UA), Luís Pereira (i3N/UA), Margarida Facão (i3N/UA), Marta Ferreira (i3N/UA), Micael Nascimento (i3N/UA), Sónia Pereira (i3N/UA), Ana Isabel Freitas (i3N/UA), Ana Rita Fernandes (CICECO/UA), Ana Sofia Assunção (i3N/UA), Diana Pereira (i3N/UA), Jorge Moura (i3N/UA), Lucca Matuck (i3N/UA), Miguel Vidal (i3N/UA), Rodrigo Rito (i3N/UA), Simone Soares (i3N and CICECO/UA), Gil Tavares (i3N/UA), João Paulo Silva (i3N/UA), Manuel M. Freitas (i3N/UA), Sandra Franco (CF-UM-UP), António Batista (CF-UM-UP), Joel Borges (CF-UM-UP), Iran Rocha Segundo (CF-UM-UP), Claver Pinheiro (CF-UM-UP), Nathalia Hammes (CF-UM-UP), Élida Margalho (CF-UM-UP and ISISE/UM), Orlando Lima Jr (CF-UM-UP and ISISE/UM)

The 7th International Conference on Applications of Optics and Photonics is set to take place in the summer of 2026 in Lisbon. With high anticipation and enthusiasm, we are looking forward to have the pleasure to welcome again our friends and colleagues from all over the world in the capital city of Lisbon for a most exciting productive and enjoyable AOP2026 conference!

Aveiro, October 7, 2024.

Manuel Filipe Costa

Carlos Alberto Ferreira Marques

D-Shaped Photonic Crystal Fiber SPR Sensor for Humidity Monitoring in Oils

Amanda F. Romeiro[1*], Hudson J. B. Rodrigues[1], Cauã C. Miranda[1], Markos P. Cardoso[1], Anderson O. Silva[2], João C. W. A. Costa[1], M. Thereza R. Giraldi[3], Jose L. Santos[4,5], Ariel Guerreiro[4,5]

[1]Federal University of Pará, Applied Eletromagnetism Laboratory, Belém Pará, Brazil
[2]Federal Center for Technological Education Celso Suckow da Fonseca, Rio de Janeiro, Brazil
[3]Military Institute of Engineering, Rio de Janeiro, Brazil
[4], INESC TEC, Porto, Portugal
[5]Faculty of Sciences, University of Porto, Porto, Portugal
*Corresponding Author: Romeiro.amanda@gmail.com

Abstract. This theoretical study presents a D-shaped photonic crystal fiber (PCF) surface plasmon resonance (SPR) based sensor designed for humidity detection in transformer oil. Humidity refers to the presence of water dissolved or suspended in the oil, which can affect its dielectric properties and, consequently, the efficiency and safety of the transformer's operation, failures in the sealing system and the phenomenon of condensation can be the main sources of this humidity. This sensor leverages the unique properties of the coupling between surface plasmons and fiber guided mode at the Au-PCF interface to enhance the sensitivity to humidity changes in the external environment. The research demonstrated the sensor's efficacy in monitoring humidity levels ranging from 0% to 100% with an average sensitivity of measured at 1106.1 nm/RIU. This high sensitivity indicates a substantial shift in the resonance wavelength corresponding to minor changes in the refractive index caused by varying humidity levels, which is critically important in the context of transformer maintenance and safety. Transformer oil serves as both an insulator and a coolant, and its humidity level is a key parameter influencing the performance and longevity of transformers. Excessive humidity can lead to insulation failure and reduced efficiency and, therefore, the ability to accurately detect and monitor humidity levels in transformer oil can significantly enhance preventive maintenance strategies, reduce downtime, and prevent potential failures, ensuring the reliable operation of electrical power systems.

1 Introduction

Although expensive, power transformers are essential components in electric power plants, and their reliable operation is critical for system stability. Replacing a damaged transformer is both costly and time-consuming, potentially leading to power outages until a new equipment is installed [1]. The reliability of a transformer largely depends on its oil, which serves as a high-quality insulating material. However, transformer oil can become contaminated with water and gas, reducing its dielectric strength and causing partial discharges [1].

Humidity in oils refers to the presence of water content, typically in the form of dissolved, emulsified, or free water, within an oil-based medium. Even small amounts of moisture in oils can significantly affect their performance, properties, and longevity. In industrial settings, such as lubrication systems, transformers, or hydraulic fluids, excessive moisture can lead to issues like oxidation, reduced lubrication efficiency, corrosion, or electrical breakdown in insulating oils [1].

The relative permittivity of a transformer oil reflects its insulating properties as well as its refractive index (RI). Therefore, changes in the oil RI indicate contamination, which makes the monitoring of this parameter crucial for assessing oil quality [1, 2]. Optical sensors whose operation relies on the excitation of surface plasmon modes are a quite promising technique to properly address the oil integrity in power transformers.

Surface plasmon resonance (SPR) sensors are becoming more popular for their real time, label-free monitoring capabilities [3]. To achieve SPR in PCF (photonic crystal fiber) sensors, a conducting material must be incorporated to the PCF structure. SPR is a coherent coupling between the incident optical beam and the electron density that vibrates collectively on the surface of the conducting material. Upon resonance condition, energy absorption increases sharply, increasing noticeably the optical mode losses [4]. A wide variety of devices have been proposed to reach this condition, many of them favouring from the properties of optical fiber guided modes to achieve plasmonic resonance. Tapered fibers [5], grating fibers [6] and D-shaped fibers [5] are leading examples.

Notably, photonic crystal fiber sensors are advantageous due to their higher effective index step from the cladding to the core, easy tailoring of large modal areas and single mode operation over a broader wavelength range [5]. They are increasingly applied to various fields, such as food safety [6], glucose monitoring [7], environmental safety [5, 7], biological and chemical material detection [6] and so on. Moreover, PCF sensors can measure physical parameters like temperature, pressure, acceleration and hydrocarbon levels [7]. A D-shaped PCF has shown to be even more interesting due to the enhanced proximity of core and surrounding media (also known as analyte) [5], easing the coupling between the fundamental core guided mode and the plasmonic mode.

This research presents a D-shaped PCF SPR based sensor designed to monitor the moisture content in transformer oil, using gold as the plasmonic material. The sensor is numerically simulated using the finite element method (FEM) based COMSOL Multiphysics software [8]. The proposed device presented high sensitivity and resolution to minor RI changes and high amplitude resonance peaks.

2 Design and modelling

Fig. 1. Provides a schematic representation of the SPR sensor's geometric configuration. The device consists of a photonic crystal fiber with a hexagonal arrangement of air holes within a silica background. The PCF is side-polished to create a D-shaped cross-section, and a plasmonic layer made of gold, which is 40nm thick and 10μm wide, placed atop the core on the flat surface. The fiber has a total diameter D = 24μm, with each air hole being 2μm apart (Λ). Numerical simulations were conducted with the computational domain enclosed by a perfectly matched layer (PML) of thickness 0.1D.

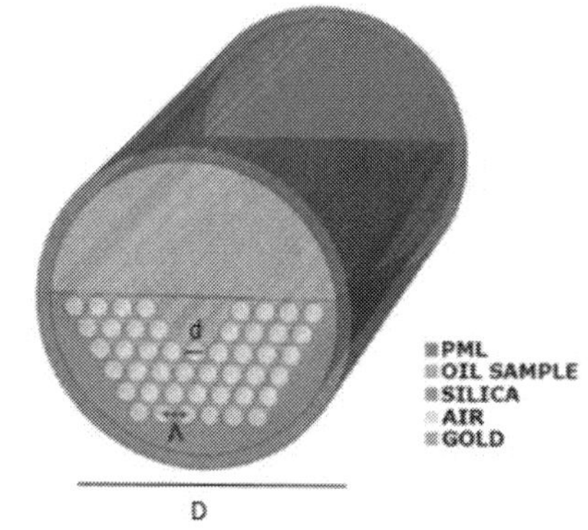

Fig. 1. Schematic of the D-shaped PCF for oil monitoring.

In the design of D-shaped photonic crystal fiber surface plasmon resonance sensors, the proximity of the core to the detected sample is a critical factor that enhances the sensor's performance [5]. By integrating a D shaped PCF design, the sensor achieves a unique configuration where a significant portion of the fiber's core is exposed to the external environment. This close proximity to the sample allows for a more substantial interaction between the surface plasmon waves and the analyte, resulting in heightened sensitivity and improved detection limits. The reduced distance between the sensing surface and the core facilitates more efficient coupling of the plasmonic fields with the target molecules, enabling the detection of lower concentrations of analytes with greater precision [7].

Moreover, the choice of gold as the plasmonic material in these sensors is particularly advantageous. Gold exhibits exceptional plasmonic properties, including a strong resonance in the visible to near-infrared spectrum, which aligns well with many common analytes and detection techniques. Gold's high biocompatibility and stability further enhance the sensor's reliability and longevity [4]. Its ability to support intense localized surface plasmon resonances makes it an ideal choice for maximizing the interaction

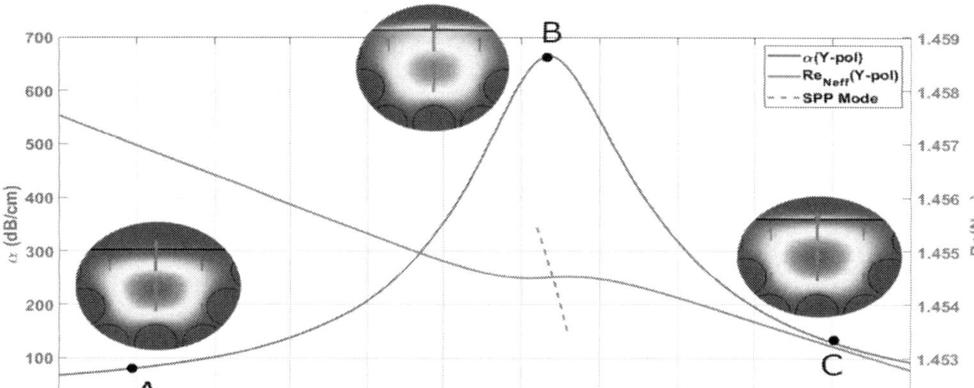

Fig. 2. Real and imaginary part (expressed by losses α) of the PCF's effective mode index and electric field distribution at (a) 550nm, (b) 604nm, (c) 640 nm.

between the plasmonic field and the analyte, thus improving the overall sensitivity and selectivity of the SPR sensor. This combination of close core-sample proximity and the superior plasmonic characteristics of gold underscores the innovative aspects of the D-shaped PCF SPR sensor design, setting it apart from conventional methods and expanding its potential applications in various fields.

To perform the simulations, we used Johnson and Cristy's constants [9] to model the optical properties of the Au layer and the permittivity of silica was obtained from the Sellmeier equation. As for the spectral response of the proposed sensor, we used a two-dimensional computational domain and the Wave Optics package in frequency domain to calculate the eigenvalues of the Helmholtz equation in angular frequency ω:

$$\nabla_{\perp}^{2} E(r_{\perp}, \omega) + k_0^2\big(\varepsilon(\omega) - n_{eff}^2\big)\, E(r_{\perp}, \omega) = 0, \quad (1)$$

Where k_0 is the magnitude of the free space wave number and ε is the complex relative permittivity, both being frequency dependent. The modal field distribution at the position $r_{\perp}$, perpendicular to the direction of light propagation, is denoted by $E(r_{\perp}, \omega)$. The complex effective index n_{eff} is related to the mode resulting from the coupling between the fundamental mode and the surface plasmon mode at the silica-gold interface. The real part of the n_{eff} refers to the mode phase and the imaginary part is related to the losses experienced by the confined mode as the coupling occurs.

3 Results

The SPR condition is reached when the phase of the fundamental fiber mode matches the phase of the plasmonic mode at the PCF-gold interface, as shown in Fig. 2. At this coupling, a pronounced peak is observed in the modal confinement losses as estimated by [7]:

$$\alpha = 8.686 \times \frac{2\pi}{\lambda} \times Im_{n_{eff}} \times 10^4 \left(\frac{dB}{cm}\right), \quad (2)$$

Where λ is the wavelength and $Im_{n_{eff}}$ is the imaginary part of the effective mode index.

Fig. 3 depicts the spectral losses computed for humidity levels of the transformer oil from 0% (ideal condition or new oil) to 100% (oil is completely saturated and there is no longer isolation).

The sensing response S is accounted by the wavelength shift that occurs when the refractive index of the analyte changes [5]:

$$S = \frac{\Delta\lambda_{peak}}{\Delta n_{analyte}} \left(\frac{nm}{RIU}\right), \quad (3)$$

Where $\Delta\lambda_{peak}$ is the wavelength resonance displacement and $\Delta n_{analyte}$ is the respective change in the analyte refractive index. Table 1 shows the humidity levels related to the refractive index change.

Table 1. Relationship between humidity levels and refractive index change [2].

Humidity level	RI	Humidity level	RI
0%	1.330	30%	1.336
5%	1.331	50%	1.337
10%	1.332	75%	1.338
15%	1.333	87.5%	1.339
20%	1.334	100%	1.340
25%	1.335		

Correspondingly, the ability of the sensor in detecting tiny changes in oil humidity levels, known as the sensor resolution R, is evaluated by [5]:

$$R = \Delta n_{analyte} \times \left(\frac{\Delta\lambda_{min}}{\Delta\lambda_{peak}}\right) (RIU) \quad (4)$$

Where $\Delta\lambda_{min}$ corresponds to an optical spectral analyzer of minimal detectable change of $\Delta\lambda_{min} = 0.1$ nm in the SPR loss spectra. The spectral losses displayed in Figure 3 indicate that the average sensitivity and resolution are 1106.1 nm/RIU and $9.04 \times 10^{-5}\ RIU$, respectively. Furthermore, the maximum sensitivity of 2136.1 nm/RIU is achieved in the range of oil humidity varying from 0% to 5%.

4 Conclusions

This work highlights the effectiveness of a D-shaped PCF SPR based sensor for the detection of humidity levels in transformer oil. The optical mode emerged from the coupling between the fundamental fiber guided mode and surface plasmons at the Au-PCF interface significantly enhances the sensitivity to fairly minute humidity changes. With calculated efficacy in monitoring humidity levels from 0% to 100%, the results show that a minimal change in the refractive index of the oil is detected by the PCF sensor, culminating in an average sensitivity of 1106.1 nm/RIU for a resolution of $9.04 \times 10^{-5}\ RIU$. This sensor provides a robust and electromagnetically compatible solution for detecting humidity variations, which is

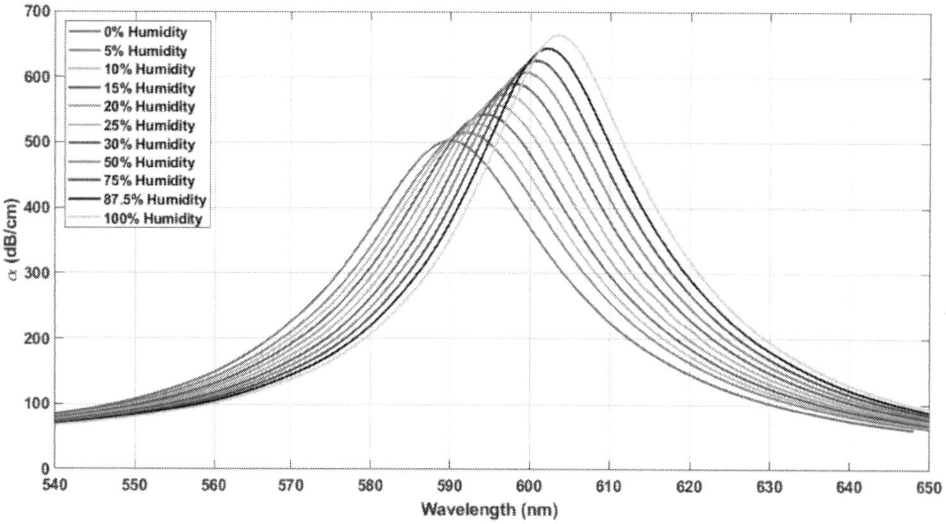

Fig. 3. Spectral losses of the proposed sensor for transformer oil humidity varying from 0% (ideal condition) to 100% (saturated oil).

critical for monitoring insulation properties of the transformer oil, as it aids to prevent failures such as harmful electrical discharges and adverts from potential damages. Consequently, the proposed sensor is promising in ensuring the continuous and reliable operation of electrical power systems.

This study was financed in part by the Coordenação de Aperfeiçoamento de Pessoal de Nível Superior – Brasil (CAPES) – Finance code 001. During the period of elaboration of this work, the authors have obtained support from the Conselho Nacional de Desenvolvimento Cientifico e Tecnologico (CNPq). This work is also financed by National Funds through the Portuguese funding agency, FCT – Fundação para a Ciência e Tecnologia, within project UIDB/50014/2020.

DOI 10.54499/UIDB/50014/2020

5 References

1. Fu Li-Hui, Dai Junfeng, Mixed Oil Detection Method Based On Tapered Fiber Spr Sensor, Optical Fiber Technology, Volume 78, 2023.
2. Alok Kumar Paul, Design And Analysis Of Photonic Crystal Fiber Plasmonic Refractive Index Sensor For Condition Monitoring Of Oil Transformer, Osa Continuum, Volume 3, 2020.
3. Maya Chauhan, Vinod Kumar Singh, Review On Recent Experimental Spr/Lspr Based Fiber Optic Analyte Sensors, Optical Fiber Technology, Volume 64, 2021.
4. A. F. Romeiro, C. C. Miranda, M. P. Cardoso, A. O. Silva, J. C. W. A. Costa, M. T. R. Giraldi, J. L. Santos, J. M. Baptista, A. Guerreiro. Plasmonic Resonances Associated With Distinct Conducting Layers Deposited On A D-Shaped Photonic Crystal Fiber: An Analysis For Sensing Purpose. International Microwave And Optoelectronics Conference, Castelldefels, Spain, 2023.
5. M. P. Cardoso, A. O. Silva, A. F. Romeiro, M. T. R. Giraldi, J. C.W.A. Costa, J. L. Santos, J. M. Baptista, A. Guerreiro, "Tunable Plasmonic Resonance Sensor Using A Metamaterial Film In A D-Shaped Photonic Crystal Fiber For Refractive Index Measurements", Applied Sciences, 2022, V. 12, P. 2153, 2022.
6. A. A. Rifat, F. Haider, R. Ahmed, G. A. Mahdiraji, F. R. Adikan And A. E. Miroshnichenko, "Highly Sensitive Selectively Coated Photonic Crystal Fiber Based Plasmonic Sensor", Optics Letters, 2018, Vol. 48, N° 4, Pp. 891-894.
7. M.P. Cardoso, A.O. Silva, A.F. Romeiro, M.T.R. Giraldi, J.C.W.A. Costa, J.L. Santos, J.M. Baptista, A. Guerreiro. "Multi-Plasmonic Resonance Based Sensor For The Characterization Of Optical Dispersion Using A D-Shaped Photonic Crystal Fiber", Ieee Instrumentation & Measurement Magazine, Vol. 24, Pp. 63-68, 2021.
8. COMSOL Website. https://Www.Comsol.com/
9. P. B. Johnson And R. W. Christy, "Optical Constants Of The Noble Metals", Physical Review B, 1972, Vol. 6, N° 1

Coil-shaped Optical Fiber Sensor for Compression Measurements

Amanda F. Romeiro[1]*, Victor H. R. Cardoso[1]*, Marcos F. C. de Souza[1], Paulo Caldas[4,5], M. Thereza R. Giraldi[2], Orlando Frazão[3,4], José L. Santos[3,4] and João C. W. A. Costa[1]
[1]Applied Electromagnetism Laboratory, Federal University of Pará, R. Augusto Corrêa, 01, Belém, Brazil
[2]Laboratory of Photonics, Military Institute of Engineering, Praça Gen. Tibúrcio, 80, Rio de Janeiro, Brazil
[3]Dept. of Physics and Astronomy, Faculty of Sciences of U.Porto, R. do Campo Alegre, 687, Porto, Portugal
[4]Inst. for Systems and Computer Engineering, Technology and Science, R. do Campo Alegre, 687, Porto, Portugal
[5]Polytech. Inst. of Viana do Castelo, R. Escola Industrial e Comercial de Nun'Álvares, Viana do Castelo, Portugal
*romeiro.amanda@gmail.com, victorcard@ufpa.br

Abstract. This study investigated the effectiveness of a coil-shaped optical fiber interferometric sensor, with a diameter of 13 mm, for measuring compression. The sensor's design utilizes the principles of interferometry to create a pattern that changes with applied pressure. This configuration significantly amplifies the sensor's sensitivity to compression due to the extended optical path length within the compact form factor. The experimental results demonstrated that even small compressive forces caused detectable alterations in the interference pattern, allowing for precise quantification of pressure changes. The 13 mm diameter proved to be particularly advantageous, providing a balance between sensitivity and practical integration into various systems, from structural health monitoring to biomedical devices. This study also highlights the sensor's robustness against electromagnetic interference and environmental variations, attributing this to the intrinsic properties of optical fiber. Overall, the findings suggest that coil-shaped optical fiber interferometric sensors are highly effective for accurate and reliable compression sensing, with potential for broad application across multiple industries.

1 Introduction

Optical fiber sensors are increasingly vital in various industrial applications due to their exceptional characteristics. These sensors utilize the unique properties of optical fibers, such as high sensitivity, immunity to electromagnetic interference, and the ability to function in harsh environments [1]. They are known for their compact size, flexibility, and durability, allowing deployment in challenging and confined spaces. Their capability to provide precise and reliable measurements makes them indispensable for monitoring structural integrity, detecting minor changes in physical parameters, and ensuring the safety and efficiency of critical systems [2].

Among the many possible configurations, interferometric optical fiber sensors are advanced devices that exploit the interference of light waves to measure various physical parameters with high precision [3]. These sensors operate based on the principle that changes in the external environment, such as pressure, temperature, or strain, can alter the optical path length of light traveling through the fiber, which makes them able to detect minute changes with remarkable sensitivity [4].

Accurate compression measurement is a significant concern across many industries as it is essential for monitoring structural integrity [5]. Detecting small pressure changes reliably can prevent catastrophic failures in bridges, buildings, and medical equipment, thereby enhancing safety and operational efficiency [6,7].

This paper explores the application of an optical fiber interferometric sensor for compression measurements. This led to the idea of using the coil to enhance interferences. By utilizing a coil configuration, we have amplified the device's sensitivity, enabling the precise detection of minor compression variations. According to the tests conducted, the 13mm diameter coil performed the best, offering a longer optical path within a compact form, thus balancing sensitivity with ease of integration into various systems. making the sensor suitable for a wide range of applications, from structural health monitoring to advanced medical devices.

2 Principle of operation and Experimental setup

In the context of a step-index single mode fiber, the macrobending loss coefficient can be denoted as [8]:

$$\alpha_c = A_c r^{-\frac{1}{2}} \exp(-Ur) \qquad (1)$$

where α_c represents the macro-bending loss per unit length, r denotes the curvature radius of the optical fiber during bending, and A_c and U are constants that correspond to the fiber types and operating conditions of the light source, respectively. Thus, it is clear that the macro-bending loss coefficient primarily depends on the curvature radius of a given single-mode optical fiber.

The operating principle of the macro-curvature interferometer relies on the interference between the core mode and cladding modes within the fiber. This occurs due to the curvature of the fiber, which allows some light to escape from the core and penetrate the cladding as it propagates through the curved section of the interferometer. Part of the mode power is reflected back into the core when the light reaches the boundary between the core and cladding. This process repeats as the light continues to travel through the curved region [9].

This repeated interaction between the core and cladding creates a modal interferometer. The interference arises due to the differences in effective refractive indexes and the optical path lengths of the light in the core and cladding modes, which are excited by the optical signal. These variations result in interference patterns that can be used for sensing or measurement purposes.

The modal interference at the output of the device results from the different effective refractive indexes and optical paths of the core and cladding modes. Since the cladding mode's effective refractive index is influenced by the surrounding medium, changes in the surrounding refractive index will alter the propagation velocity, leading to a spectral shift in the transmission spectrum. Assuming only one cladding mode is excited in the curved region of the balloon-shaped interferometer, the transmission intensity after light passes through this region can be described as:

$$I = I_{core} + I_{cladding} + 2\sqrt{I_{core}I_{cladding}}\cos(\Delta\phi) \quad (2)$$

Here, I_{core} and $I_{cladding}$ represent the intensities of the core and cladding modes, and $\Delta\phi$ is the phase difference between them. The phase difference is expressed as:

$$\Delta\phi = \frac{2\pi L(n_{core} - n_{cladding})}{\lambda} = \frac{2\pi\Delta nL}{\lambda} \quad (3)$$

where λ is the wavelength, L is the effective length of the curved region of the balloon-shaped interferometer, and n_{core} and $n_{cladding}$ are the effective refractive indexes of the core and cladding modes, respectively. Δn is the difference in these refractive indexes. When $\Delta\phi = (2m+1)\pi$, where m is an integer, the interference pattern reaches a minimum. The resonant wavelength λ_d is then given by:

$$\lambda_d = \frac{2\Delta nL}{2m+1} \quad (4)$$

As demonstrated in [10], the radius r of a fiber balloon-like interferometer is defined as half the distance between the centers of two symmetrical fibers. Due to the stress-optical effect, the refractive index of the outer layers of the bent single-mode fiber increases, while that of the inner layers decreases compared to the original straight fiber [8]. This phenomenon is evident in the balloon-like interferometer. To describe the refractive index profile, we can use the equivalent modal refractive index as expressed by the following equation [8, 9]:

$$n'(x) = n(x)\left(1 + \frac{x}{r}\right) \quad (5)$$

In this equation, r represents the radius of curvature, and x denotes the perpendicular distance from the axis of curvature of the fiber. The refractive index profiles, $n'(x)$ and $n(x)$, correspond to the curved and straight states of the fiber, respectively.

In this experiment, a 1525 nm laser (SLD - Superluminescent Diode, for THORLABS) served as the light source to illuminate a single-mode optical fiber sensor enclosed within a capillary tube. The fiber was carefully coiled into a single turn with a diameter of 13 mm [8] and a height of 2 mm, optimizing its sensitivity to mechanical deformations. Connected to an OSA (Optical Spectrum Analyzer, for YOKOGAWA) with 1 nm resolution, the setup monitored changes in the light spectrum transmitted through the coiled fiber. These spectral variations were indicative of the mechanical compression applied to the fiber, which was precisely measured using a pachymeter. This instrument allowed for accurate quantification of the coil's deformation, ensuring a clear correlation between the physical compression and the optical responses detected by the OSA.

Fig. 1. in the schematic representation illustrates the experimental configuration, depicting the placement of

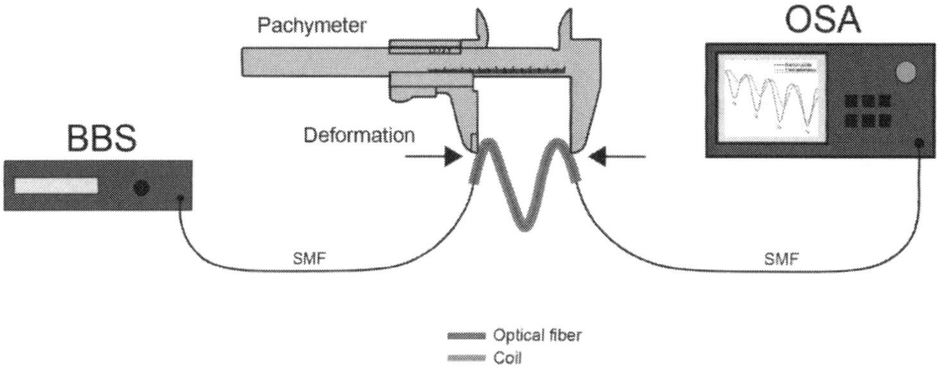

Fig. 1. Experimental setup.

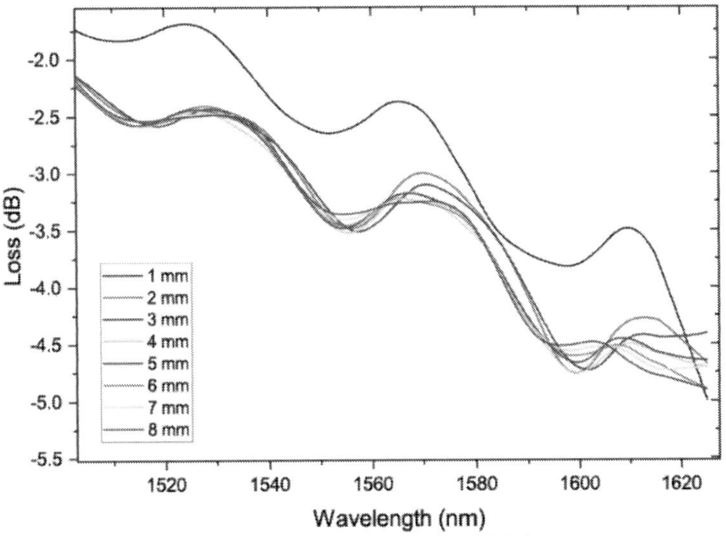

Fig. 2. Spectral losses for each coil height.

the light source, the coiled optical fiber sensor within the capillary tube, the connection to the OSA for spectral analysis, and the integration of the pachymeter for compression measurement. This setup facilitated a detailed investigation into the sensor's sensitivity and performance under controlled mechanical stress conditions, offering insights into its potential applications in precise deformation sensing technologies.

3 Results

This section details the behavior analysis of an optical fiber sensor utilizing a spring-shaped structure to monitor micro-displacements. The coil-shaped configuration of an optical fiber sensor significantly enhances its sensitivity to compression by increasing the optical path length. When the fiber is wound in a helical shape, the light traveling through it covers a longer path compared to a straight fiber. This extended optical path allows the sensor to experience more pronounced changes in the transmission of light, such as phase shifts, intensity modulation, or wavelength variations. As a result, even small compressive forces cause noticeable changes in the optical signals, making the sensor highly sensitive.

Moreover, the helical configuration increases the strain sensitivity of the fiber. The coiled structure allows more of the fiber to be exposed to external forces during compression, which amplifies the sensor's response to deformation. The distribution of compressive forces along the coil also enhances the sensor's ability to measure small changes in pressure or force. In this way, the coil-shaped optical fiber sensor's sensitivity is improved because the extended optical path magnifies the impact of compression on the fiber, allowing for more precise measurements.

The sensor operates within a spectral range of 1450 to 1600 nm. Before the deformation analysis, the height of the coil was studied for determining the best configuration for the sensor. Fig. 2. depicts the spectral response for the analysed heights.

Fig. 2. shows that the spectral response for the 2mm coil height exhibited the most well-defined wavelength dips, making it the chosen configuration for this work. Fig. 3. presents the spectral response for a 1mm deformation in the coil-shaped sensor.

To investigate the sensor's response to micro-displacement variations, it was subjected to incremental changes in displacement in 1-milimiter steps. As these variations were applied, a noticeable wavelength shift occurred, highlighting the sensor's sensitivity to micro-displacement changes. Sensitivity analysis for the four selected wavelengths revealed the following observations: the first wavelength had a sensitivity of 7 nm/mm, the second wavelength exhibited a sensitivity of 10 nm/mm, the third wavelength showed sensitivity at 4 nm/mm and the fourth wavelength presented a sensitivity of –2 nm/mm.

These results demonstrate the optical fiber sensor's capability to accurately detect and measure micro-displacements. The selected wavelengths and their corresponding sensitivities provide valuable insights into the sensor's performance, making it suitable for precise measurements in various applications. Additionally, the broad spectral range from 1450 to 1600 nm allows for versatile monitoring across different scenarios, ensuring compatibility with diverse light sources and signal processing equipment.

4 Conclusion

In summary, the optical fiber sensor with a spring-shaped configuration has demonstrated remarkable efficacy in detecting and tracking micro-displacements. By utilizing a detailed wavelength analysis, the sensor exhibited notable sensitivity levels for the designated wavelengths, ensuring precise and reliable measurements. Specifically, the incremental displacement changes in 1-milimeter steps revealed clear shifts in wavelength, indicating the sensor's high

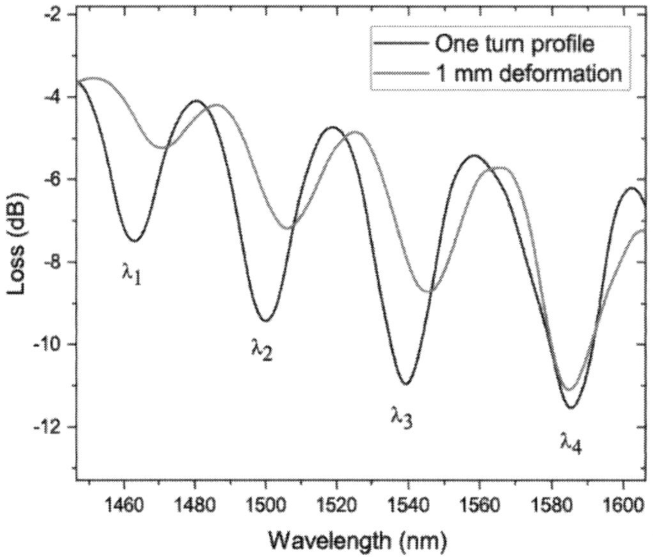

Fig. 3. Spectral loss for the coil-shaped sensor and a 1mm deformation.

sensitivity to micro-displacement variations. This high level of sensitivity was quantified for four specific wavelengths, each showing distinct sensitivity values, thereby providing valuable insights into the sensor's performance. Moreover, the broad wavelength range from 1450 to 1600 nm further enhances the sensor's versatility, allowing it to be a valuable tool for a wide array of applications that require precise micro-displacement monitoring. This broad spectral range not only facilitates versatile monitoring but also ensures compatibility with diverse light sources and signal processing equipment, making the sensor adaptable to various scenarios and enhancing its overall utility in different fields.

This study was financed in part by the Coordenação de Aperfeiçoamento de Pessoal de Nível Superior – Brasil (CAPES) – Finance code 001. During the period of elaboration of this work, the authors have obtained support from the Conselho Nacional de Desenvolvimento Cientifico e Tecnologico (CNPq). This work is also financed by National Funds through the Portuguese funding agency, FCT – Fundação para a Ciência e Tecnologia, within project UIDB/50014/2020.

DOI 10.54499/UIDB/50014/2020

5 References

1. K.T.V. Grattan And B.T. Meggitt. Optical Fiber Sensor Technology Fundamentals. Springer Science+Business Media, Llc. (Pag. 11).
2. Qingbin Li, Farhad Ansari. "Circumferential Strain Measurement Of High Strength Concrete In Triaxial Compression By Fiber Optic Sensor" International Journal Of Solids And Structures 38 (2001) 7607—7625.
3. Yue Wu, Shiying Xiao, Yao Xu, Ya Shen, Youchao Jiang, Wenxing Jin, Yuguang Yang, Shuisheng Jian. Highly Sensitive Force Sensor Based On Balloon-Like Interferometer. Optics And Laser Technology 103 (2018) 17–21.
4. Grattan, K.T.V.; Meggitt, B.T. Optical Fiber Sensor Technology; Chapman & Hall: Orange, Ca, Usa, 1995.
5. Ostachowicz, W.; Soman, R.; Malinowski, P. Optimization Of Sensor Placement For Structural Health Monitoring: A Review. Struct. Health Monit. 2019, 18, 963–988.
6. Zhihang Deng, Minshui Huang, Neng Wan And Jianwei Zhang. The Current Development Of Structural Health Monitoring For Bridges: A Review. School Of Civil Engineering And Architecture, Wuhan Institute Of Technology, Wuhan 430073, China; Dengzhihang@Stu.Wit.Edu.Cn (Z.D.).
7. Yunzhu Chen And Xingwei Xue. Advances In The Structural Health Monitoring Of Bridges Using Piezoelectric Transducers. Sensors Mdpi.
8. Cardoso, Victor Hr, Et Al. Optical Fiber Sensor For Micro Displacement Monitoring Based On A Balloon-Like Interferometer With A Spring-Shaped Structure. In: 2023 Sbmo/Ieee Mtt-S International Microwave And Optoelectronics Conference (Imoc). Ieee, 2023. P. 4-6.
9. Y. Zhao, X. Liu, R.-Q. Lv, And Q. Wang, "Simultaneous Measurement Of Ri And Temperature Based On The Combination Of Sagnac Loop Mirror And Balloon-Like Interferometer," Sensors And Actuators B: Chemical, Vol. 243, Pp. 800–805, 2017.
10. Victor Cardoso, "Development of Optical Sensors for Measuring Diameter Variation in Cylindrical Structures: Applications in Monitoring Plant Species and Gas Pipelines", Thesis, 2023, pp. 53-78.

Enzymatic Plasmonic Optical Fiber Biosensor for Uric Acid Detection

Ana Sofia Assunção[1], *Mariana Sebastião*[1], *António J. S. Fernandes*[1], *António J. S. Teixeira*[2], *Florinda M. Costa*[1], *Cátia Leitão*[1], *Sónia O. Pereira*[1*]

1 i3N, Department of Physics, University of Aveiro, 3810-193 Aveiro, Portugal
2 IEETA, DETI, LASI, University of Aveiro, 3810-193 Aveiro, Portugal
*Corresponding author: sonia.pereira@ua.pt

Abstract. Uric acid plays important roles in the body, acting as an antioxidant, contributing to the elimination of metabolic waste, and helping to combat some infections. Alterations in uric acid concentration can be important indicators of disease and their real-time and point-of-care monitoring is of extreme importance, for instance in primary care facilities. Optical fibre biosensors are a promising option, allowing real-time detection, quantification, and monitoring of uric acid concentrations in blood plasma or urine. In this work, enzymatic plasmonic sensors were developed for the detection of uric acid using multimode optical fibres composed of a 400 µm silica core and a polymer cladding. To prepare the sensors, the coating and cladding were removed from a section at the fibre tip, and nano film of gold was deposited on their surface, obtaining gold-coated sensors at the fibre tips (Au-tips). Initially, the sensor was characterized to bulk refractive index, exhibiting a sensitivity of 1420.7 ± 59.8 nm/RIU. The Au-tip was biofunctionalized with the enzyme uricase and detection tests were carried out. Uric acid detection tests revealed a response consistent with the Michaelis-Menten function, which was employed to assess the enzymatic kinetics of uricase immobilized on the Au-tips.

1 Introduction

Uric acid is a biologically active compound found in human fluids such as blood plasma and urine [1]. Monitoring uric acid concentrations in both blood plasma and urine is crucial for diseases diagnosis and follow-up. In blood, uric acid is generally found in the range from 1.5 to 6.0 mg/dL for women and 2.5 to 7.0 mg/dL for men in adulthood [2,3], whereas in urine, varies between 25 and 74 mg/dL [4].

In recent decades, there has been significant progress in developing biosensors for uric acid detection. This progress is due to extensive research in the field driven by the growing interest in precise and rapid detection. Various principles have been explored, including techniques like fluorescence, electrochemical, and surface plasmon resonance (SPR). These techniques offer distinct advantages over traditional methods in terms of miniaturization, response time, and ease of use, contributing to diagnosis and patient monitoring at point-of-care [5].

The high specificity of enzyme-analyte interactions and the their ability to perform catalytic functions over time without depletion contribute to the efficiency, stability, and reproducibility of the results, making this type of biosensor attractive [6]. On the other hand, optical biosensors have significant advantages over traditional analytical techniques, allowing label-free and real-time detection of various biological and chemical substances, with high specificity, sensitivity, compact size, and cost-effectiveness [7]. In the scope of optical biosensing, SPR technique is one of the most explored. SPR is a phenomenon where free electrons on a metal surface, in contact with a dielectric medium, oscillate collectively in resonance with an incident light wave. Surface plasmons corresponds to electron oscillations occurring at the interface of two media with opposite dielectric constants, usually a metal-dielectric interface [8]. Gold or silver, with a negative dielectric constant ($\varepsilon < 0$), are the most commonly used metals for plasmonic applications, while the dielectric media usually is a liquid solution or air, having a positive dielectric constant ($\varepsilon > 0$) [9]. For resonance to occur, certain conditions must be met that depend on the properties of the metal, the dielectric medium, and the light.

To excite surface plasmons, coupling elements like prisms, waveguides, and/or diffraction gratings are necessary. The most conventional approach is the Kretschmann method, which uses a prism coupler and a thin metal layer (10 to 100 nm thick) at the base of the prism [8]. A *p*-polarized light beam strikes through the coupling element–metal interface at an angle equal or greater than the total internal reflection (TIR) critical angle. During TIR, an evanescent wave is generated at the interface, enabling surface plasmon excitation at the metal-dielectric interface [8]. When the propagation constant of the evanescent wave matches the one of the surface plasmon wave, resonance occurs [10].

Optical fibres utilize the TIR principle, for light propagation, which occurs at the core-cladding interface [11,12]. The propagation of light in an optical fibre occurs in two modes: through the core in guided modes and as an evanescent field with exponential decay at the core-cladding interface. Therefore, it is crucial to ensure adequate exposure of the evanescent field to the gold nanofilm and external medium to optimize the sensor's effectiveness [13].

Gupta *et al.* [14], reported a SPR silver-silicon biosensor using polyacrylamide gel as immobilization agent of uricase in an uncladded fibre used in transmission. Besides promising results, the transmission approach difficult miniaturization and therefore its application in point-of-care contexts. This work explores gold SPR based optical fibre biosensors working in reflection as reported by Assunção *et al.* [15], but, here an enzyme (uricase) is used as biorecognition element for specific detection of uric acid.

2 Materials and methods

2.1 Production of Au-tips

The sensors were produced following the procedure described in [15]. Here, the sensors were produced with multimode silica core (diameter 400 μm) fibres with plastic cladding (FT-400-URT, Thorlabs, USA). First, a 15 cm piece of optical fibre was cut, and then, with the help of a blade, the coating and the cladding were removed from each end: 3 cm from one end and 1 cm from the other end. After this, the stripped area was cleaned with a cloth soaked in acetone to remove any cladding residues. The ends were polished using polishing sheets to ensure that both locations had a smooth surface. Then, to evaluate the quality of the fibre cleaving, the ends were observed under an optical microscope using 10x and 40x objectives. This allowed for the evaluation of the polishing quality, flatness, and cross-sectional shape. After the fibres were cleaved and cleaned, a thin film of gold was deposited on the 1 cm fibre tip, corresponding to the sensing zone. For this purpose, the previously cut and cleaved optical fibres were aligned in a holder, and the gold film was deposited on the fibres by sputtering. In the vacuum chamber, the process starts by placing the aligned tips in a parallel position with the gold target at a distance of 5 cm. After applying vacuum, Argon is introduced, and the coating process is performed with a pressure of $\sim 10^{-2}$ Torr, a current of 10.25 mA, a voltage of 1.2 kV, and a sputtering time of 60 seconds. After this time, the fibre holder was rotated 180° and the entire process was repeated to ensure that the entire surface of the fibres was coated. Finally, the Au-tips were subjected to an annealing process in a thermal chamber for 2 hours at 180° C. The produced Au-tips have a gold coating thickness of around 55 nm. The produced Au-tips have a gold coating thickness of around 55 nm.

2.2 Data acquisition setup

The experimental setup comprised a white light source (LS-W7, Sarspec, Portugal), which has an emission range of 380-3000 nm, allowing light ejection into the optical fibre through a bifurcated fibre (FCB-UVIR200-1, Avantes, Netherlands). A spectrometer (FLAME-UVvis, Ocean Optics, USA) with a detection range of 180-890 nm and a resolution of 0.3 nm was used to acquire the light reflected by the Au-tip. The spectra were monitored and acquired in real-time using a computer with the OceanView 2.0.8 software (Ocean Optics, USA), which is represented in **Fig. 1**.

For the acquisition of reflected light spectra from the fibres, an integration time of 240 ms was used, with an average of 10 samples and a bandwidth filter of 10. All the spectra were first normalized to the range [0, 1]. The reflectance spectra were obtained by subtracting the reference spectrum, acquired with the sensing zone exposed to air, from the spectra obtained in each studied solution. Finally, a Savitzky-Golay (SG) filter with a 20-point window was used to smooth the spectra. The resulted reflectance spectra were plotted, and the minimum reflectance was considered the SPR wavelength (λ_{SPR}). All acquired data were processed with Origin 2018 64 Bit software (OriginLab, USA).

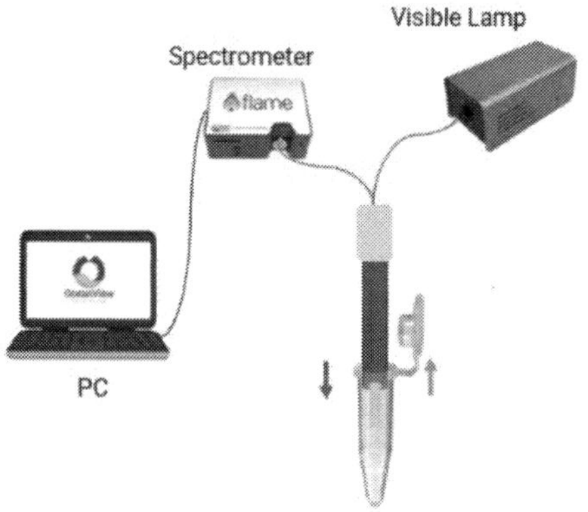

Fig. 1. Data acquisition instrumentation for the Au-tips in reflection mode.

2.3 Characterization of Au-tips to RI

The produced Au-tips were characterized to RI, and their sensitivity was determined. For this purpose, several glucose aqueous solutions were prepared with the concentrations ranging from 0.1 to 30 %(w/v), which corresponds to a RI range from 1.3331 to 1.3692 RIU (refractive index units). The RI value of each solution was measured using a refractometer (Abbemat 200, Anton Paar). First, the Au-tip was submerged in deionised (DI) water (1.3330 RIU) for 30 minutes, and spectra were taken every 2 minutes. The same procedure was repeated for each glucose solution in increasing order of concentration. At the end of the spectral acquisition, the surface of each Au-tip was sequentially cleaned in an aqueous solution containing 10 %(v/v) ethanol and then in DI water.

2.4 Biofuncionalization and detection tests

The biofunctionalization protocol of the Au-tip was adapted from [16]. The first step of biofunctionalization involves immersing the Au-tip in an ethanolic 11-mercaptoundecanoic acid (MUA) solution (0.5 mM) during ~16 h (overnight). After that time, the unbound MUA was removed by washing the tip three times in ethanol and then three times in DI water. Next, the MUA-Au-tip was immersed for 30 minutes in a 500 μL solution containing 250 μL of 1-ethyl-3-(3 dimethylaminopropyl) carbodiimide (EDC, 200 mM) and 250 μL of N-hydroxysuccinimide (NHS, 50 mM), both prepared in phosphate-buffered saline (PBS, 10 mM, pH 7.4) solution. After 30 minutes, the Au-tip is removed and directly immersed in 500 μL of uricase enzyme solution (0.1 mg/ml in PBS) for two hours. After these two hours, the fibre is removed from the solution, washed three times in PBS. Finally, the Au-tip surface was passivated in an ethanolamine solution (1 mM) prepared in PBS, for 30 minutes. The entire biofunctionalization experimental protocol was performed in a controlled environment at room temperature. The biofunctionalization procedure was monitored over time and the respective steps are illustrated in **Fig. 2**.

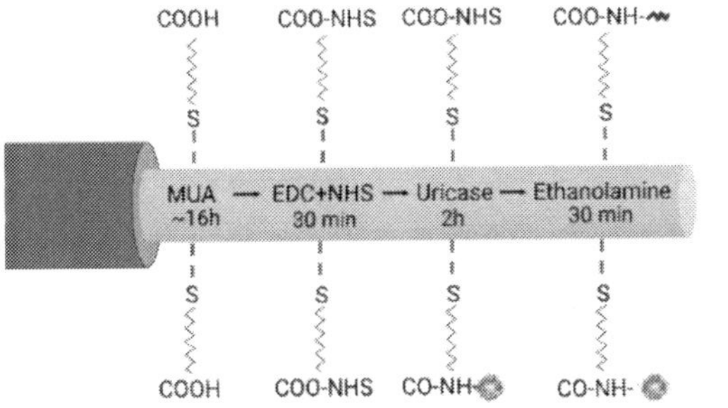

Fig. 2. Schematic representation of the biofunctionalization steps.

After biofunctionalization, the Au-tips were ready to be tested with different concentrations of uric acid to evaluate their sensing performance. A stock solution of uric acid (20 mM) in NaOH (0.01 M) was prepared for this purpose. The biofunctionalized Au-tip was immersed in 10 mL of PBS under stirring, and the first spectrum (0 mg/dL of uric acid) was collected. Then, small volumes of the uric acid solution were added under stirring to ensure homogeneity of the solution before collecting the optical spectra, corresponding to a tested range concentration from 0.2 to 42.0 mg/dL (0.01 to 2.50 mM). For each amount added, a period of 20 minutes was allowed for the response to stabilize and to observe the reaction kinetics. The RI of the solutions containing the different concentration of uric acid were measured (1.3348 RIU) and it was confirmed that with the increasing of uric acid concentration the RI of the solutions was kept constant.

3 Results

3.1 Characterization to RI

The produced Au-tip was firstly characterized to RI using glucose solutions with different RIs. To verify the time response and stability of the sensor's response over time, the Au-tip remained immersed in each solution for 30 minutes. It was noted that the Au-tip response was almost instantaneous and constant over time, meaning there were no significant changes of the SPR wavelength during the 30 minutes. Therefore, the λ_{SPR} for each of the 15 spectra acquired, during the 30 min, for each concentration was taken and the average and respective standard deviation for each refractive index were determined, as shown in **Table 1**.

As an example, **Fig. 3**a presents the reflectivity spectra of the Au-tip after 30 minutes immersed in each solution with varying RI. From the observations of the obtained spectra, it is possible to conclude that the position of the SPR signature shifts to higher wavelength values, with increasing RI values, as expected, attaining a total red shift of 52.7 nm. **Fig. 3**b shows the linear relationship between the average λ_{SPR} shift and the RI. RI sensitivity can thus be determined by the slope of the obtained linear regression, which is 1420.7 ± 59.7 nm/RIU.

Table 1. λ_{SPR} average of the 15 spectra acquired through the time (30 min) and respective standard deviation, for different refractive indices.

Refractive index	Average λ_{SPR} (nm)	Standard deviation (nm)
1.3330	624.73	0.18
1.3331	626.49	0.12
1.3338	627.69	0.21
1.3405	633.73	0.15
1.3467	640.52	0.24
1.3597	663.95	0.14
1.3692	677.27	0.25

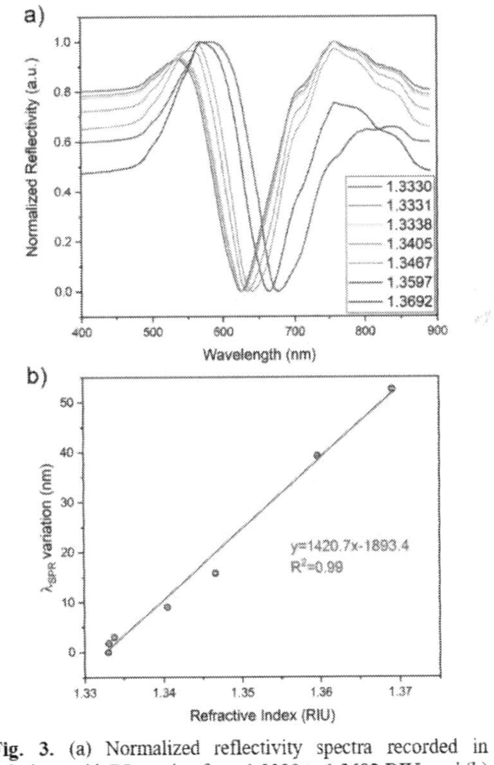

Fig. 3. (a) Normalized reflectivity spectra recorded in solutions with RI ranging from 1.3330 to 1.3692 RIU, and (b) variation of the λ_{SPR} as a function of RI, and the respective linear fitting (red line).

3.2 Biofunctionalization

The effectiveness of the biofunctionalization steps was evaluated by monitoring spectral variations after each step. The spectra acquisition was always performed under the same conditions, in PBS and after washing the fibre, to prevent unbound molecules on the Au-tip from interfering and causing changes in the surface RI. In **Fig. 4** is clearly noticeable the shift in wavelength to the red region caused by each step of biofunctionalization for each Au-tip.

The first step of biofunctionalization involves leaving the Au-tip immersed in MUA for approximately 16 hours to allow the fibre surface to acquire carboxyl groups. A wavelength shift to the red region of the spectrum of 7.27 nm occurred. In the next step, uricase was covalently linked to the carboxyl groups using the EDC/NHS chemistry, in this case a red shift of 8.23 nm was registered. In the passivation step, there was also an increase in wavelength of 1.97 nm. At the end of the biofunctionalization process, the total wavelength shift was 17.47 nm.

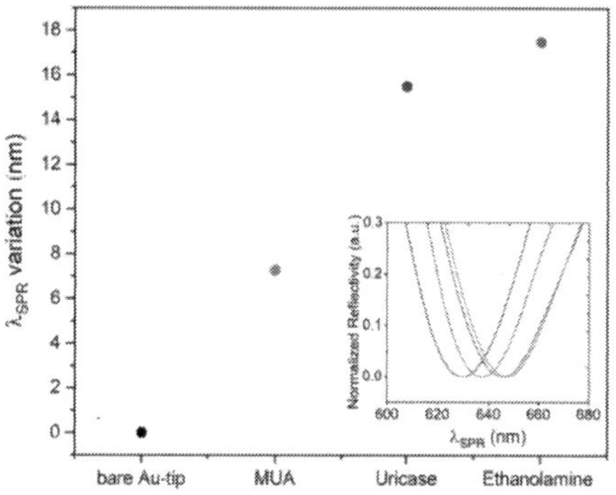

Fig. 4. Variation of the SPR wavelength obtained in PBS solution after each biofunctionalization step. *Inset* shows the spectra of each functionalization step.

3.3 Uric Acid Detection

Two tests for the detection of uric acid were performed using the same biofunctionalized Au-tip. In both cases, a red shift in the spectrum and a decrease in reflectivity occurred. However, in the first detection, the reflectivity of the last two concentrations increased compared to the previous ones. Thus, the maximum variation in wavelength shift was 1.31 nm, for the first detection and 1.97 nm for the second one.

The response of the biofunctionalized Au-tip was assessed by plotting the average of the SPR wavelength variation in function to the uric acid concentration, as displayed in **Fig. 5**. The Michaelis-Menten equation (Eq. 1) [17] was used to fit the data and to evaluate the enzymatic kinetics of the uricase enzyme immobilized on the Au-tips.

$$y = \frac{V_{max} x}{K_m + x} \qquad (1)$$

Where V_{max} corresponds to the maximum reaction rate when the enzyme is saturated with uric acid, and K_m is the Michaelis-Menten constant. By applying the Michaelis-Menten function, a V_{max} of 2.57 nm and a K_m of 1.35 mM were obtained

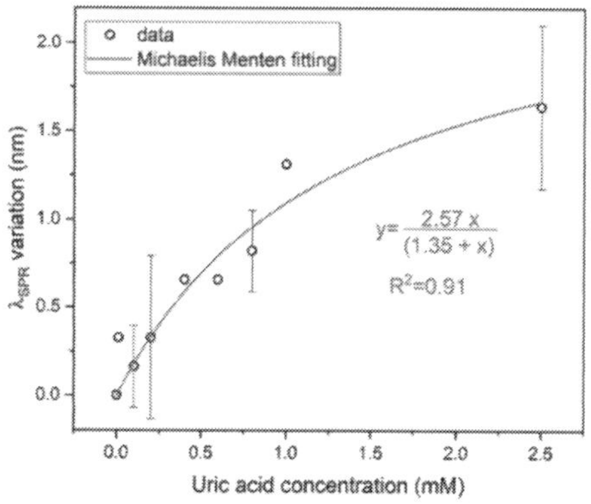

Fig. 5. Evaluation of the biofunctionalized Au-tip response to uric acid by following the average variation of the SPR wavelength, and the respective fitting using the Michaelis-Menten function (n=2).

4 Conclusion

This work involved the development of an optical fiber-based enzymatic biosensor using SPR for uric acid detection, using a 400 µm core fiber. This gold SPR sensor works in a reflective configuration, and for that is referred to as Au-tip.

The Au-tip demonstrated a RI bulk sensitivity of 1420.7 ± 59.8 nm/RIU. The Au-tips were subjected to the biofunctionalization process using MUA, EDC, NHS, uricase, and ethanolamine.

In response to different concentrations of uric acid, the Au-tip exhibited a red shift of the λ_{SPR}, thus allowing the use of the Michaelis-Menten function to evaluate the enzymatic kinetics of the uricase enzyme immobilized on the Au-tips. Consequently, a V_{max} of 2.57 nm and a K_m of 1.35 mM was obtained for the average wavelength shift.

This work presents preliminary results in the development of plasmonic optical fiber enzymatic biosensors for uric acid detection. Therefore, for future work, repeating detections would be useful to verify the reproducibility of biofunctionalization processes and the responses of the Au-tip. However, for reported enzymatic biosensors using SPR phenomenon, it is not clear neither demonstrated what is contributing for the variations in the SPR wavelength, therefore future work should investigate if enzymes are a reliable bioreceptor for SPR-based sensors due to its catalytic nature, where there is not a real immobilization of the analyte suitable for RI alterations at the sensors' surface. Hence, at this time the question arises as to whether enzymatic biosensors are the best choice to use with the SPR principle.

Acknowledgements

The present study was supported by project i3N (UIDB/50025/2020, UIDP/50025/2020 and LA/P/0037/2020) funded by national funds through the FCT and MEC of Portugal, and developed in the scope of the Project "Agenda ILLIANCE" [C644919832-00000035 | Project n° 46] – WP4 OLI Health, financed by PRR – Plano de Recuperação e Resiliência under the Next Generation EU from the European

Union. S.O. Pereira and A. S. Assunção thank the Researcher contract and the PhD Research Grant BI/UI64/10684/2023, respectively, within the scope of the ILLIANCE Project. Cátia Leitão thanks the CEEC contract (2020.00154.CEECIND/CP1589/CT0019, https://doi.org/10.54499/2020.00154.CEECIND/CP1589/CT0019), funded by national funds through the FCT.

References

1. S. Aafria, P. Kumari, S. Sharma, S. Yadav, B. Batra, J. S. Rana, and M. Sharma, "Electrochemical biosensing of uric acid: A review," Microchem. J. **182**, 107945 (2022).
2. S. Durgapal, A. Jantwal, J. Upadhyay, T. Joshi, and A. Kumar, "Uric acid," in *Antioxidants Effects in Health* (Elsevier, 2022), pp. 505–516.
3. J. Maiuolo, F. Oppedisano, S. Gratteri, C. Muscoli, and V. Mollace, "Regulation of uric acid metabolism and excretion," Int. J. Cardiol. **213**, 8–14 (2016).
4. C. Ma, N. Jiang, X. Sun, L. Kong, T. Liang, X. Wei, and P. Wang, "Progress in optical sensors-based uric acid detection," Biosens. Bioelectron. **237**, 115495 (2023).
5. M. Amjadi, T. Hallaj, and Z. Kouhi, "An enzyme-free fluorescent probe based on carbon dots – MnO2 nanosheets for determination of uric acid," J. Photochem. Photobiol. A Chem. **356**, 603–609 (2018).
6. H. H. Nguyen, S. H. Lee, U. J. Lee, C. D. Fermin, and M. Kim, "Immobilized Enzymes in Biosensor Applications," Materials (Basel). **12**(1), 121 (2019).
7. P. Damborský, J. Švitel, and J. Katrlík, "Optical biosensors," Essays Biochem. **60**(1), 91–100 (2016).
8. A. K. Sharma, R. Jha, and B. D. Gupta, "Fiber-optic sensors based on surface plasmon resonance: A comprehensive review," IEEE Sens. J. **7**(8), 1118–1129 (2007).
9. T. Haque and H. K. Rouf, "A Performance Improved Kretschmann Configuration Based Surface Plasmon Resonance (SPR) Sensor," in *2019 1st International Conference on Advances in Science, Engineering and Robotics Technology (ICASERT)* (IEEE, 2019), pp. 1–4.
10. C. L. Wong and M. Olivo, "Surface Plasmon Resonance Imaging Sensors: A Review," Plasmonics **9**(4), 809–824 (2014).
11. B. J. Elliott, "Optical cable technology — optical fibre," in *Cable Engineering for Local Area Networks* (Elsevier, 2000), pp. 161–185.
12. A. Leung, P. M. Shankar, and R. Mutharasan, "A review of fiber-optic biosensors," Sensors Actuators B Chem. **125**(2), 688–703 (2007).
13. B. D. Gupta and R. Kant, "[INVITED] Recent advances in surface plasmon resonance based fiber optic chemical and biosensors utilizing bulk and nanostructures," Opt. Laser Technol. **101**, 144–161 (2018).
14. R. Kant, R. Tabassum, and B. D. Gupta, "Fiber Optic SPR-Based Uric Acid Biosensor Using Uricase Entrapped Polyacrylamide Gel," IEEE Photonics Technol. Lett. **28**(19), 2050–2053 (2016).
15. A. S. Assunção, M. Vidal, M. J. Martins, A. V. Girão, M. Loyez, C. Caucheteur, J. Mesquita-Bastos, F. M. Costa, S. O. Pereira, and C. Leitão, "Detection of NT-proBNP Using Optical Fiber Back-Reflection Plasmonic Biosensors," Biosensors **14**(4), 173 (2024).
16. L. Singh, G. Zhu, R. Singh, B. Zhang, W. Wang, B. K. Kaushik, and S. Kumar, "Gold Nanoparticles and Uricase Functionalized Tapered Fiber Sensor for Uric Acid Detection," IEEE Sens. J. **20**(1), 219–226 (2020).
17. D. L. Nelson, M. M. Cox, and A. A. Hoskins, *Lehninger Principles of Biochemistry* (W.H. Freeman & Co. Ltd, 2021).

Vegetation health evaluation using cost-effective aerial reflectance measurements

Anaísa Lucena [1*], *Rodrigo* Rendeiro[1], and *Carlos* Marques[1]

[1]CICECO - Aveiro Institute of Materials, Physics department, University of Aveiro, 3810-193 Aveiro, Portugal.

Abstract. With the increase of wildfires, posing a significant threat to ecosystems, human life, and infrastructure, early detection and management of vegetation becomes crucial. The Green-Red Vegetation Index (GRVI) and the Excess Green (ExG) index are often used as a health indicator for plants. While the ExG index excels in identifying vegetation, the GRVI can distinguish between healthy and dead vegetation by calculating the difference between reflected green and red light. This study utilized a cost-effective commercial-off-the-shelf (COTS) drone (DJI MAVIC 2 PRO) and in-house software to retrieve the forementioned GRVI and ExG indices. These enable the classification of vegetation in rural areas to identify biomass fuel that requires cleaning to prevent the spread of fires. Validation through comparison tests with field samples demonstrated that the drone and software can effectively identify and categorize vegetation types, as well as estimate biomass fuel.

1 Introduction

Wildfires pose a significant and escalating threat to ecosystems, human life, and property worldwide. In 2021 alone, 9 million hectares were lost, an area the size of Portugal [1]. In Portugal, 110 thousand hectares were lost in 2022, surpassing the annual totals of the previous four years [2], and in 2023, over 15 million hectors were lost in Canada alone [3].

Wildfires data collected from 2008-2018 by the *Instituto Conservação da Natureza e da Floresta* (Institute for the Conservation of Nature and Forest - ICNF) revealed a reduction in ignitions due to imputable person, fire restarts, fireplaces, pasture management and burning of trash/agriculture/forest leftovers [4]. However, there has been an increase in other accidental, less controlled, ignitions caused by machinery use, transport and communications and lightning strikes. The fires caused by lightning and electricity infrastructures have been associated with more severe consequences [5]. Therefore, the constant monitorization of areas with tendency for the build-up of biomass fuel, the main fuel for the spread of fire, and its cleaning, is required.

One crucial aspect of fire prevention is the early identification of vegetation health, as dead or unhealthy vegetation is more susceptible to ignition and can fuel the rapid spread of fires. Traditional methods of monitoring vegetation health, such as ground surveys and satellite imaging, are limited by being labor-intensive, time-consuming, and often lacking the resolution and frequency needed for precise, timely assessments.

Unmanned Aerial Vehicles (UAVs) present a promising solution to these challenges. UAVs (colloquially drones) can cover large areas quickly, access remote and difficult terrains, and capture high-resolution images, providing detailed and timely data on vegetation health.

Exploiting the photogrammetry capabilities of the drone, a commercially available DJI MAVIC PRO 2 (2018), images of several urban and rural areas containing different types of landscapes were taken and analyzed using vegetation indices based on the absorption and reflectance of the vegetation. These images were later processed to categorize and highlight critical areas of the photographed scenes.

By employing computer vision algorithms and vegetation indices, it is possible to distinguish between healthy and dry vegetation, enabling early identification of areas at high risk of fire, as seen in Fig.1.

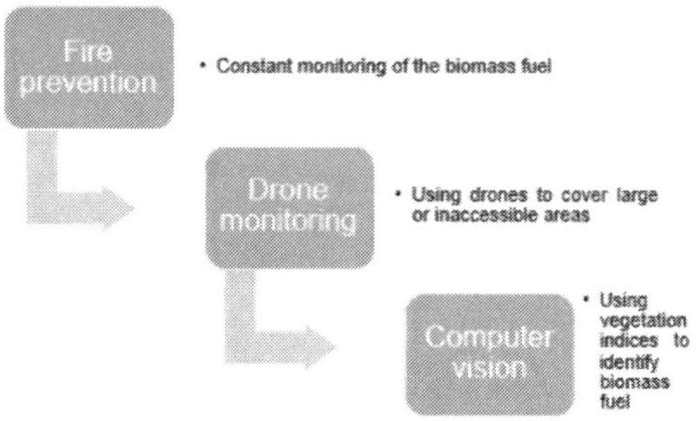

Fig. 1. Diagram illustrating the workflow of the proposed vegetation classification method using a drone and vegetation indices.

---

* Corresponding author: anaisalucena@ua.pt

2 Methodology

The drone has a maximum flight time of 30 minutes per battery and, when flying at an altitude of 100 meters, covers over 1 hectare of terrain per shot. Besides the mentioned, it has a native autonomous mode where it can fly through predetermined waypoints, dramatically decreasing the effort required to monitor large areas.

Using computer vision, it was possible to do an estimation of the areas with a higher amount of biomass fuel.

Moreover, this drone captures images in RGB format using a Hasselblad L1D-20c, which can capture visible and near-infrared light, but has a natively integrated 800 nm cut-off filter. Since it does not have a thermal sensor or a specific near-infrared sensor, the analysis focused on RGB-based vegetation indices. While it is possible to detect near-infrared radiation, in order to use the Normalized Difference Vegetation Index (NDVI) [6], a commonly used metric to estimate vegetation, this requires taking two pictures per scenario, one regular and another with a near infrared (>700 nm) filter over the camera lens. This dual-image method causes some problems, since even when programming the flight coordinates, the drone may be on a different angle, making it difficult to perfectly align the pixels, which increases the complexity of the analysis.

Therefore, the focus was on RGB indices. There are multiple such indices [7], which were tested and analyzed to categorize three types of areas: no vegetation (i.e. soil, houses, concrete roads, etc.), dry or dead vegetation (biomass fuel), and healthy vegetation.

Most of the indices faced some issues, such as difficulty with some areas, or separating dry and healthy vegetation. Therefore, a combination of two indices was chosen: the Green-Red Vegetation Index (GRVI) [8] and the Excess Green (ExG) index [9]. Typically, vegetation reflects more green light than red, due to absorption of red wavelengths in the chlorophyll. Both GRVI and ExG harness this characteristic to segment the different vegetation states, where GRVI is normalized and ExG is unnormalized and amplifies the green component of the image. While the GRVI (Eq.1) struggled with the color cyan and some shadows, it excelled in distinguishing between dry and healthy vegetation, and whereas the ExG (Eq. 2) was capable of effectively separating vegetation from the non-vegetation, it was outperformed by the GRVI in distinguishing vegetation types.

$$GRVI = \frac{g - r}{g + r} \tag{1}$$

$$ExG = 2 \times g - b - r \tag{2}$$

where g, b and r are the normalized (divided by the maximum) pixel intensities of the green, blue and red channels, respectively.

2.1 Finding the thresholds

Furthermore, to accurately classify the three types of rural areas, it is essential to determine appropriate thresholds for the indices used. The first step involved distinguishing vegetation from the absence of vegetation, which was done using the ExG index.

In Fig.2, this process is illustrated, where it is possible to observe an image of a rural area, on (a), as well the respective ExG heat map, on (b), showcasing the intensity of the ExG index for each pixel. This heat map aids in visualizing areas with varying levels of vegetation.

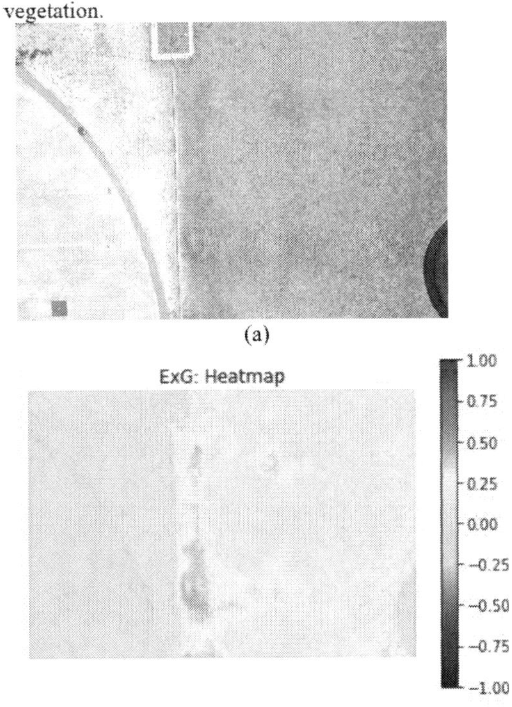

Fig. 2. RGB image, 20 m high containing both concrete and vegetation (a), as well as the heatmaps of the ExG index on (b).

This specific image, in Fig.2(a) was chosen for this test since the vegetation and non-vegetation areas are fairly balanced. By examining the heatmap, it is possible to understand that the threshold lies somewhere between 0 and 0.15. As such, to verify this, and to obtain a more concrete value, a histogram of the ExG for each pixel in the image was made, as seen in Fig.3.

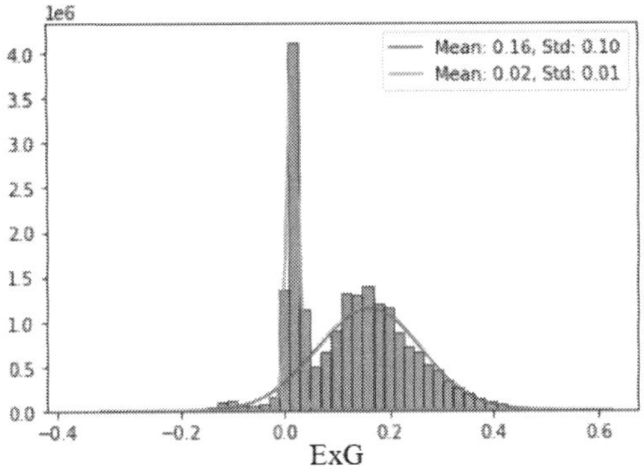

Fig. 3. Histogram of the ExG value of each pixel of Fig.2(a).

Looking at the histogram in Fig.3, it is possible to observe two peaks. These peaks represent the sum of two normal distributions, as represented in the figure: the left one, in orange, can be interpreted as the area that does not contain vegetation, and the right one, in blue, as corresponding to the vegetated areas. As such, the local minima between these two distributions, 0.06, was selected as the threshold value, meaning that every pixel with an ExG index over 0.06 was considered to be vegetation.

Finally, regarding the separation between healthy and dry vegetation using the GRVI, it was used 0 as the threshold, meaning that vegetation with $GRVI > 0$ was identified as healthy and vegetation with $GRVI \leq 0$ was identified as dry [10]. In summary:

- $ExG \leq 0.6 \rightarrow$ absence of vegetation
- $ExG > 0.06 \cap GRVI > 0 \rightarrow$ healthy vegetation
- $ExG > 0.06 \cap GRVI \leq 0 \rightarrow$ dry vegetation

3 Results

After defining these limits, it was possible to individually represent the three types of vegetation, with Fig.4 clearly illustrating the separation between the proposed categories. Based on these categories, the percentage of the area, or pixels, corresponding to each vegetation type can be calculated. In this case, this particular rural area consists of approximately 15% of healthy vegetation, represented by the green pixels, and 47% of dry vegetation, represented by the orange pixels.

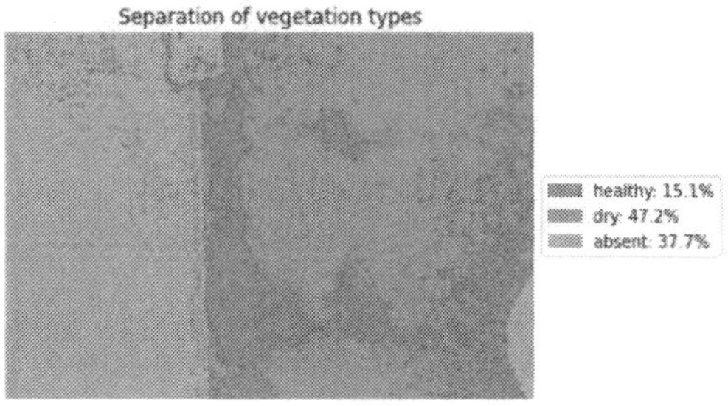

Fig. 4. Separation of vegetation types for each pixel of Fig.2(a), as well as the percentage of each type, where green indicates healthy vegetation, orange dry vegetation, and gray the absence of vegetation.

Furthermore, Fig.5 through 7 provide three further examples of rural areas and their respective vegetation classification. Since the example on Fig.2(a) and Fig.4 is mainly composed by dead vegetation, and to ensure a wide range of rural areas, these figures illustrate various compositions of vegetation: in Fig.5 it is shown a rural area composed of mainly healthy vegetation, in Fig.6 an area with few vegetation, and in Fig.7 a mixed area with similar percentages of each category.

(a)

(b)

Fig. 5. Original RGB image (a) and respective separation of vegetation types (b) in a rural area, 50 m high, composed mainly of healthy vegetation.

(a)

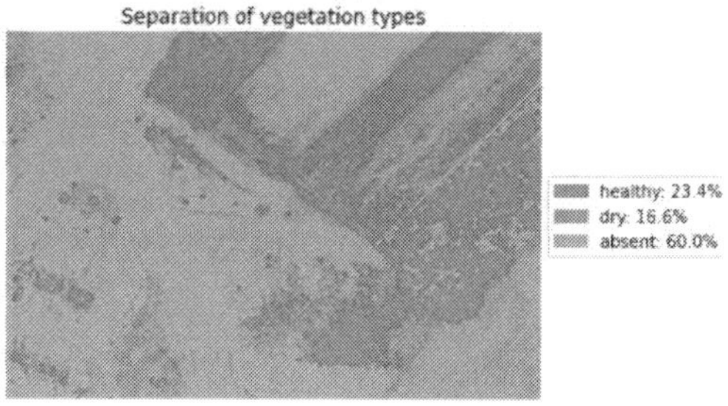

(b)

Fig. 6. Original RGB image (a) and respective separation of vegetation types (b) in a rural area, 100 m high, with limited vegetation.

(a)

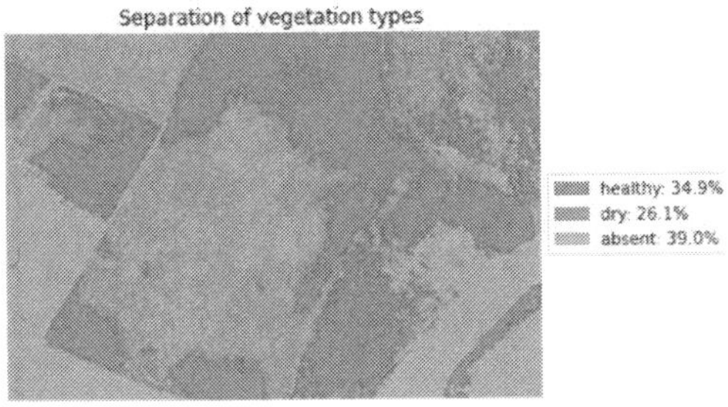

(b)

Fig. 7. Original RGB image (a) and respective separation of vegetation types (b) in a rural area, 80 m high, with balanced amounts of healthy, dead and absence of vegetation.

As it is possible to observe, the algorithm was capable to effectively identify and categorize the rural areas, as well as give an estimation of biomass fuel. For example, in the Fig.8, an alternative representation is provided, where the real photo and the tri-colored map are overlapped.

Fig. 8. RGB image of a rural area, 100 m high, overlapped with the tri-colored map of vegetation types.

4 Conclusion and Future Work

The aerial view provided by the drone allows for a simple, low-cost solution to obtain large area samples. The analysis method based on computer vision and the ExG and GRVI indices proved effective at categorizing healthy vegetation, dry vegetation and the absence of vegetation. Moreover, using the categorized data, the distance from the drone to the terrain and mathematical relations, it is possible to estimate the area covered with biomass fuel.

In conclusion, drone photogrammetry can be used to identify areas with a high risk of fire due to biomass fuel, with the benefit of the easy adoption by private individuals, institutions or industries due to the low-cost associated with the maintenance of such systems. Furthermore, these routine monitoring flights can be integrated with remote AI tools and/or other computer vision methods for a real-time tracking of eminent fires [11].

This work was developed in the framework of DigiAqua Project PTDC/EEI-EEE/0415/2021. We acknowledge the financial support of CICECO (UIDP/50011/2020, UIDB/50011/2020) funded by national funds through the FCT.

References

[1] F. Carta, C. Zidda, M. Putzu, D. Loru, M. Anedda, and D. Giusto, "Advancements in Forest Fire Prevention: A Comprehensive Survey," Jul. 01, 2023, *Multidisciplinary Digital Publishing Institute (MDPI)*. doi: 10.3390/s23146635.

[2] Agência Portuguesa do Ambiente, "Riscos Ambientais: Incêndios Rurais," *Portal do Estado do Ambiente Portugal*, 2024.

[3] P. Jain *et al.*, "Canada Under Fire – Drivers and Impacts of the Record-Breaking 2023 Wildfire Season," Feb. 2024, doi: 10.22541/essoar.170914412.27504349/v1.

[4] Rui Moreira, "Padrões de incêndios em Portugaç Continental e modelos de combustível," Instituto para a Conservaçao da Natureza e da Floresta.

[5] C. Miller *et al.*, "Electrically caused wildfires in Victoria, Australia are over-represented when fire danger is elevated," *Landsc Urban Plan*, vol. 167, pp. 267–274, Nov. 2017, doi: 10.1016/j.landurbplan.2017.06.016.

[6] A. A. Gitelson, Y. J. Kaufman, R. Stark, and D. Rundquist, "Novel algorithms for remote estimation of vegetation fraction," *Remote Sens Environ*, vol. 80, no. 1, pp. 76–87, 2002, doi: https://doi.org/10.1016/S0034-4257(01)00289-9.

[7] A. Agapiou, "Vegetation Extraction Using Visible-Bands from Openly Licensed Unmanned Aerial Vehicle Imagery," *Drones*, vol. 4, no. 2, 2020, doi: 10.3390/drones4020027.

[8] H. Ramdan, "Drone technology for identification of healing forest spot at Kampung Cisamaya Mount Ciremai National Park," *IOP Conf Ser Earth Environ Sci*, vol. 918, no. 1, p. 12040, Nov. 2021, doi: 10.1088/1755-1315/918/1/012040.

[9] D. M. Woebbecke, G. E. Meyer, K. Von Bargen, and D. A. Mortensen, "Color Indices for Weed Identification Under Various Soil, Residue, and Lighting Conditions," *Transactions of the ASAE*, vol. 38, no. 1, pp. 259–269, 1995, doi: https://doi.org/10.13031/2013.27838.

[10] T. Motohka, K. N. Nasahara, H. Oguma, and S. Tsuchida, "Applicability of Green-Red Vegetation Index for Remote Sensing of Vegetation Phenology," *Remote Sens (Basel)*, vol. 2, no. 10, pp. 2369–2387, 2010, doi: 10.3390/rs2102369.

[11] H. Cruz, M. Eckert, J. Meneses, and J.-F. Martínez, "Efficient Forest Fire Detection Index for Application in Unmanned Aerial Systems (UASs)," *Sensors*, vol. 16, no. 6, 2016, doi: 10.3390/s16060893.

Manipulation of Microparticles in Optofluidic Devices Fabricated by Femtosecond Laser Micromachining

Carolina Cameira[1,2*], *João M. Maia*[2], *and P. V. S. Marques*[1,2]

[1]Department of Physics and Astronomy, Faculty of Sciences of University of Porto, 4169-007 Porto, Portugal
[2]CAP – Centre for Applied Photonics, INESC TEC, 4150-179 Porto, Portugal

Abstract. This study reports the fabrication of three-dimensional microfluidic channels in fused silica, using femtosecond laser micromachining, to achieve two-dimensional hydrodynamic flow focusing in either the horizontal or the vertical directions. Spatial focusing of 3 μm polystyrene particles was successfully demonstrated, showing the ability of the fabricated devices to confine microparticles within a 6 μm layer over a channel width of 420 μm and within a 5 μm layer over a channel height of 260 μm. Integration of laser-direct written optical waveguides inside a microfluidic chip and orthogonal to the channel also enabled the implementation of a dual-beam optical trap, with trapping of polystyrene microparticles using a 1550 nm beam being demonstrated.

1 Introduction

In the past decade, LOC (lab-on-a-chip) devices have proven to be a powerful and reliable biosensing technology for medical and healthcare applications, offering benefits such as point-of-care diagnostics, faster analysis times, reduced reagent costs, and less chemical waste [1]. LOC technology integrates multiple functions (fluidic, optical, electrochemical, ...) onto a single chip with mm^2 footprint, enhancing the automation, compactness, and portability of these devices.

LOC devices have traditionally been fabricated by planar processes, such as photolithography and soft lithography. However, the manufacturing of 3D structures through these microfabrication techniques is considerably complex and time-consuming, given that these require stacking of multiple 2D structures. Another promising method for fabricating LOC devices is femtosecond (fs) laser micromachining. This technique, which can be applied to a wide range of materials transparent to the laser wavelength, relies on non-linear absorption of the pulsed laser beam and relaxation processes to alter the material properties at the laser beam's focal point [2]. For instance, fs-laser exposure in fused silica can induce a refractive index increase or higher etching rate in HF acid [3]. Besides its capability for 3D machining, this technique also permits the integration of both optical and fluidic functions on a single chip [4]. It enables the development of devices leading to processing of fluidic samples—such as cell imaging [5], fluid mixing [6], particle separation or focusing [7]—as well as their analysis through optical trapping [8], fluorescence [9], and backscattered light [10].

Here, fs-laser micromachining is used to fabricate optical waveguides and 3D microfluidic channels in a single chip to develop a flow cytometry device for optical manipulation and analysis of microparticles and cells. To achieve single-particle level sensing, it is fundamental to have precise control over particle movement and positioning within the channels [11]. For this reason, we design microfluidic channels with a three-dimensional geometry capable of controlling the distribution of microparticles both horizontally and vertically through hydrodynamic flow focusing (HFF), by adjusting the ratio between sheath and sample flow rates. On-chip integration of optical waveguides towards microparticle/cell optical sensing is also demonstrated through the creation of a 1550 nm dual-beam optical trap.

2 Experimental Methods

2.1 Fabrication

A fiber amplified diode pumped ultrafast laser (Amplitude Systèmes Satsuma HP) operating at 515 nm and emitting 250 fs long pulses at 500 kHz was used to either induce an increase of the refractive index or of the chemical etching rate in 30×25×1 mm^3 fused silica (Suprasil 1) substrates. The linearly polarized laser beam was focused inside the substrate via a 0.42 NA objective (Mitutoyo M Plan Apo NIR 50x). The substrate is held in place by vacuum on a stainless-steel sample holder, which in turn is set on top of Aerotech direct-drive translation stages for controlled movement along the XY plane (ANT130XY-110 PLUS) and along the Z direction (ANT130V-5 PLUS). Stage movement,

*Corresponding author: carolina.cameira@inesctec.pt

as well as pulse energy and polarisation, are all controlled with the Workshop of Photonics workstation's SCA software.

Optical waveguides are directly written by translating the substrate along the XY plane relative to the laser focus, with the laser beam polarisation parallel to the scanning direction.

The microfluidic channel fabrication is more intricate since it requires the programming of more complex 3D geometries. These are directly programmed in the SCA software interface, with the 3D object being sliced and hatched into a set of linearly spaced laser-tracks that are scanned sequentially from bottom-to-top with the laser beam polarization always perpendicular to the scanning direction for a more effective chemical etching rate increase. Chemical etching is performed by immersing the substrate in hydrofluoric acid (HF) at either a 10% or 20% volume concentration in an ultrasonic bath inside an ISO-6/7 clean room facility. Typical surface roughness values of the fabricated microchannels are around 90 nm. This result was obtained in a separate study using the same laser direct writing and chemical etching systems, similar materials and processing parameters [12]. The surface roughness can be reduced through thermal annealing [12], oxyhydrogen flame heating [13] or CO2 laser irradiation [14]. In between steps, the substrate side facets are mechanically polished to suppress the waveguide tapered end region arising due to the laser failing to converge to a single focal spot at the interface.

Lastly, a polydimethylsiloxane (PDMS) layer is fabricated, pierced with a blunt-end steel syringe needle to fit the capillary tubes (Darwin Microfluidics, PTFE Tubing 1/16" OD - 0.5mm) used for fluid injection or withdrawal, and plasma-bonded to the fused silica chip to seal the microfluidic device.

2.1.1 Horizontal HFF Device

A scheme of the device is depicted in figure 1. The microfluidic channel has three inlets, one for the sample and two for the sheath solutions, and one outlet. The sample and sheath flows intersect at a 90° angle. The main channel has a square cross-section measuring 400×250 μm^2 (width by height), and the diameter of the inlets/outlets is 1.2 mm. The other dimensions are indicated in fig. 1.

The 3D channel was sliced into 5 µm spaced layers, which were then hatched into linear tracks with a period of 10 µm. Each track was written with a pulse energy of 125 nJ at 5 mm/s. The etching step for the channel fabrication took two hours and fifteen minutes with a 20% HF volume concentration. Microfluidic channel width and height increased by 20 and 10 µm, respectively, during chemical etching.

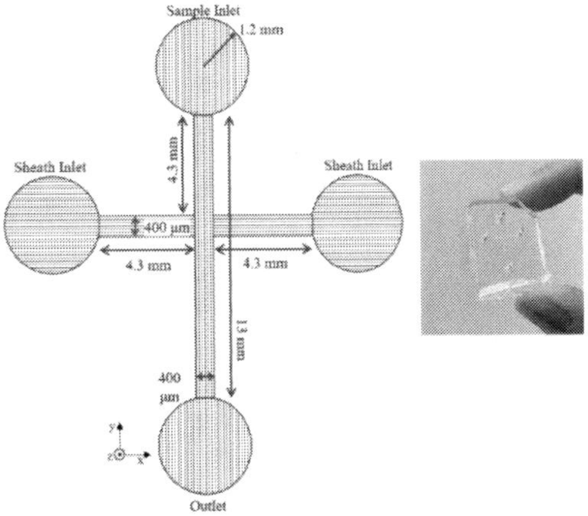

Fig. 1. Top view schematic of the fabricated device to test horizontal HFF (left) and an image of the device (right). Schematic not to scale.

2.1.2 Vertical HFF Device

The device geometry is shown in figure 2. The microfluidic channel contains two inlets, for sample and sheath flows, and an outlet. The sheath flow divides into two subchannels, one at the surface of the chip and another inside it. The sample flow is enclosed by these two subchannels and runs co-directional to them. All dimensions are indicated in fig. 2. During initial trials, it was observed that, despite the symmetric design of the two sheath flow subchannels, these became asymmetric after the etching process. To address this issue, the programmed geometry was adjusted to include slight differences in the dimensions of the subchannels, as illustrated in figure 2. Although this modification mitigated the problem, it did not completely resolve it. Further discussion on the extent of this attenuation will be provided later in the experimental results section.All subchannels have a common width of 400 µm, the sheath subchannels have heights of 52 and 36 µm, the sample subchannel narrows down to a height of 36 µm right before the intersection, and the channel depth is 262 µm after the intersection.

The 3D channel was sliced into 5 µm spaced layers, which were then hatched into linear tracks with a period of 4 and 25 µm for the main channel and inlet/outlet, respectively. Each track was written with a pulse energy of 60 nJ at 5 mm/s. The etching step for the channel fabrication took two hours with a 20% HF volume concentration. During chemical etching, the channel width and height increased by 20 and 10 µm, respectively, and the sheath subchannels became asymmetric, as will be discussed further ahead.

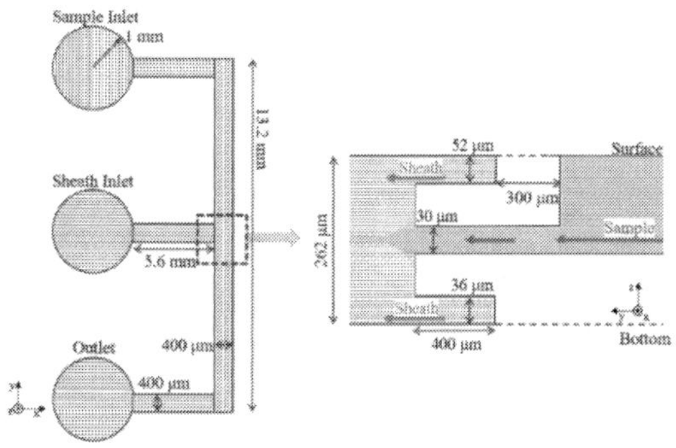

Fig. 2. Top view (left) and side view (right) schematic representation of fabricated device to test vertical HFF. Schematic not to scale.

2.1.3 Dual-Beam Optical Trap Optofluidic Device

The device, shown in figure 3, is composed of a microfluidic channel with one inlet and an outlet for fluid flow with several waveguides crossing the main channel. The microfluidic channel has a depth of 100 µm, the diameter of the circular inlet/outlet is 1.5 mm, and the main channel is 14 mm long and 36 µm wide. The 3D channel was sliced into 5 µm spaced layers, which were then hatched into linear tracks with a period of 2.5 and 5 µm for the main channel and inlet/outlet, respectively. Each track was written with a pulse energy of 60 nJ at 2 mm/s. The etching step for the channel fabrication took one hour and a half with a 10% HF volume concentration.

The optical waveguides were laser-written after chemical etching with a pulse energy of 100 nJ at 200 µm/s. These were written at varying depths covering the channel depth, from 90 to 30 µm with a pitch of 5 µm. Afterwards, the substrate facets are mechanically polished and a V-groove assembly containing an SMF28 fiber is aligned with the 50-µm deep laser written waveguide and bonded to the substrate using UV-cured Norland 81 optical adhesive.

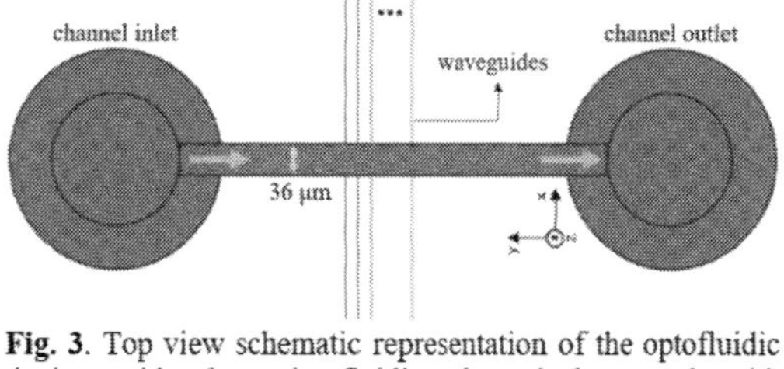

Fig. 3. Top view schematic representation of the optofluidic device with the microfluidic channel integrated with perpendicular optical waveguides. Despite several waveguides being written, only one was used for testing the device. Schematic not to scale.

2.2 Device Characterization

All devices were characterized using red food dye and/or non-functionalized 3-µm spherical polystyrene (PS) particles (Phosphorex, PolySciences) packaged in a 2.7% solids (w/v) aqueous suspension and subsequently diluted in deionized water. The flow rates were controlled by syringe pumps (SP210iw Syringe Pump; AL-4000 Two-Channel Syringe Pump), where the syringes (3 mL plastic B-D; 0.5 and 1 mL glass Hamilton Gastight) were placed and connected to the capillary tubes. Choice of the syringes depends on the used flow rate; lower-volume syringes provide more stability at lower flow rates. A homemade microscope system (employing either one of the following objectives 0.40 NA 20x Olympus ULWD, 0.25 NA 10x Nikon, and 0.10 NA 4x Nikon) was used to record live images of the samples passing through the microfluidic channels via a CCD camera (Thorlabs CS165CU/M Zelux).

3 Experimental Results and Discussion

3.1 Hydrodynamic Flow Focusing

3.1.1 Horizontal Focusing

The device to test horizontal HFF (fig. 1) was firstly tested using deionized water in the two sheath inlets and red food dye diluted in deionized water in the sample inlet, making it possible to observe flow focusing of the sample fluid. Both sheath inlets were connected to the same syringe pump, with 3 mL syringes, so that the sheath's flow rate, Q, was always the same for both inlets, since the channels are symmetrical. On the other hand, the sample fluid flow rate, q, is controlled by another syringe pump, with a 1 mL syringe. This allows the control of the ratio between the sample and sheath flow rate, $f = q/2Q$, which in turn determines the ratio between the width of the focused sample stream, Wf, and the total width of the channel, W (fig. 4).

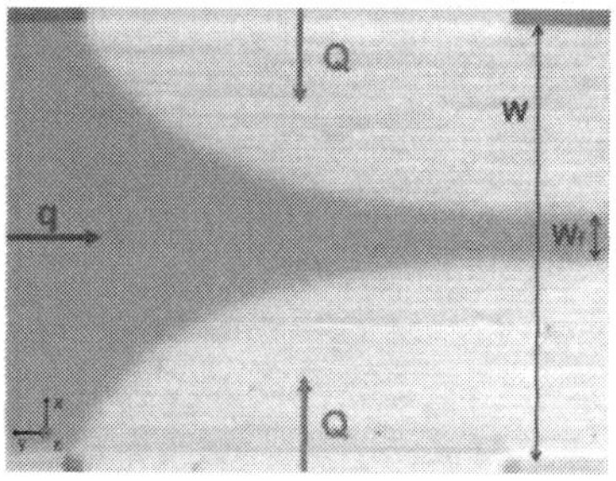

Fig. 4. Image of 2D hydrodynamic focusing on the channel junction of the fabricated device. The total flow rate after the intersection is q+2Q. Image was taken for Q = 20 µL/min and q=5µL/min (f=0.125).

The flow rates Q and q were varied from 10 to 40 µL/min and 1 to 80 µL/min, respectively. Images of the channel were taken at distances ranging from 280 to 380 µm from the centre of the junction and analysed using ImageJ to calculate the ratio between the width of the focused sample stream and the total channel width (Wf/W). This region was selected because the hydrodynamic entrance length for laminar flow, which is the distance at which the fluid's velocity profile becomes fully developed [15], ranges from 6.6 to 89.2 µm under the conditions of this study, thus ensuring that the fluid is fully developed in the measurement region.

Additionally, longer distances were avoided to minimize potential issues arising from diffusion effects. The experimental results are depicted in figure 5. Wf was measured 3 times for each experimental condition, with the resulting standard deviation depicted as an error bar. The dashed lines represent the theoretical results obtained from Lee *et al* [16].

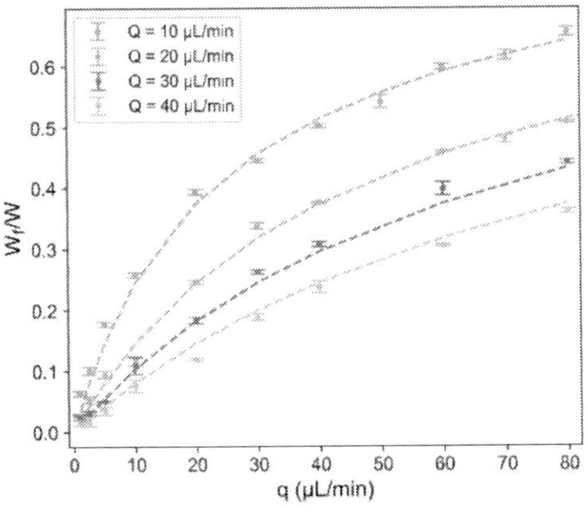

Fig. 5. Experimental results of the variation of the ratio between the focused sample stream width and the total width of the channel, Wf/W, with the sample flow rate, q, for constant sheath flow rates, Q. The dashed lines represent the corresponding theoretical results.

The obtained results follow the expected theoretical behaviour. For a constant sheath flow rate Q, the width of the focused sample stream, Wf, increases as the sample flow rate q is increased. In the extreme condition where Q=0 (no sheath flow) the food dye fills the entire channel. On the other hand, for a constant q, a decrease on the flow rate Q increases the width of the focused sample stream.

Tests with microparticles diluted in water were also performed. Since the diffusion coefficient of PS particles suspended in water is approximately three orders of magnitude lower than that of the food dye [17, 18], it was possible to use lower flow rates, decreasing the particles' flow velocity and thereby facilitating the analysis. In order to test the spatial focusing efficiency of microparticles in the device, their position distribution along the channel was obtained through image analysis using ImageJ. The results for two different flow rate ratios are depicted in figure 6.

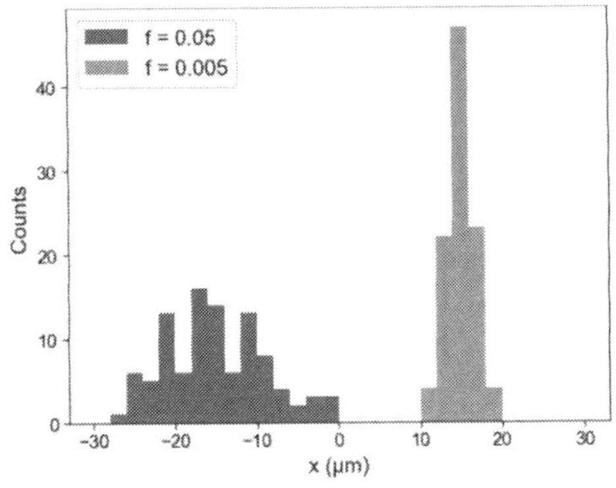

Fig. 6. Histogram of the position of the PS particles along the channel's width for f=0.05 and f=0.005, centered at x = -15 and 15 µm, respectively. The particle solution flow rate was q = 0.05 µL/min for both histograms.

When the sample flow rate, q, is kept constant while the sheath flow rate, Q, is varied, resulting in a change in the flow rate ratio, f, the focused stream width, Wf clearly increases as f increases. Furthermore, considering the width confining 90% of the measured particle position, it was shown that by employing a flow rate ratio of 0.005 it was possible to spatially confine the PS particles within a 6 µm thick layer. Smaller flow rate ratios result in a narrower focal spot, whereas smaller ratios lead to a more dispersed microparticle solution. In the context of this work, the more spatially focused the particles are, the easier it will be to use light to interact with the microparticles and to perform single-particle analysis.

The presented study proved the capability of the fabricated device for horizontal two-dimensional HFF. However, here the focusing was only performed in one direction meaning that the particles can still freely move vertically within the channel. To meticulously achieve single-particle measurements it is crucial to confine the particles in all directions, both in horizontal and vertical directions.

3.1.2 Vertical Focusing

Similarly to the tests performed to test the horizontal HFF, the microfluidic device to test vertical HFF (fig. 2) was firstly tested using deionized water in the sheath inlet and red food dye diluted in deionized water in the sample inlet, which makes it possible to observe flow focusing of the sample fluid. The sheath inlet was connected to a syringe pump, with 1 mL glass syringes, allowing the control of the sheath's flow rate, Q_1+Q_2. On the other hand, the sample fluid flow rate, q, was controlled by another syringe pump, with an identical 1 mL syringe. This allows the control of the ratio between the sample and sheath flow rate, $f = q/(Q_1+Q_2)$, which in turn determines the ratio between the height of the focused sample stream, Hf, and the total height of the channel, H (fig. 7).

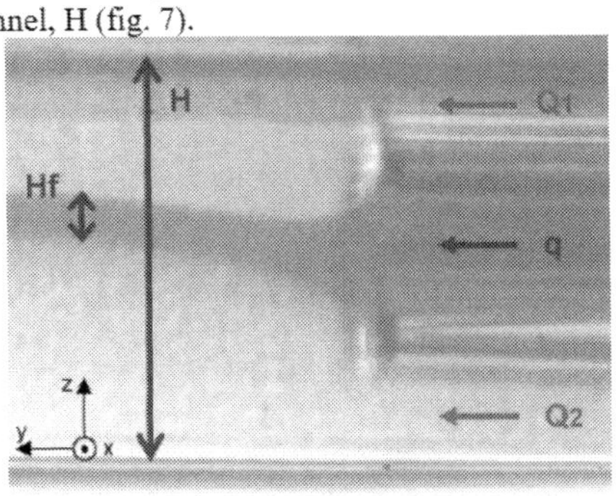

Fig. 7. Image of 2D vertical hydrodynamic focusing on the fabricated device. The total flow rate after the junction is $q+(Q_1+Q_2)$. Image was taken for $(Q_1+Q_2) = 100$ µL/min and q = 10 µL/min (f=0.1).

As shown in figure 7, the sample fluid is vertically displaced to the centre. This derives from a limited etching selectivity, which means that the subchannel at the glass surface is etched for a longer period than the bottom subchannel, resulting in two sheath subchannels of different heights and, in turn, different flow rates. This asymmetry affects the overall performance of the device, since the total flow rate (Q_1+Q_2) is not equally distributed between the two sheath subchannels. Consequently, the height of the focused stream can vary based on its position across the channel height, since the stream-wise velocity profile may not be uniform. In order to study the vertical flow focusing, the obtained results were analysed and compared to the asymmetric hydrodynamic flow focusing model, developed by Lee *et al* [16]. This model considers the position of the focused stream, giving rise to more accurate analysis, even though it was developed for a T-junction channel.

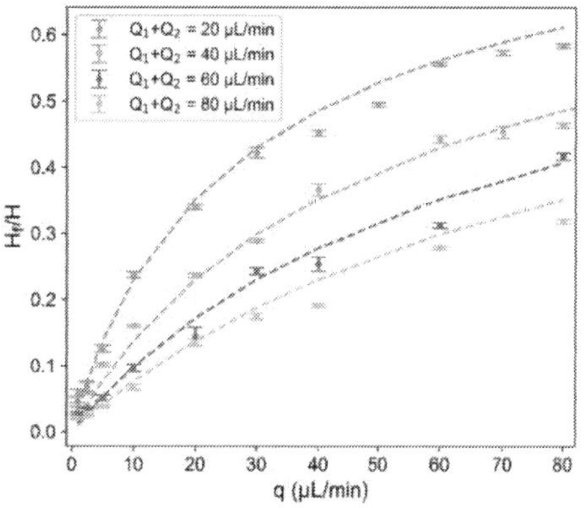

Fig. 8. Experimental results of the variation of the ratio between the focused sample stream height and the total height of the channel, Hf/H, with the sample flow rate, q, for constant sheath flow rates, Q_1+Q_2. The dashed lines represent the corresponding analytical results predicted by Lee's model.

Similarly to the analysis for the horizontal 2D flow focusing, images of the channel were taken and subsequently analysed using the software ImageJ. The experimental results are depicted in figure 8. Hf was measured 3 times for each experimental condition, with the resulting standard deviation depicted as an error bar. The dashed lines represent the analytical results calculated using Lee's asymmetric hydrodynamic flow focusing model, following a similar approach as was used for the study of horizontal HFF.

Again, the tendency of the obtained results follows the expected theoretical behaviour. When the sheath flow rate is kept constant, the height of the focused sample stream, Hf, increases as the sample flow rate is increased. For a constant q, a decrease on the sheath flow rate increases the height of the focused sample stream. It is also worth noting that here, in comparison to the results obtained for the horizontal flow focusing, the experimental errors are considerably higher. This occurs because the etching step significantly impacts the device's geometry, particularly in the vertical direction, which in turn affects flow focusing. Besides, here the angle between the channels was 0°, unlike the 90° angle used in the horizontal focusing device. This difference also contributes to deviations in focusing [19].

Further tests were conducted with PS microparticles. The particle positions along the channel were determined, and the resulting histograms are shown in figure 9. The presented results confirm the capability of the fabricated device to spatially focus particles on the vertical direction inside the channel, with the height of the focused sample stream depending on the flow rate ratio. Particularly, considering the height confining 90% of the measured particle position, by employing a flow rate ratio of 0.005 it is possible to spatially confine the microparticles to a height of around 5 µm. Since the aim is using light to manipulate particles, their position should be restrained to the illuminated volume, and therefore the width/height of the layer containing particles should be similar to the modal field diameter of the optical waveguide guiding light which will exert the optical gradient/scattering force. Consequently, a flow rate ratio of 0.005 was used (in both the horizontal and vertical HFF devices) as it focuses the particles down to a width/height of 5-10 µm. Higher focusing ratios would increase the width/height of the focal spot, whereas smaller focusing ratios are not possible due to limitations imposed by the syringe pump. Further, to obtain a flow rate ratio of 0.005, flow rates of 0.05 µL/min (sample flow) and 10 µL/min (total sheath flow) were used; larger sample and sheath flow rates would increase the particle velocity, making imaging of the particles, and eventually optical trapping, more difficult, the latter due to higher Stokes drag force. Still, it should be noted that the ideal flow rate ratio depends on the geometry and dimensions of the microfluidic channel, and on the viscosity of the sample and sheath fluids.

Lastly, we highlight that the designed 3D geometry is difficult to fabricate through conventional lithographic techniques given the requirement to machine the bulk of the material.

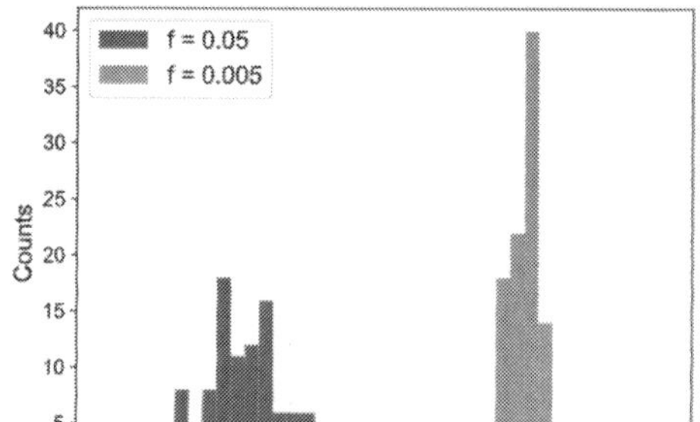

Fig. 9. Histogram of the position of the PS particles along the channel's height for f=0.05 and f=0.005, centered at z = -15 and 15 µm, respectively. The particle solution flow rate was q = 0.05 µL/min for both histograms.

3.2 Dual-Beam Optical Trap

The optical trapping process is made using the capabilities of fs-laser micromachining to integrate both optical waveguides and microfluidic channels. Since the waveguides can be written perpendicular to the fabricated channel, it is possible to create a dual-counter-propagating-Gaussian beam. To understand the origin of the optical trap, we first discuss the effect of a single waveguide on the microparticle. Given the diameter of the microparticles (3 μm) is larger than the light wavelength (1.55 μm), a particle in the path of one light beam experiences a gradient force pulling it toward the beam axis and a scattering force pushing it away from the light source. Here, with two opposing and identical divergent beams, a microparticle becomes trapped at the point of symmetry (point where resultant scattering force is null), with any displacement from this centre leading to a restoring force. Consequently, inside the channel, a trapping region is created due to the two beams coming out of the integrated optical waveguides. It should also be highlighted that the particle velocity should also be small (≈10 μm/s), otherwise the Stokes drag force can be stronger than the exerted optical force, preventing optical trapping.

The optofluidic device shown in fig. 3 was tested using 3 μm PS particles. Figure 10 shows the obtained results. First with light coming from only one of the waveguides, it was possible to see that a displacement of the particles was induced due to the radiation pressure of a 1550 nm beam coming out of the integrated waveguide. Afterwards, with light coming from both waveguides, a dual-beam trap (1550 nm) derived from the two counter-propagating Gaussian beams coming out of the integrated waveguides was created [20]. It was possible to perform optical trapping with this technique for particles flowing around the same height as the waveguides (fig. 10).

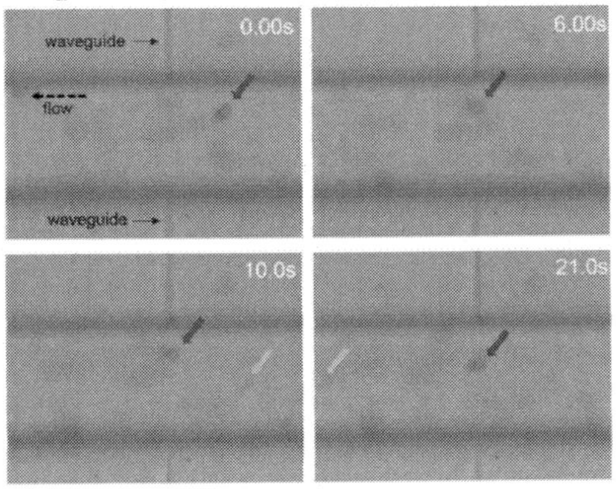

Fig. 10. Snapshots illustrating the effect of the two counter propagating Gaussian beams creating a dual-beam trap for two in-focus PS particles (red arrow). Out-of-focus particles' trajectory (light blue arrow) is not affected by the beams.

These results show our capability of performing both optical trapping and displacement due to pressure radiation on PS microparticles in an integrated monolithic device, which is crucial for the development of cell sorters and stretchers. However, this preliminary study has also proved that with such small channel widths (~36 μm) the flow rate needs to be around 1 nL/min in order to image the particles, which places several restrictions in the syringe pumps stability. Further, even though the channel was small, particles were still randomly distributed over it. Therefore, some of the particles were unaffected by the propagating light beams, which prevents their analysis. A more efficient solution would be to replace this channel by the previously described flow focusing geometries; this is currently under study.

4 Conclusion

On-silica chip 2D HFF devices, in both vertical and horizontal directions, were fabricated through fs-laser direct writing followed by chemical etching. The first tests, performed with red food dye, show that the dimensions of the focused sample stream follow the expected behaviour predicted by Lee *et al,* and can be controlled by adjusting the sample and sheath flow rates. Further tests with 3 μm PS particles show that, with a flow rate ratio of 0.005, the particles could be focused within a 6 μm layer across a 420 μm wide channel, and in a 5 μm layer across a 260 μm high channel.

Light induced movement through pressure radiation and optical trapping of microparticles was also demonstrated by integrating optical waveguides within the microfluidic chip. With this technology, further optical functions can be included for cell sensing by fluorescence detection or light scattering analysis. Combination with the previous HFF geometry can be used for single-line particle focusing near the optical waveguide depth, which paves the way for efficient single particle/cell analysis.

This project is co-financed by Component 5 – Capitalization and Business Innovation, integrated in the Resilience Dimension of the Recovery and Resiliency Plan within the scope of the Recovery and Resilience Mechanism (MRR) of the European Union (EU), framed in the Next Generation EU, for the period 2021-2026 within the scope of the HfPT project with reference 4.

References

1. S. K. Vashist, P. B. Luppa, L. Y. Yeo, A. Ozcan, J. H. Luong, Emerging Technologies for Next-Generation Point-of-Care Testing. Trends in Biotechnology, **33**, 692-705 (2015).
2. D. Tan, K. N. Sharafudeen, Y. Yue, J. Qiu, Femtosecond laser induced phenomena in transparent solid materials: Fundamentals and applications. Progress in Materials Science, **76**, 154-228 (2016).
3. R. Osellame, Femtosecond laser micromachining: photonic and microfluidic devices in transparent materials (Berlin: Springer, 2012)
4. G. Testa, G. Persichetti, R. Bernini, Optofluidic biosensing: devices, strategies, and applications. TrAC Trends in Analytical Chemistry, 117865 (2024).
5. P. Paiè, *et al*, Structured-light-sheet imaging in an integrated optofluidic platform. Lab on a Chip, **24**, 34-46 (2024).
6. N. Bellini, *et al*, Femtosecond laser fabricated monolithic chip for optical trapping and stretching of single cells. Optics Express, **18**, 4679-4688 (2010).

7. J. Qi, *et al*, A microfluidic mixer of high throughput fabricated in glass using femtosecond laser micromachining combined with glass bonding. Micromachines, **11**, 213 (2020).
8. A. A. Al-Halhouli, W. Al-Faqheri, B. Alhamarneh, L. Hecht, A. Dietzel, Spiral microchannels with trapezoidal cross section fabricated by femtosecond laser ablation in glass for the inertial separation of microparticles. Micromachines, **9**, 171 (2018).
9. F. Bragheri, *et al*, Optofluidic integrated cell sorter fabricated by femtosecond lasers. Lab on a Chip, **12**, 3779-3784 (2012).
10. P. Paiè, F. Bragheri, R. M. Vazquez, R. Osellame, Straightforward 3D hydrodynamic focusing in femtosecond laser fabricated microfluidic channels. Lab on a Chip, **14**, 1826-1833 (2014).
11. F. Storti, S. Bonfadini, L. Criante, Simplified 3D hydrodynamic flow focusing for lab-on-chip single particle study. Scientific Reports, **13**, 14671 (2023).
12. J. M. Maia, V. A. Amorim, D. Viveiros, P. V. S. Marques, Femtosecond laser micromachining of an optofluidics-based monolithic whispering-gallery mode resonator coupled to a suspended waveguide. Scientific reports, **11**, 9128 (2021).
13. F. He, Y. Cheng, Z. Xu, Y. Liao, J. Xu, H. Sun, H., X. Chen, Direct fabrication of homogeneous microfluidic channels embedded in fused silica using a femtosecond laser. Optics letters, **35**, 282-284 (2010).
14. J. Drs, T. Kishi, Y. Bellouard, Laser-assisted morphing of complex three dimensional objects. Optics express, **23**, 17355-17366 (2015).
15. D. Ting, Thermofluids: from nature to engineering (Academic Press, 2022)
16. G. B. Lee, C. C. Chang, S. B. Huang and R. J. Yang, The hydrodynamic focusing effect inside rectangular microchannels. Journal of Micromechanics and Microengineering, **16**, 1024 (2006).
17. S. Kendig, Use of microchannels to determine diffusion of food coloring in water. Microelectronics Processing Technology Fall Term (2003).
18. D. Rusu, D. Genoe, P. van Puyvelde, E. Peuvrel-Disdier, P. Navard, G. G. Fuller, Dynamic light scattering during shear: measurements of diffusion coefficients. Polymer, **40**, 1353-1357 (1999).
19. C. Kunstmann-Olsen, J. D. Hoyland, H. G. Rubahn, Influence of geometry on hydrodynamic focusing and long-range fluid behavior in PDMS microfluidic chips. Microfluidics and nanofluidics, **12**, 795-803 (2012).
20. J. M. Taylor, Counter-Propagating Gaussian Beam Traps. Optical Binding Phenomena: Observations and Mechanisms (Springer Theses, 2011).

Design and Development of a Color Picker System to Integrate in POC Device Systems.

Caterina Serafinelli [1,2,3,4], *Alessandro* Fantoni [1,3], *Miguel* Tavares Fernandes [1,3], *Elisabete* C. B. A. Alegria [1,2] and *Manuela* Vieira [1,3,4]

[1] ISEL—Instituto Superior de Engenharia de Lisboa, 1949-014 Lisboa, Portugal;

c.serafinelli@campus.fct.unl.pt (C.S.)

[2] Centro de Química Estrutural, Institute of Molecular Sciences, Instituto Superior Tecnico IST, Universidade de Lisboa, 1049-001 Lisboa, Portugal

[3] CTS—Centre of Technology and Systems and Associated Lab of Intelligent Systems (LASI), 2829-516 Caparica, Portugal

[4] DEE-FCT-UNL, Department of Electrotechnical and Computer Engineering of the Faculty of Science and Technology of the Universidade NOVA de Lisboa, 2829-516 Caparica, Portugal

Abstract: Nowadays is increasing the demand for miniaturized, user-friendly, automated, and portable sensing systems able to provide a fast and reliable response. In this context, colorimetric detection has emerged for its intrinsic advantages, such as simplicity and rapidity, but also for the outstanding development of novel materials, such as plasmonic nanoparticles, and new technologies. Here, the Color Picker system, a system reproducing in the ba has been developed and tested on a plasmonic paper. The aim is to provide a tool for a colorimetric detection that can be successively integrated in next generation diagnostic devices for real world applications.

Keywords: Color Sensor, RGB Display, AuNPs, Plasmonic Paper.

1 Introduction

During the last few years an exponentially growing attention has been focused towards the easy and fast identification of many biomarkers for several diseases so that many lives can be saved thanks to the detection of a disease at an early stage as well as facilitating the management of more personalized and efficient therapies, aiming to develop and improve personalized medicine. With this in mind, fervent activity is directed towards the advancement of Point-of-Care (POC) devices [1]: these tests enable analysis using very small sample volumes with no or little preparation, providing results in shorter times compared to routine lab tests. They offer the additional advantage of being interpreted and implemented without any special equipment or training, while they are administered externally to hospitals and clinics practice, in a home environment [2]. Furthermore, they are cost-effective and can be easily transported. In this scenario, also stimulated by the development of a large number of novel materials, colorimetric detection has blossomed due to its advantages: sophisticated or expensive instrumentation is not required because the color changes can be read by naked eyes, so they are characterized by simplicity, practicality and low cost [3]. The key issue in colorimetric detection is to convert the event of detection into a color change so that several smart materials, such as for example noble metal nanoparticles, 2D nanomaterials as graphene or reduced graphene, and hybrids composed by plasmonic particles and bidimensional material, have been developed to be exploited as sensing platform in colorimetric assays [4]. Since the earliest days of nanoparticle synthesis, the engineering of metal nanoparticles has progressed dramatically, with significant contributions in multiple fields, such as chemistry, physics, engineering, medicine and biology but including also completely new fields as microelectronics and boosting their applications [5]. The growing interest is originated not only from the bright, visible colors that can be obtained, but also because it is possible to control these colors by the fine tuning of various parameters, such as particle shape and size, state of aggregation, or the nature of the surrounding medium. Nanoparticles of noble metals, such as gold, silver, and copper display strong localized surface plasmon resonances, (LSPRs) that are oscillations of d-electrons cloud driven by an electromagnetic radiation [6] and are the source of their unusual optical properties. The resonance frequency strongly depends also on the dielectric constant of the medium around the metal nanoparticles, so that according to the changes of the refractive index the position of the LSPR is displaced. In this scenario, the sensitivity towards the variations in the refractive index is used to identify biorecognition events occurring onto the particle surface, providing a great potential for novel sensing platform [7]. Anyway, the detection of molecules using the LSPR of metal nanoparticles is commonly based on spectroscopic measurements, requiring bulky instrumentation and trained personnel. To transfer colorimetric detection based on metal nanoparticles to applications in routine diagnostic, there is still a strong request of user-friendly devices. In the present work a system able to reproduce the color detected by the TCS34725 Color Sensor in the backlight of an I2C RGB Display, the Color Picker system, has been developed with the aim to provide a tool that can be integrated in POC devices for the next generation of sensing platforms enabling a real-time, highly sensitive and accurate diagnostics. The Color Picker system, in addition, has been tested to measure the changes in the colors of a plasmonic paper (paper with AuNPs incorporated in its fiber network), arising from the different refractive index around the particle. In a further step, the sensing platform based on those novel materials, will be integrated in POC systems relying in new technologies such as lab-on-chip (LOC) hardware and remote connections based on Internet-of-Things (IoT) infrastructure.

2 Materials

2.1 Color Picker

The principal components of the Color Picker are the Color sensor I2C V2 (TCS34725FN) – Seeed and I2C 16x2 Arduino LCD with RGB Backlight Display V2.0 from DFRobot. The TCS34725 sensor and the RGB backlight display are connected by means of a breadboard and controlled by an Arduino Uno Microcontroller. The TCS34725 is a digital Light-to Digital Converter, meaning it converts light from the visible spectrum to a digital signal that is read by an external microcontroller. One of the main components is a 3x4 photodiode array, composed of red-filtered, green-filtered, blue-filtered, and clear (unfiltered) photodiodes that are coated with an IR blocking filter minimizing the IR spectral component of the incoming light and allows color measurements to be made accurately. Also, the amplified photodiode currents are simultaneously converted to a 16-bit digital value by four integrating analog-to-digital converters (ADC). Upon completion of a conversion cycle, the results are transferred to the data registers, which are double - buffered to ensure the integrity of the data. All the internal timing, as well as the low - power wait state, is controlled by the state machine. Communication of the TCS3472 data is accomplished over a fast, up to 400 kHz, two-wire I²C serial bus. The LCD1602 RGB Backlight Module is a 16x2 LCD screen with an adjustable RGB backlight color, up to 16M (2563) backlight colors in theory. That display uses an I2C communication interface meaning that it only needs 2 communication lines for the communication and backlight control and only 4 pins for the connection with a microcontroller: VCC, GND, SDA, SCL. Also, it is provided with an onboard AiP31068L LCD driver chip (LCD controller) and a PCA9633 RGB control chip (RGB controller), can display 2x16 characters and support functions like scrolling-displaying, cursor movement and backlight color adjustment. By programming the LED brightness with a PWM controller, it is possible to tune the color of the backlight display. The functional block diagrams of TCS34725 color sensor and DFRobot 1602 RGB backlight have been reported in Fig. 1A and 1B.

A)

B)

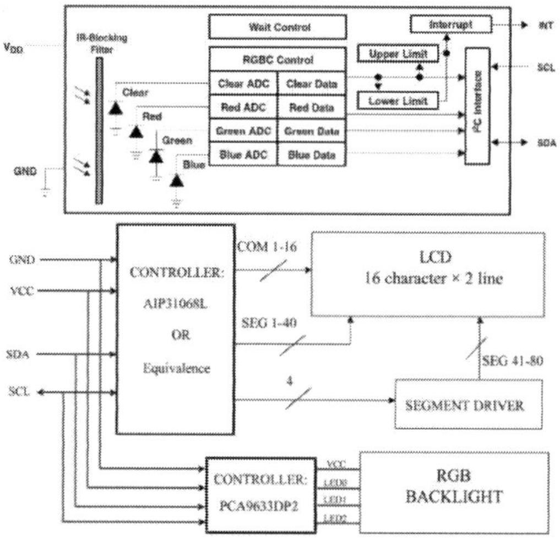

Fig. 1: A) Functional block diagram of TCS34725 sensor; B) Functional Block diagram of DFRobot 1602 Backlight display.

Also, the calibration and the tests have been performed using the Spyder Checker 24 from Datacolor and using MATLAB software to calculate the Moon Penrose inverse and the transformation matrix.

2.2 AuNPs and Plasmonic Paper

All the chemicals and reagents used in this study were analytical grade and used as received without further purification. Tetrachloroauric acid ($HAuCl_4.3H_2O$), trisodium citrate dihydrate and glycerol were purchased from Aldrich. The citAuNPs are adsorbed on laboratory filter paper (Whatman grade No. 1, 180 µm thick).

3 Methods

The Color Picker has been tested for first measuring the RGB values of the Sunflower patch of the Spider Checkr 24 at different integration Times. In this case, the values returned from the sensor are normalized to the clear. Successively, a further test has been carried out following the TCS 34725 calibration. The calibrated

RGB values returned from the sensor have been calculated according to the following formula 1

$$\begin{pmatrix} R \\ G \\ B \end{pmatrix} = (M) \times \begin{pmatrix} R' \\ G' \\ B' \end{pmatrix} \quad (1)$$

Where:

-R, G, B are the RGB values in the sRGB Color Space

-M is the transformation matrix

-R' G' B' are the responses from the RGB channel of the sensor normalized to the clear.

The transformation matrix is the following:

$$(M) = \begin{pmatrix} R_r^* & R_g^* & R_b^* \\ G_r^* & G_g^* & G_b^* \\ B_r^* & B_g^* & B_b^* \end{pmatrix} \times \begin{pmatrix} R_r' & R_g' & R_b' \\ G_r' & G_g' & G_b' \\ B_r' & B_g' & B_b' \end{pmatrix}^{-1} \quad (2)$$

Where:

$$R_r^* \quad R_g^* \quad R_b^*, G_r^* \quad G_g^* \quad G_b^* \text{ and } B_r^* \quad B_g^* \quad B_b^*$$

are the reference sRGB values for red, green, and blue Color Checker Patches,

$$R_r' \quad R_g' \quad R_b', \quad G_r' \quad G_g' \quad G_b' \text{ and } B_r' \quad B_g' \quad B_b'$$

are the measured values of the red, green, and blue Color Checker Patches normalized to the clear, and the matrix

$$\begin{pmatrix} R_r' & R_g' & R_b' \\ G_r' & G_g' & G_b' \\ B_r' & B_g' & B_b' \end{pmatrix}^{-1} \quad (3)$$

is the Moon Penrose pseudoinverse.

After the calibration, the RGB values of the Sunflower patch of the Color checker have been measured for different Integration Times (IT).

3.1 Synthesis of AuNPs

The AuNPs have been prepared by heating 98 mL of milliQ water under reflux while stirring, and at the boiling point 2 mL of trisodium citrate dihydrate solution (1 wt%) have been added. After 5 minutes, a tetrachloroauric(III) acid trihydrate ($HAuCl_4.3H_2O$) solution has been added with a final concentration in the system of 1mM and left boiling for 5 minutes. Finally, the solution turned into a red color, indicating the presence of AuNPs (CitAuNPs) and cooled down.

3.2 Plasmonic Paper preparation

A common filter paper for laboratory use has been used for CitAuNps absorption. One strip of paper (about 20x20 mm) has been immersed in the solution of the pre-synthesized AuNPs and left overnight at 4 °C, as proposed in the work of Singameni [8]. The paper strip was taken out and washed with MilliQ water. As final step, the paper has been dried for about 45 minutes at 44 °C into the oven.

3.3 Test of refractive index sensitivity

To test the sensitivity towards variations in the refractive index, the plasmonic paper has been wetted in water and glycerol and then the RGB values measured with the Color Picker system.

4 Results and Discussions

4.1 citAuNPs Characterizaction

The citAuNPs have characterized by UV-Vis spectroscopy (Fig. 2):

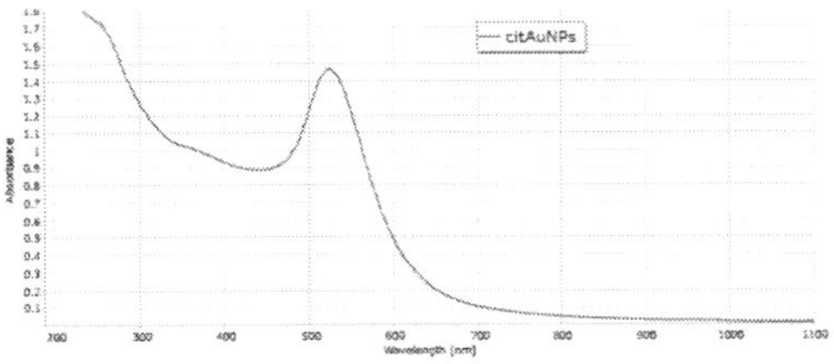

Fig. 2: UV-Vis. Spectra of the citAuNPs used to prepare the plasmonic paper.

The LSPR is centered at about 524 nm and from spectral data a size of 15-20 nm is evaluated for citAuNPs.

4.2 TCS34725 Sensor calibration

The data obtained from the first test have been reported in table 1:

Table 1. RGB values measured for the Sunflower patch of the Color Checker normalized to the clear.

	Red	Green	Blue	Clear
IT 24 ms	110	91	46	125
IT 120 ms	110	93	46	622
IT 614 ms	110	93	46	3188

In this test, the RGB data returned by the sensor have been normalized to the clear, so that, as it is possible to observe, they are independent from the Integration Times and are almost the same. The color space of the RGB data normalized to the clear is not defined, so that is not possible a comparison with the reference values that are defined in the L*a*b*, sRGB and Adobe RGB color spaces, losing some information. To acquire a more accurate estimate of the RGB values and to define the color space of the data returned from the sensor, a calibration of the TCS34725 Color Sensor has been performed so that the RGB output values are in the sRGB color space. The data obtained in this test have been reported in table 2:

Table 2. RGB data obtained with the TCS34725 calibration.

	Red	Green	Blue	Clear
Sunflower patch Color Checker (sRGB Color Space)	**238**	**158**	**25**	
IT 24 ms	148	81	35	125
IT 120 ms	150	81	36	623
IT 614 ms	151	81	35	3218

From the data reported in table 2 it is possible to observe that at increasing of the Integration Time, the difference between the measured values in the sRGB color space is quite small. Also, the measured values are different from the reference values for the Sunflower patch of the Color Checker. The tiny difference in the measured values for different Integration times may be motivated by the normalization to the clear. In fact, the data used in Eq. 1 for the calibration, are normalized to the clear, so that they are independent from the Integration time. Also, the difference with the reference values of the Color Checker may be justified with the different illuminant used to measure the references for the Color Checker and the color of the Sunflower patch in the experiment. In the case of the Color Checker, the reference values have been measured using a D50/2 degrees illuminant, whereas the illuminant used in the experiment to measure the sRGB values of the Sunflower patch is the led integrated in the TCS34725 sensor. In Fig. 3 are reported A) the spectral power distribution for CIE D50 light source (line 1) [9], and B) the measured spectrum of the LED integrated into the TCS34725 sensor.

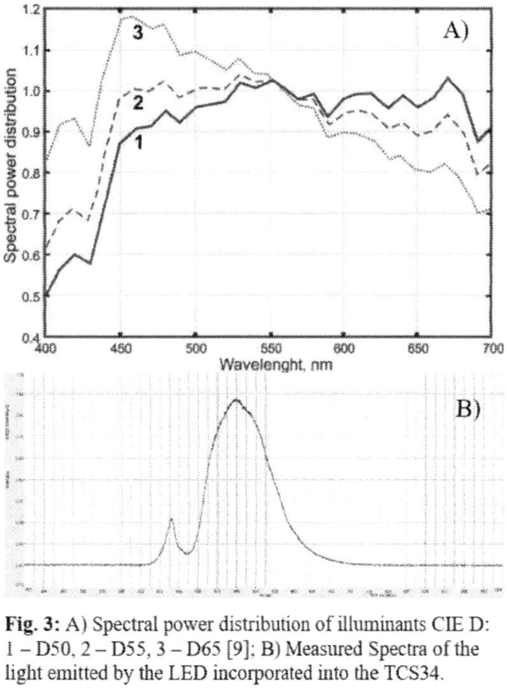

Fig. 3: A) Spectral power distribution of illuminants CIE D: 1 – D50, 2 – D55, 3 – D65 [9]; B) Measured Spectra of the light emitted by the LED incorporated into the TCS34.

It is possible to observe that whereas the spectral power distribution for the CIE D50 illuminant is a broad band, the spectrum of the LED presents two peaks, whose characteristics are reported in table 3:

Table 3. Data of the spectrum emitted from the LED integrated into the TCS 34725 sensor

	CENTROID WAVELENGTH [nm(AIR)]	PEAK WAVELENGTH [nm(AIR)]	PEAK LEVEL [INTENSITY]	FWHM [nm(AIR)]	ΔPEAK POSITION [nm(AIR)]	OFFSET PEAK POSITION [nm(AIR)]
FIRST PEAK	451.321	451.464	0.275	15.529	0.000	0.000
SECOND PEAK	566.961	562.959	0.972	107.819	111.495	111.495

Currently the majority of LEDs use phosphor conversion (phosphor-converted LEDs or pcLED). White light is generated by mixing a portion of the blue light emitted directly from an LED with a Gallium Nitride (GaN) substrate (smaller peak at 451 nm) with the light converted to yellow emitted by the phosphors (broad peak at 567 nm). The phosphors are located on the emitting surface of the LED, inside the encapsulant, or away from the LED (remote phosphors). The differences in the spectrum are then reflected in the different sRGB values detected. Although the output values are different from the reference sRGB, after calibration the sensor acquired an increased precision producing the same output for the same input: measuring for example the Sunflower patch (but also other objects such as, for example, the cover of the notebook) the sensor always returned the same values even after multiple measurements. In Fig. 4A the picture of the overall Color Picker system has been reported, whereas in Fig. 4B is shown the picture of the Color Picker while sensing the blue cover of a notebook. It is possible to note how the backlight of the DFRobot RGB display turned blue as the cover. Finally, in Fig. 4C the Spyder Checkr24 picture with the patches used for the calibration and the test has been provided.

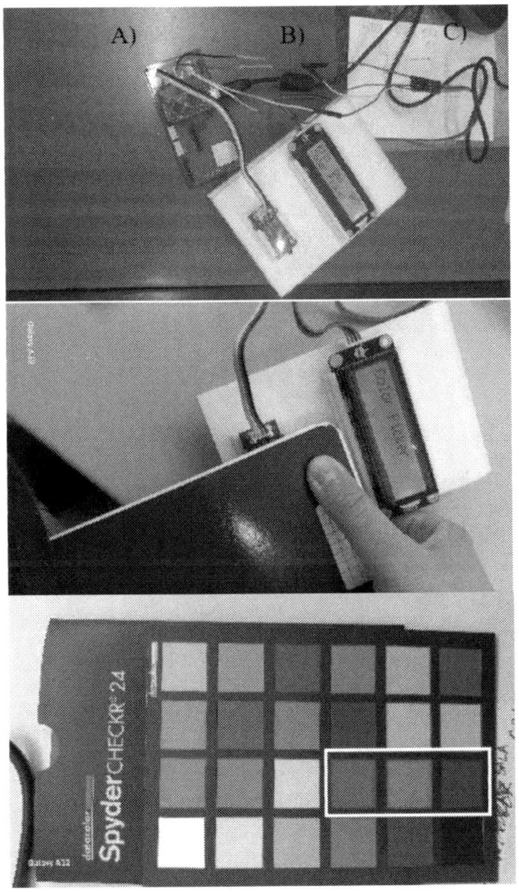

Fig. 4. A) picture of the Color Picker system with the TCS34725 sensor, the RGB backlight DFRobot display, and the microcontroller Arduino Nano assembled onto a board; B) esting the Color Picker with the blue cover of a notebook; C) the Spyder Checkr 24 used to calibrate and to test the Color Picker system. In the picture the primary red, the primary green and the primary blue patches used for the calibration along with the Sunflower patch used to test the device have been indicated.

4.3 Test on Plasmonic Paper

The Color Picker device with the calibrated TCS34725 sensor successively has been employed to test the changes in the color of a plasmonic paper strip arising from the different local dielectric environment. In figure 5 the picture of the plasmonic paper A) in its dry state and wet with B) water and C) glycerol is shown. In figure 6 is illustrated the display of the Color Picker device with the color of plasmonic paper in the backlight.

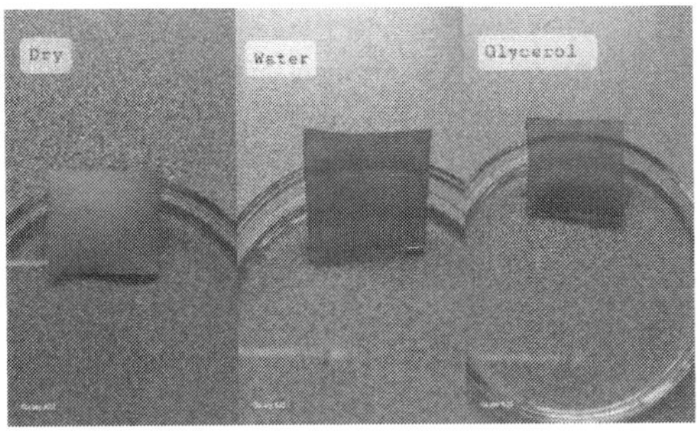

Fig. 5. Plasmonic paper A) dry B) wet with water C) wet with Glycerol

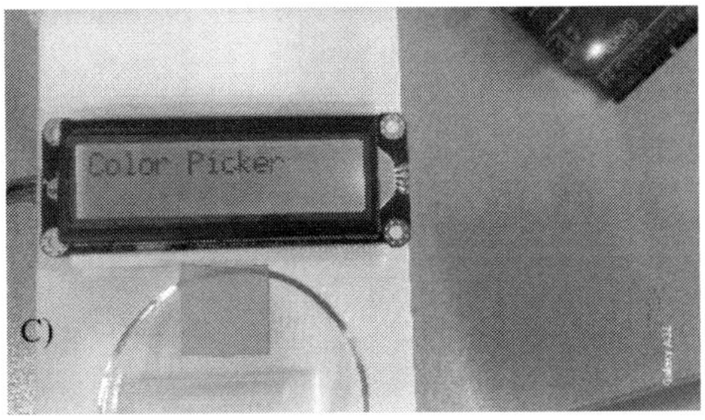

Fig. 6. Backlight of the I2C 16x2 LCD Display reproducing the color of the plasmonic paper detected by the TCS34725 color sensor.

The RGB values obtained from the sensor, along with the refractive index of the solvent embedding the plasmonic paper have been reported in table 4.

Table 4. RGB values for plasmonic paper

	Refractive Index of the embedding medium	R	G	B
Dry Plasmonic Paper	Air (1)	85	83	78
Paper in Water	1.33	90	80	79
Paper in Glycerol	1.47	93	81	78

From the values measured by the sensor it is possible to observe a different trend for the separated RGB channels. Whereas the values of the red channel increase when the refractive index of the medium increases, the values of the green and blue channels do not show a specific trend, are almost identical for the different refractive indices. On this basis, it has been possible to demonstrate that the red channel is responsive to the changes in the refractive index of the local dielectric environment around the surface of a plasmonic AuNPs grafted onto the paper structure.

5 Conclusions

In summary, a device able to reproduce in the backlight of a RGB display the color detected by the TCS34725 color sensor, the Color Picker system, has been developed aiming to create an output tool to be integrated into advanced sensing platforms for a more accurate and rapid diagnostic. The Color system has been tested before and after calibration, finding an improved precision after calibration providing the same values of the sRGB values even after numerous measurements. In addition, the TCS34725 sensor calibration offers the additional advantage to define the Color Space of the output values (sRGB). Differences in the measured sRGB values with the reference of the color checker have been found, depending on the different light source used for the measurements. After calibration, the Color Picker System has been tested to measure the differences in the color of plasmonic paper when immersed in solvents with different refractive index. Whereas the values of the red channel increase at increasing of the refractive index, the values of the green and blue channel are almost identical, thus proving that the Color Picker system is able to detect changes in color of plasmonic substrates arising from variations in the dielectric environment around plasmonic nanoparticles. In further step, the Color Picker device will be employed to detect differences in the color of plasmonic nanoparticles arrays arising from binding events between analytes and receptor localized onto the nanoparticle surface for a colorimetric detection, with the aim to be incorporated into POC devices for an easy and real time diagnostic.

Acknowledgments

This research was supported by Portuguese national funds provided by FCT – Fundação para a Ciência e Tecnologia, through the Portuguese FCT program, Center of Technology and Systems (CTS) UIDB/00066/2020 / UIDP/00066/2020, and by project IPL/IDI&CA2023/LUMINA_ISEL. The authors are grateful to Carlo Fantoni (FCUL-Lisboa) for his helpful support in the practical implementation of *Color Picker system*.

References

1. Purohit Buddhadev, Vernekar Pramod R., Shetti Nagaraj P., Chandra Pranjal. Biosensor nanoengineering: Design, operation, and implementation for biomolecular analysis. Sens. Int. 1, 100040 (2020)
2. Maria Soler, Cesar S. Huertas & Laura M. Lechuga. Label-free plasmonic biosensors for point-of-care diagnostics: a review. Expert Review of Molecular Diagnostics. 19, 71-81 (2019)
3. Yujun Song, Weili Wei, Xiaogang Qu. Colorimetric Biosensing Using Smart Materials. Adv. Mat. 23, 4215-4236 (2011)
4. Caterina Serafinelli, Alessandro Fantoni, Elisabete C. B. A. Alegria and Manuela Vieira. Hybrid Nanocomposites of Plasmonic Metal Nanostructures and 2D Nanomaterials for Improved Colorimetric Detection. Chemosen. 10, 237 (2022)
5. Lu Wang, Morteza Hasanzadeh Kafshgari, Michel Meunier. Optical Properties and Applications of Plasmonic-Metal Nanoparticles. Adv. Funct. Mat. 30, 2005400 (2020)
6. Ben Tadgell and Luis M. Liz-Marzán. Probing Interactions between Chiral Plasmonic Nanoparticles and Biomolecules. Chem. an Eur. Jour. 29, e202301691 (2023)
7. Tianping Xu and Zhaoxin Geng. Strategies to improve performances of LSPR biosensing: Structure, materials, and interface modification. Bios. and Bioelec. 174, 112850 (2021)
8. Sirimuvva Tadepalli, Zhifeng Kuang, Qisheng Jiang, Keng-Ku Liu, Marilee A. Fisher, Jeremiah J. Morrissey, Evan D. Kharasch, Joseph M. Slocik, Rajesh R. Naik and Srikanth Singamaneni, Scientific Reports, 2015, Vol. 5, Article number: 16206: Peptide Functionalized Gold Nanorods for the Sensitive Detection of a Cardiac Biomarker Using Plasmonic Paper Devices.
9. Alexander P. Sergeev, Dmitry A. Tarasov, Sergey Y. Arapov, Svetlana P. Arapova. Qualimetric Researches of Educational Resources: Standardizing of Light Conditions in the Light Booth. Procedia - Social and Behavioral Sciences. 174, 1285-1291 (2015)

Fully Connected Feedforward Neural Network for the Prediction of Amorphous Silicon Grating Couplers Efficiency

Daniel Almeida [1,2,3*], *Alessandro* Fantoni[1,2], *João* Costa[1,2], *Manuela* Vieira[1,2,3], and *José* Fonseca[2,3]

[1]ISEL-Instituto Superior de Engenharia de Lisboa, Instituto Politécnico de Lisboa, Portugal
[2]Center of Technology and Systems (UNINOVA-CTS) and Associated Lab of Intelligent Systems (LASI), 2829-516 Caparica, Portugal
[3]NOVA School of Science and Technology, NOVA University of Lisbon, 2829-516, Caparica, Portugal

Abstract. Photonic circuits are an enabling technology for the development of novel solutions in different fields such as healthcare, quantum computing, neural networks, communications, and manufacturing. Interconnections between devices and systems require low-loss light coupling strategies. Grating couplers are a promising solution to couple light between photonic circuits and optical fibres due to their off-plane coupling capabilities. Hydrogenated amorphous silicon (a-Si:H), which can be deposited by PECVD over a substrate of silica or glass, is a suitable low-cost solution for the production of such light coupling devices. In this work we developed, trained and tested a fully connected feedforward neural network for coupling efficiency prediction in a-Si:H grating couplers. The light coupling gratings were simulated by two-dimensional finite-difference time-domain (FDTD) analysis and field distributions were analysed with the Finite Element Method (FEM). Simulated gratings include non-apodized, linear and quadratic refractive index variation designs featuring full or partial etching, operating at 1550 nm. Not featuring any type of bottom reflector, the couplers exhibit coupling efficiencies up to about 40 % (~ -4 dB). The neural network multiclass grating coupler efficiency classifier was trained with over 3000 simulation results, reaching an accuracy over 85%, for coupling efficiencies between 0 and 30%+.

1 Introduction

Silicon photonic circuits are basic building blocks of several devices and systems finding applications in the fields of healthcare [1], communications [2], quantum computing [3], neural networks [4] and manufacturing. The design, simulation and optimization of photonic devices are expensive and time-consuming tasks, typically requiring high computational power. Of the three tasks, optimization can become the most challenging, due to the fact it usually requires the two previous steps to be redone. One of the simplest approaches consists in varying design parameters using an optimizer, which can be used to narrow down the possibilities and isolate the variables which are critical to achieve the best possible device performance, usually under human supervision.

The high computational power of modern graphics processing units (GPUs) and central processing units (CPUs) have the potential for being applied in highly demanding tasks, which are time and resource consuming, like photonic circuits' optimization. Artificial intelligence (AI), machine learning (ML) and particularly neural networks (NNs) are interesting tools for the research and development of photonic devices [5, 6], due to their power in the prediction of the outcomes of the typical workflow or in a reverse approach [7]. In other words, NNs can be used to either predict the performance of a given design or to start with an intended result and generate a possible candidate design. As examples, we have the works of Hammond and Camacho (2019) [8], which have successfully employed neural networks for modelling strip waveguides and chirped Bragg gratings and Miyatake et al. (2020) [9], that explored the utilization of NNs and the covariance matrix adaptation evolution strategy (CMA-ES), for the optimization of silicon-on-insulator (SOI) grating couplers (GCs). The particle swarm optimization method was successfully applied by Vitali et al. (2022) [10], for the optimization of dual-level silicon nitride grating couplers, based on two gratings (fully and partially etched) featuring opposite sign apodization.

Grating couplers, due to their off-plane coupling nature and high efficiency on near-infrared (NIR) wavelengths [11], constitute a flexible solution for photonic devices' interconnections. In this study, we trained a feedforward fully-connected neural network, for the prediction of the coupling efficiency of grating couplers composed of hydrogenated amorphous silicon (a-Si:H), deposited over a silicon substrate by low temperature plasma-enhanced chemical vapor deposition (PECVD), a low cost fabrication technique [12]. The grating couplers operate at the 1550 nm wavelength, both fully etched, and partially etched designs were simulated by 2D-FDTD. This work intends to demonstrate the applicability of artificial intelligence

* Corresponding author: dalmeida@deetc.isel.ipl.pt

(AI) and machine learning (ML), for the prediction of photonic circuits behaviour based on their design and setup parameters. By using neural networks to predict the behaviour of photonic components based on their design and configuration parameters it is possible to optimize and simplify the design procedure of new photonic solutions that can achieve better efficiency and performance.

2 Theory and Methodology

2.1 Amorphous Silicon Grating Couplers

The simulated grating couplers consist of a 220 nm thick hydrogenated amorphous silicon layer over a silica substrate, no bottom reflectors of any kind are included. Three different types of grating couplers are studied in this report: non-apodized gratings with fill-factors greater than 50 % (80, 85 and 90 %), apodized gratings featuring linear (negative slope) refractive index variation, and apodized gratings based on the quadratic variation of the refractive index [13]. All grating couplers and access waveguides have the same length, of 20.8 and 12 μm, respectively. Etching ratio was defined as 1 and 0.5 (100 % and 50 %), resulting in etch depths of 220 and 110 nm.

A total of 4950 bidimensional (2D) FDTD simulations were performed using the RSoft software package (Synopsys, Inc.), via University Donation Program [14]. Six input variables were swept, and the resulting coupling efficiency was measured, considering waveguide loss.

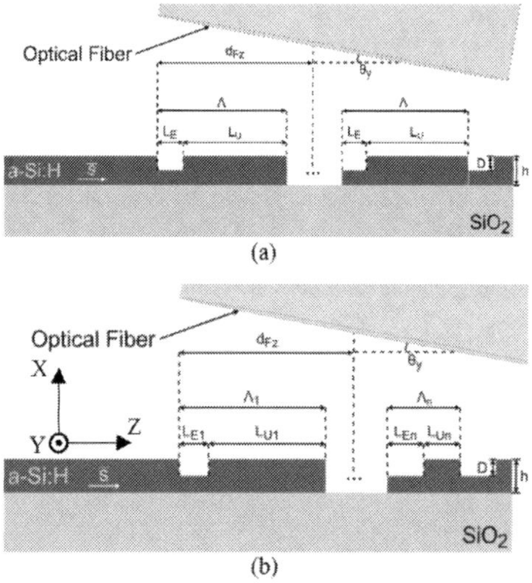

Fig. 1. Representation of the access waveguides, initial and final sections of the simulated grating couplers, showing geometrical design parameters and optical fibre positioning. Two types of grating coupler are represented: (a) Non apodized grating coupler with fill factor greater than 50 %; (b) Grating coupler with apodization, the fill factor and period (and consequently the refractive index) vary over the grating's length. The representation depicts coupler devices with a etch ratio of 50 % (110 nm etch depth).

Table 1. List of simulation variables and respective variation.

Parameter	Min.	Max.	Step
Coupling Angle (θ_y)	5°	15°	1°
Min. Fill Factor (F_{min})	50 %	70 %	5 %
Max. Fill Factor (F_{max})	80 %	90 %	5 %
Fiber Distance (d_{Fz})	3.2 μm	7.2 μm	1 μm
Variation Order (k)	0	2	1
Etch Ratio ($r = D/h$)	0.5	1	0.5

Grating coupler design and setup parameters that are fed to the NN are given as a span of integer and floating-point values, comprising start value, step and final value, as shown in Table 1. There are 7 input parameters in total. Grating coupler design parameters are diffraction angle (integer), maximum fill-factor (integer), minimum-fill factor (integer), effective refractive index variation order (integer), etch depth (float) and grating coupler height (float). The last two parameters are used in the calculation of the etch ratio (r). A setup parameter, the optical fibre offset along the grating coupler's length is also included as an input parameter, expressed as a floating-point value. Grating coupler and coupling configuration parameters are represented in Fig. 1, for designs with (b) and without (a) apodization. In Table 1, the simulation variables are presented along with the corresponding variations. The coupling angle (θ_y), or diffraction angle, corresponds to the angle formed between the plane of deposition and the coupling facet of the optical fibre, around the y axis, measured in units of degree (°). The fill factor is given as a ratio (between 0 and 1). Minimum fill factor (F_{min}) is only applicable to apodized gratings, maximum fill factor (F_{max}) is relevant for all grating designs. The fibre distance (d_{Fz}), corresponds to the distance measured between the grating's starting point (access waveguide's end) and the center of the optical fibre coupling facet, along the z axis, represented in units of micrometre (μm). Variation order (k) is the order of the grating coupler apodization along the z axis (0 means no apodization, 1 stands for linear refractive index variation and 2 for quadratic refractive index variation). The etch ratio (r), corresponds to the fraction of the etch depth (D) over the grating's height (h) and can take values between 0 and 1. The generic expression of the effective refractive index of the grating coupler along the z axis is given in (1).

$$n_{eff}(z) = \begin{cases} n_{eff}(0), & k = 0 \\ m \times z + n_{eff}(0), & k = 1 \\ a \times z^2 + b \times z + n_{eff}(0), & k = 2 \end{cases} \quad (1)$$

In expression (1), k is the variation order, $n_{eff}(0)$ corresponds to the grating's effective refractive index at its starting point, m is the slope in the linear variation of the refractive index, in this work it always has negative sign, a and b are the coefficients of the quadratic function. Since the gratings covered in this work have decreasing refractive index along the z axis, a is always positive and b is always negative. The independent variable, z, is the distance along the z axis, in nanometer (nm), measured from the grating coupler's starting point.

2.1.1 Non-apodized Grating Couplers

The simplest grating coupler studied in this work is the non-apodized grating coupler with fill factor greater than 0.5 (or 50 %). The coupler device is constructed with the given values of fill factor (F_{max}) and coupling angle (θ_y), in this case, the variation order, k is defined as 0. The effective refractive index of the grating coupler is given by (2) [11]:

$$n_{eff} = F \times n_U + (1 - F) \times n_E \tag{2}$$

In expression (2), F, corresponds to the fill-factor, n_U is the effective refractive index of the unetched grating segment, which was calculated using FemSIM from RSoft (*Synopsys, Inc.*), as approximately 2.8209, the same value applies for the access waveguide. The parameter n_E corresponds to the effective refractive index of the etched grating segment, which was also obtained using FEM simulation, having a value of 2.1610, for a partially etched GC, with a etch depth of 110 nm (0.5 etch ratio), or 1.0000 for a fully etched GC, with an etch depth of 220 nm (1.0 etch ratio). The effective refractive index values were calculated considering the hydrogenated amorphous silicon GUTL model, as proposed by Fantoni *et al.* (2017) [15]. With the effective refractive index it is now possible to obtain the grating coupler's period from its effective refractive index, coupling angle and operating wavelength (in vacuum), expression (3) [11]:

$$\Lambda = \frac{\lambda_0}{n_{eff} - \sin\theta} \tag{3}$$

Where λ_0 is the wavelength in vacuum in units of nanometer (nm), θ (or θ_y in this case) is the coupling angle in degree (°) and Λ is the grating period in the same units as the wavelength (nm). The etched and unetched segment lengths are given by (4) and (5):

$$\mathrm{L}_U = F \times \Lambda \tag{4}$$
$$\mathrm{L}_E = (1 - F) \times \Lambda \tag{5}$$

Having all geometric design parameters of the non-apodized grating coupler, it can now be constructed, see Fig. 1(a).

2.1.2 Grating Couplers with Apodization

For grating couplers featuring apodization, expressions (2) and (3), must be modified to take into account the refractive index variation along the grating's length:

$$n_{eff}(z) = F(z) \times n_U + (1 - F(z)) \times n_E \tag{6}$$
$$\Lambda(\mathrm{z}) = \frac{\lambda_0}{n_{eff}(z) - \sin\theta} \tag{7}$$

Now, the effective refractive index (n_{eff}), the fill-factor (F), and the period Λ, vary over the grating's length. In (6) and (7), z corresponds to the distance between an arbitrary point along the grating's length and the grating's starting point, measured in nanometer (nm). The other variables have the same values as described in the previous subchapter. Since the refractive index decreases over the coupler's length, and $n_U > n_E$, the maximum fill-factor (F_{max}) corresponds to the fill-factor of the first grating period ($F(0)$). Likewise, the minimum fill-factor (F_{min}) is the fill-factor of the last grating period ($F(L_{GC})$), where L_{GC} corresponds to the length of the grating coupler. To obtain the period (Λ(z)) and the fill-factor ($F(z)$) for the remaining grating periods, expression (1) is used, with $k = 1$, for 1st order (linear) refractive index variation, and $k = 2$, for 2nd order (quadratic) refractive index variation. Expression (8) is employed to obtain the linear variation slope ($k = 1$):

$$\mathrm{m} = \frac{n_{eff}(L_{GC}) - n_{eff}(0)}{L_{GC}} \tag{8}$$

Where $n_{eff}(L_{GC})$, corresponds to the effective refractive index at the grating coupler's end, $n_{eff}(0)$ is the effective refractive index at the grating's starting point and L_{GC} is the length of the grating coupler in nanometer (nm). The slope of the GC effective refractive index function (m) is given in nm^{-1}. For the quadratic refractive index variation ($k = 2$), the following system of linear equations is defined (9):

$$\begin{cases} n_{eff}(L_{GC}) - n_{eff}(0) = a \times L_{GC}^2 + b \times L_{GC} \\ L_{GC} = -\dfrac{b}{2a} \end{cases} \tag{9}$$

Solving the system of linear equations to a and b, the following expressions result (10):

$$\begin{cases} a = \dfrac{n_{eff}(0) - n_{eff}(L_{GC})}{L_{GC}^2} \\ b = -2a \times L_{GC} \end{cases} \tag{10}$$

The coefficient a is given in units of nm^{-2} and the coefficient b is represented in units of nm^{-1}. With the coefficients of the quadratic function or the slope of the linear function it is possible to obtain $n_{eff}(z)$ from expression (1), which can be used to obtain all grating periods, by using expression (7) and setting z to the starting point of the nth grating period. The fill-factor can be obtained from (11), the unetched and etched GC segment lengths from (4) and (5), respectively:

$$F(z) = \frac{n_{eff}(z) - n_E}{n_U - n_E} \tag{11}$$

Having all geometric design parameters of the apodized grating couplers, these can now be constructed, refer to Fig. 1(b).

2.1.3 Optical Fiber and Photonic Waveguide

The simulated optical fibre employed in the grating couplers' efficiency performance tests is an SMF-28, which is commonly employed on the optical C-band

(1530-1565 nm), due to its ultra-low propagation loss, which is typically under 0.3 dB/km [16].

Typical core refractive index is approximately 1.4682 at 1550 nm, with a core-cladding index difference of 0.36% [16], which results in a cladding refractive index of about 1.4629 at 1550 nm. The SMF-28 has an average core diameter of 8.2 µm and a typical cladding diameter of 125 µm, the mean field diameter (MFD) is typically 10.4 µm at 1550 nm [16]. Due to this fact the grating length was set as twice the optical fibre MFD, or 20.8 µm (20800 nm) and the average fibre facet centre distance along the z axis, measured from the GC's starting point was set at one half MFD, or 5.2 µm (5200 nm).

The grating coupler efficiency corresponds to the overlap between the diffracted field at the fibre facet and the fundamental transverse electric (TE) mode of the optical fibre, expressed as a ratio (or percentage).

The effective refractive index of the fundamental transverse electric mode (TE_{00}) of the optical fibre, operating at a wavelength of 1550 nm is approximately 1.4669, the electric (E_y) and magnetic (H_x) field distributions are represented in Fig. 2(a) and (b), respectively.

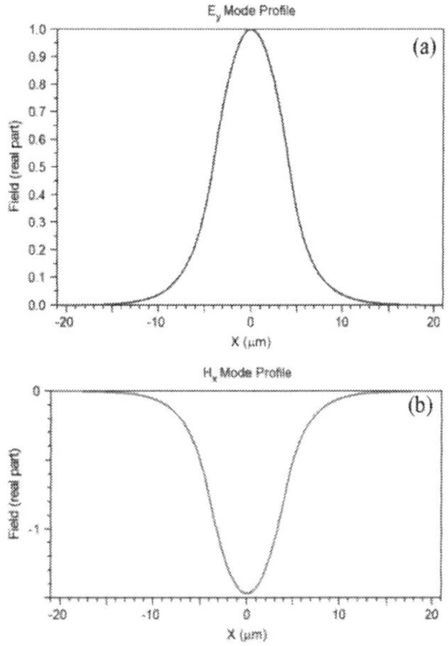

Fig. 2. Real component of the electric (a) and magnetic (b) fields of the simulated SMF-28 optical fibre, for the fundamental TE mode. In the horizontal axes, zero represents the centre of the optical fibre. Field amplitude, having its scale represented on the vertical axes is represented in arbitrary units.

The effective refractive index of the 220 nm a-Si:H photonic access waveguide was also calculated by the Finite Element Method (FEM), its value is approximately 2.8209, with an extinction coefficient of about 3.085×10^{-8}, for the quasi-TE fundamental mode (TE_{00}). The distributions of the electric (E_y) and magnetic (H_x) fields on the waveguide are represented in Fig. 3(a) and (b), respectively.

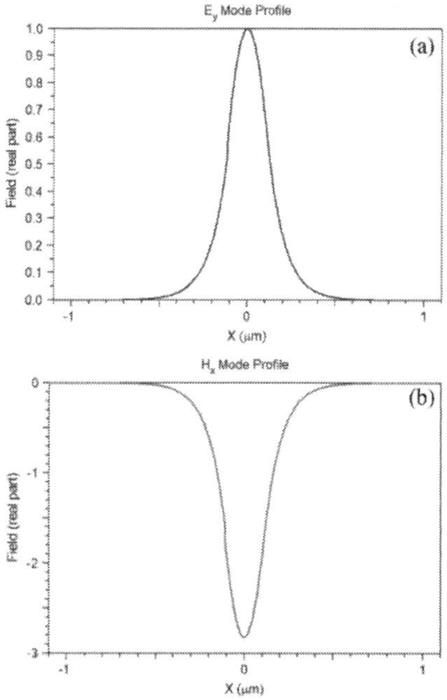

Fig. 3. Real component of the electric (a) and magnetic (b) field components of the photonic waveguide, for the fundamental quasi-TE mode. Amplitude, represented on the vertical axis is in arbitrary units. The horizontal scale represents the position, relative to the centre of the waveguide (zero), the a-Si:H layer is deposited from the right-side of the graphs over the silica layer, which starts at -0.11 µm, the air (cover) starts at +0.11 µm.

2.2 Neural Network

A feedforward fully connected neural network (NN) was trained with the input datasets from Table 1 and labeled with the efficiency results from the 2D-FDTD simulations. Input values were converted to floating point and normalized between 0 and 1, to prevent neuron saturation. The neural network implements a multiclass classifier, the labels were divided into 7 classes and fed to the NN as integer values between 0 and 6, having into account the GCs efficiency range:

- 0 to 5% (exc.),
- 5 to 10% (exc.),
- 10 to 15% (exc.),
- 15 to 20% (exc.),
- 20 to 25% (exc.),
- 25 to 30% (exc.),
- 30%+

The neural network features 2 layers with rectified linear unit (ReLU) activation functions, 2 with 2048 output neurons (Layer 1 and Layer 2) and a final layer with 7 output neurons (Layer 3), one per class, implementing a softmax activation function, which gives a statistical probability per each class, Fig. 4. The optimizer which yielded better implements the Adam (adaptive moment estimation) algorithm. For this classification problem the loss function of choice was defined as categorical crossentropy (i.e. the outputs are given as a fraction, whose sum equals 1). The learning rate was kept at the Keras' default value, of 0.001. Batch size was set at 512, the power of two which resulted in the best training performance. The number of hidden layers was defined as 2, based on testing, each layer as 2048 neurons, a value corresponding to the power of two which yields the best neural network training and validation results, without excessively increasing complexity and training time.

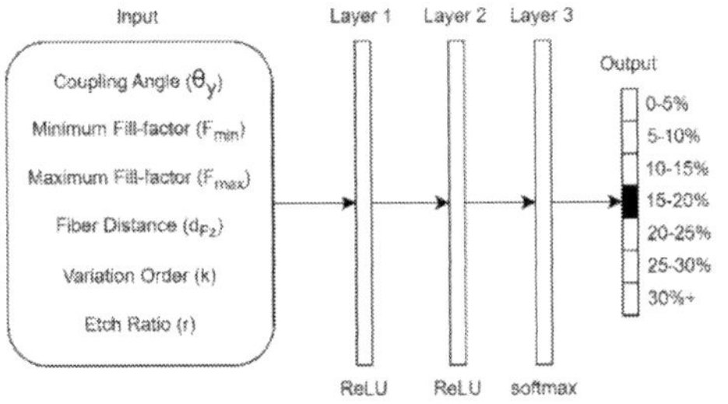

Fig. 4. Simplified diagram of the neural network used for grating coupler efficiency estimation based on design and setup parameters.

The simple feedforward neural network (NN) configuration was chosen to achieve the best computational cost vs. performance compromise. Due to its simplicity this NN is easier to replicate than more complex architectures. The NN was able to achieve a training accuracy in excess of 95 % and over 90 % validation accuracy. The NN is expected to run in modern central processing units (CPUs) and graphics processing units (GPUs), in machines with *TensorFlow* and *Keras* open-source machine learning platforms, running on Python environments. It can be trained for 100 epochs in under 8 seconds with a GeForce RTX 2080 Ti (*Nvidia Corp.*) GPU (with approximately 8.73 GiB of usable graphics memory and *TensorFlow* device compute capability 7.5) and in less than 22 seconds in an older GeForce GTX 1050 Ti (*Nvidia Corp.*) GPU (with around 2.7 GiB of usable graphics memory and *TensorFlow* device compute capability 6.1). To run *TensorFlow GPU 2.10.0* in *Nvidia Corp.* GPUs, the system must be configured with the Compute Unified Device Architecture (CUDA) API and CUDA Deep Neural Network (cuDNN). The NN architecture was tested with CUDA (*Nvidia Corp, Inc.*) 11.2.0, cuDNN (*Nvidia Corp, Inc.*) 8.1.0.77, Visual Studio Community 2019 (*Microsoft Corp.*) and Python 3.9.19.

3 Results and Discussion

3.1 Simulation Results

A total of 4950 bi-dimensional finite-difference time-domain (FDTD) simulations were performed, using the FullWAVE software tool from RSoft package (*Synopsis, Inc.*). A script written in Python language was necessary to dynamically design the grating couplers from the input parameters (Table 1). To avoid performance issues and bugs a *batch* script was written to perform variable sweep operations, instead of using the optimizer from the RSoft package (*Synopsis, Inc.*).

The simulation results are given as a fraction of input (launch) power, a positive floating-point value between 0 and 1, which can be expressed as a percentage. Two performance thresholds for apodized GC efficiency were defined for fully and partially etched gratings, of -3.9 dB (≈ 40.74 %) and -4.32 dB (≈ 36.98 %), respectively. Having into account such baselines, 7 fully etched and 5 partially etched designs were selected as achieving the top performance of the simulations' batch, Table 2.

Table 2. List of inputs which result in the best efficiencies for the apodized grating couplers, grey filled lines represent partially etched GCs, and white filed lines fully etched GCs. The meanings of the simulation/NN inputs are represented in Table 3.

Sim/NN Inputs						Efficiency
θ_y	F_{min}	F_{max}	d_{Fz}	k	r	
11°	55%	90%	5.2 µm	2	0.5	-4.311 dB
12°	50%	90%	5.2 µm	2	0.5	-4.309 dB
12°	55%	90%	5.2 µm	2	0.5	-4.312 dB
14°	55%	90%	6.2 µm	2	0.5	-4.297 dB
14°	60%	90%	6.2 µm	2	0.5	-4.316 dB
15°	55%	90%	6.2 µm	1	1	-3.827 dB
15°	60%	90%	6.2 µm	1	1	-3.751 dB
15°	60%	90%	7.2 µm	1	1	-3.887 dB
15°	65%	90%	6.2 µm	1	1	-3.759 dB
15°	65%	90%	7.2 µm	1	1	-3.691 dB
15°	70%	90%	6.2 µm	2	1	-3.879 dB
15°	70%	90%	7.2 µm	1	1	-3.777 dB

Table 3. List of simulation and neural network inputs (variables), represented in Table 2, associated with their respective meaning.

Sim./NN Input	Meaning
θ_y	Grating coupler diffraction angle
F_{min}	Minimum fill-factor
F_{max}	Maximum fill-factor
d_{Fz}	Fibre distance (from GC start)
k	GC refractive index variation: $k = 1$, linear $k = 2$, quadratic
r	Etch ratio ($r = D/h$)

Following a similar approach, two thresholds were set for non-apodized grating couplers. For fully etched grating couplers the limit was set at -4.8 dB (≈ 33.11 %), for partially etched designs the baseline was defined as -5.3 dB (≈ 29.51 %). Three designs of each etch ratio met such conditions, Table 4.

Table 4. List of inputs which result in the best coupling efficiencies for the design of non-apodized (constant refractive index) grating couplers. Partially etched GCs are represented by the grey lines and fully etched GCs by the white lines. The meanings of the simulation/NN inputs are represented in Table 5.

Sim/NN Inputs				Efficiency
θ_y	F	d_{Fz}	r	
14°	80%	4.2 μm	0.5	-5.296 dB
14°	80%	5.2 μm	0.5	-5.200 dB
15°	80%	5.2 μm	0.5	-5.235 dB
14°	85%	5.2 μm	1	-4.765 dB
14°	85%	6.2 μm	1	-4.757 dB
15°	85%	6.2 μm	1	-4.724 dB

Table 5. List of simulation and neural network inputs (variables), represented in Table 4, associated with their respective meaning.

Sim./NN Input	Meaning
θ_y	Grating coupler diffraction angle
F	Fill-factor
d_{Fz}	Fibre distance (from GC start)
r	Etch ratio ($r = D/h$)

As can be observed, with this set of design parameters and optical fibre positioning, fully etched grating couplers exhibit better performance regarding efficiency, Table 3. The best apodized GC design is fully etched, with linear refractive index apodization, and achieves a coupling efficiency of approximately -3.691 dB (≈ 42.75 %). In contrast, the best apodized partially etched design, reaches an efficiency of around -4.297 dB (≈ 37.18 %), and features quadratic refractive index variation. The same is true for non-apodized grating couplers, Table 5, the best fully etched design achieves a coupling efficiency of about -4.724 dB (≈ 33.70 %). The best partially etched non-apodized GC exhibits an efficiency of approximately -5.2 dB (≈ 30.20 %).

Grating couplers with efficiencies over 30 % (≈ -5.229 dB), account for 872 designs, or about 17.62 % of all simulated devices. Having into account the percentage of simulation inputs and outputs reserved for NN training (72 % of all data), only around 244 designs remain for test and validation purposes. Due to this fact the highest classification which was attributed to the best performing devices was 30 %+. Of the 4950 simulations, 3564 are reserved for training, 396 for validation and 990 for test purposes. Data split ratio is 72 % (training), 8 % (validation) and 20 % (testing), this ratio is similar to that found in the textbook *Deep Learning with Python* (Chollet, 2021) [17]. The simulations are shuffled each time the neural network is trained, to assure better dispersion and to avoid training, validating or testing with a predefined data distribution, which could lead to poor training accuracy for the datasets with fewer occurrences, or incur in improper testing and/or validation.

3.2 Neural Network Training and Predictions

To establish the most appropriate number of training epochs, assuring an appropriate fit and avoiding overfitting or underfitting, training and validation loss and accuracy plots were drawn, Fig. 5(a) and (b). The traces were drawn until overfitting was evident, the neural network was trained for 540 epochs to obtain these performance metrics.

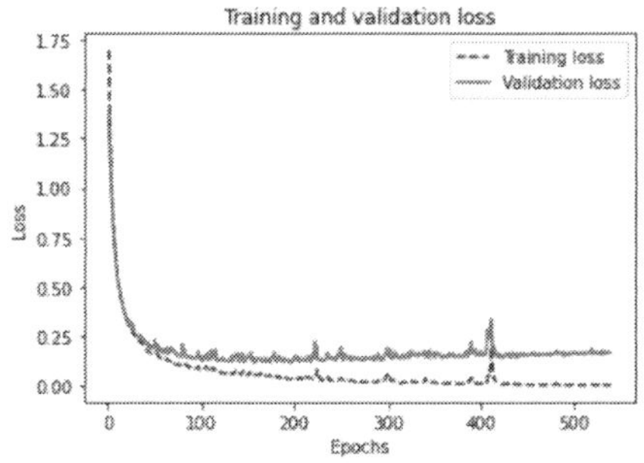

(a)

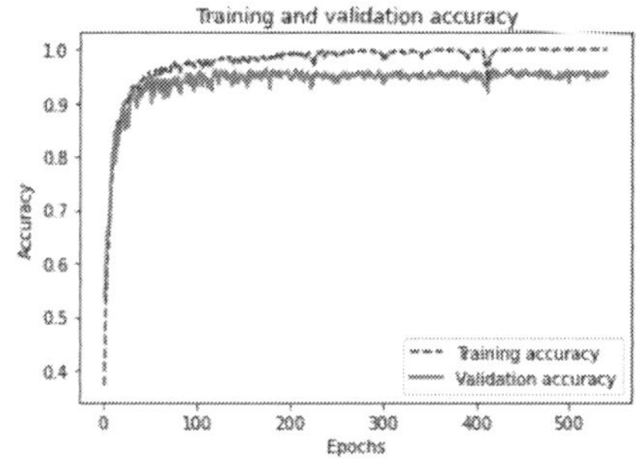

(b)

Fig. 5. Neural network training (dashed blue line) and validation (red line) loss (a) and accuracy (b) versus number of training epochs.

It can be noticed from the analysis of Fig. 5(a), that validation loss does not improve after about 100 training epochs, moreover, the line seems to slowly increase and diverge from the training loss line, this features mark the beginning of overfitting. Under about 50 training epochs both lines show a steep negative slope which clearly indicates underfitting. The same conclusions can be drawn from Fig. 5(b), where the validation accuracy does not (significantly) improve past the 100 epochs mark, and the training and validation lines start to diverge, showing overfitting. Under 50 training epochs the accuracy is clearly increasing, demonstrating underfitting.

In our perspective, the best fit occurs for 93 training epochs, or the average number of training epochs obtained with early stopping targeting validation loss with a patience of 10. For this number of training epochs, no non-adjacent target class classification errors were found, and test accuracy reaches an average value of 94.28 %, the average worst case classification accuracy (for the class with the lowest percentage of NN hits) is approximately 85.96 %. Numbers of training epochs below 93, demonstrated underfitting with the average measured test accuracy below 94 % and with an average worst case classification accuracy under 85 %, non-adjacent class classification errors were found. Further training, for more than 134 epochs, results in overfitting, the training accuracy increases, but test accuracy is maintained slightly over the 94 % mark, on average. With an increase in the number of training epochs, a higher number of non-adjacent class classification errors were found.

Maximum learning accuracy is achieved between 444 and 540 epochs (on average), the first number was obtained by defining early stopping with training accuracy as target, and by setting the patience to a (very high) 100 epochs. The second number, 540 epochs, was the value which was found to give a higher percentage of 100 % training accuracy and the highest average for the same metric, which was found to be approximately 99.83 %. In fact, it is difficult to find the exact point where the NN reaches the highest training accuracy, since on each training cycle, the initial weights would be different, and the behaviour of the optimizer (Adam), may also differ slightly. Overfitting the model to the limit does not seem to bring any benefits.

A confusion matrix, displaying the "hits" and "misses" of the neural network, when trained for 93 epochs, is represented on Fig. 6.

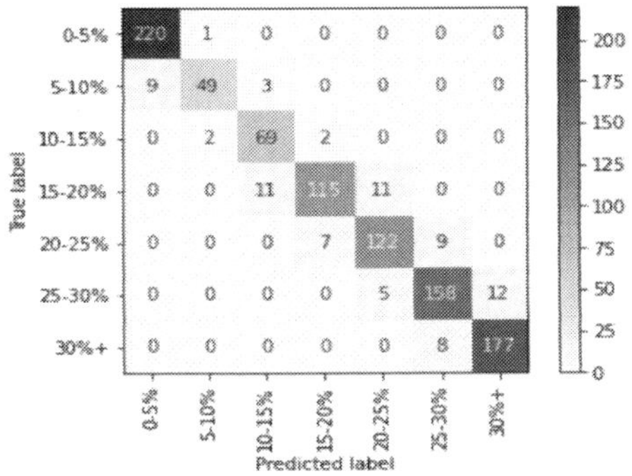

Fig. 6. Confusion matrix for an NN trained for 93 epochs, the test data has 990 occurrences. In this case test accuracy is around 95.35 % and the worst-case accuracy occurs for the 10-15% class, in which the NN have an accuracy of about 90.77 %.

The confusion matrix validates that a test accuracy of over 94 % and a worst-case accuracy of over 85 % are achievable, with a neural network properly dimensioned and trained to a near perfect fit.

4 Conclusion

A feedforward fully connected neural network, was successfully trained for the prediction of hydrogenated amorphous silicon (a-Si:H) grating couplers' efficiency, based on some of its setup and design parameters. The bidimensional (2D) finite-difference time-domain (FDTD) simulations, show that fully etched apodized grating couplers can achieve over -3.9 dB ($\approx$ 40.74 %) coupling efficiency, and partially etched designs values above -4.4 dB ($\approx$ 36.31 %). Non-apodized couplers can reach efficiencies over -4.8 dB ($\approx$ 33.11 %), in fully etched designs and in excess of -5.3 dB ($\approx$ 29.51 %), in the case of partially etched grating couplers. All devices are made by depositing a-Si:H over a silica substrate, which can be deposited by plasma-enhanced chemical vapor deposition (PECVD), at low temperatures, significantly reducing production costs. None of the devices feature any bottom reflector or mirror of any kind, nor any buried silicon layer.

The neural network, composed of two rectifier linear unit (ReLU) layers with 2048 output neurons and a final 7 neurons softmax output layer, when trained for 93 epochs, can achieve a classification accuracy of over 94 % and a worst-case accuracy above 85 %, while being compatible with older graphics processing units (GPUs). These results attest the relevance of neural networks for the prediction of grating couplers' efficiency, showing its potential for behaviour prediction of generic photonic circuits. The usage of machine learning for the predictability of grating coupler efficiencies, constitutes an important step for the design of more efficient photonic devices which can achieve better performance.

A downside of this study was the relatively low number of occurrences of grating couplers with coupling efficiencies over 30 %, which restricted the highest efficiency neural network class to 30 %+, and the number of classes to 7, instead of having the possibility of using 9 classes and two more classification intervals 30 % - 35%, 35 % - 40 % and 40 %+.

Acknowledgements

This research was supported by Portuguese national funds provided by FCT – Fundação para a Ciência e a Tecnologia, through grant SFRH/BD/07792/2021 (https://doi.org/10.54499/2021.07792.BD), funded by the Portuguese FCT program, Center of Technology and Systems (CTS) UIDB/00066/2020 / UIDP/00066/2020 and by FCT project ASER-META 2022.07694.PTDC.

References

1. Dhote, C., Singh, A., Kumar, S.: Silicon Photonics Sensors for Biophotonic Applications - A Review. IEEE Sens. J. 22, 18228–18239 (2022). https://doi.org/10.1109/JSEN.2022.3199663
2. Bernabé, S., Wilmart, Q., Hasharoni, K., Hassan, K., Thonnart, Y., Tissier, P., Désières, Y., Olivier, S., Tekin, T., Szelag, B.: Silicon photonics for terabit/s communication in data centers and exascale computers. Solid. State. Electron. 179, (2021). https://doi.org/10.1016/j.sse.2020.107928
3. Adcock, J.C., Bao, J., Chi, Y., Chen, X., Bacco, D., Gong, Q., Oxenløwe, L.K., Wang, J., Ding, Y.: Advances in Silicon Quantum Photonics. IEEE J. Sel. Top. Quantum Electron. 27, (2021). https://doi.org/10.1109/JSTQE.2020.3025737
4. Shastri, B.J., Huang, C., Tait, A., Ferreira de Lima, T., Prucnal, P.R.: Silicon photonic neural network applications and prospects. 52

(2022). https://doi.org/10.1117/12.2614865
5. Ma, W., Liu, Z., Kudyshev, Z.A., Boltasseva, A., Cai, W., Liu, Y.: Deep learning for the design of photonic structures. Nat. Photonics. 15, 77–90 (2021). https://doi.org/10.1038/s41566-020-0685-y
6. Jiang, J., Chen, M., Fan, J.A.: Deep neural networks for the evaluation and design of photonic devices. Nat. Rev. Mater. 6, 679–700 (2021). https://doi.org/10.1038/s41578-020-00260-1
7. Mao, S., Cheng, L., Zhao, C., Khan, F.N., Li, Q., Fu, H.Y.: Inverse design for silicon photonics: From iterative optimization algorithms to deep neural networks. Appl. Sci. 11, (2021). https://doi.org/10.3390/app11093822
8. Hammond, A.M., Camacho, R.M.: Designing integrated photonic devices using artificial neural networks. Opt. Express. 27, 29620–29638 (2019). https://doi.org/10.1364/oe.27.029620
9. Miyatake, Y., Sekine, N., Toprasertpong, K., Takagi, S., Takenaka, M.: Computational design of efficient grating couplers using artificial intelligence. Jpn. J. Appl. Phys. 59, (2020)
10. Vitali, V., Lacava, C., Domínguez Bucio, T., Gardes, F.Y., Petropoulos, P.: Highly efficient dual-level grating couplers for silicon nitride photonics. Sci. Rep. 12, 1–12 (2022). https://doi.org/10.1038/s41598-022-19352-9
11. Marchetti, R., Lacava, C., Khokhar, A., Chen, X., Cristiani, I., Richardson, D.J., Reed, G.T., Petropoulos, P., Minzioni, P.: High-efficiency grating-couplers: Demonstration of a new design strategy. Sci. Rep. 7, 1–8 (2017). https://doi.org/10.1038/s41598-017-16505-z
12. Fantoni, A., Costa, J., Lourenço, P., Vieira, M.: Computer simulation study about the dependence of amorphous silicon photonic waveguides efficiency on the material quality. EPJ Appl. Phys. 90, 1–10 (2020). https://doi.org/10.1051/epjap/2020190250
13. Almeida, D., Rossi, M., Lourenço, P.J.P.S., Fantoni, A., Costa, J., Vieira, M.: Amorphous silicon grating couplers based on random and quadratic variation of the refractive index. In: Proc. SPIE 12880, Physics and Simulation of Optoelectronic Devices XXXII, 128800J (11 March 2024). p. 38 (2024)
14. Synopsys Inc.: RSoft Photonic Device Tools, https://www.synopsys.com/photonic-solutions/rsoft-photonic-device-tools/rsoft-products.html
15. Fantoni, A., Lourenco, P., Vieira, M.: A model for the refractive index of amorphous silicon for FDTD simulation of photonics waveguides. Proc. Int. Conf. Numer. Simul. Optoelectron. Devices, NUSOD. 167–168 (2017). https://doi.org/10.1109/NUSOD.2017.8010044
16. Corning Inc.: Corning® SMF-28TM Optical Fiber, http://www.photonics.byu.edu/FiberOpticConnectors.parts/images/smf28.pdf
17. Chollet, F.: Deep Learning with Python. Manning Publications Co. (2021)

Use of laser technology for the postural classification of bedridden people

David S. Canzobre[1], *Pablo Pardiñas Torrado*[1], *Javier Lamas Vigo*[1], *Alberto Ramil Rego*[1]

[1]Laboratorio de Aplicaciones Industriales del Láser (LAIL), Campus Industrial de Ferrol, Universidade da Coruña, Mendizábal s/n 15403 Ferrol

Abstract. This work presents an innovative method for the automated classification of postures of bedridden people using laser technology. With the aim of improving the quality of medical care and facilitating the continuous monitoring of patients in clinical and home environments, a system is proposed that uses laser sensors to capture the three-dimensional geometry of body postures. Through advanced computer vision processing techniques, a robust classification algorithm is developed capable of identifying and categorizing lateral decubitus postures, also known as lateral safety position, which are commonly used to lay down patients who are permanently bedridden. Experimental results, performed so far, show sufficient accuracy in posture classification, suggesting the potential of this technology to improve the monitoring and care of bedridden patients, while providing periodic warnings to medical staff when a patient has exceeded the recommended time in the same posture, to avoid the appearance of skin ulcers.

1 Introduction

According to the report "The aging of the population and its effects on the labor market" [1], presented by the Spanish Confederation of Business Organizations (CEOE), between 2015 and 2050 the number of people over 65 years of age could double in our country. On the other hand, the study "The Elderly in Spain" [2], prepared by the Ministry of Health and Social Policy and the General Secretariat of Social Policy, through the IMSERSO, indicates that people over 65 years of age represent 16.7% of the population of our country, and of these, 27.8% are over 80 years of age, which means that the number of elderly has multiplied by 8 with respect to 1900. The report also points out that "the aging of the population is a generalized phenomenon in advanced economies, although in Spain it will be more intense". The most obvious consequence of this demographic change is that in 2050 there will be four employed people for every three retired people, which could translate into a considerable increase in the needs of the social and healthcare market due to generational aging.

One of the consequences of aging is that there will be an increasing number of elderly dependents, with almost half of the elderly population expected to be octogenarians by 2060 (6,891,590 people). As mentioned, this will make the supply of innovative solutions to the various situations of dependency an indispensable element of the society and market of the future, and already of the present.

In this context, where social and health care is an important niche market opportunity for companies that offer innovative solutions in the field of telecare, arises the collaboration among the Laboratory of Industrial Laser Applications of the University of A Coruña, Anta Norte that is a small company in the province of A Coruña dedicated to the development of technological products for the field of social and health care sector and, a Anta Norte's client, Raiola Residencial that it a company that manages accommodation for the elderly, under a service model oriented to the well-being of the residents.. As a result of this cooperation, different ideas were born that will become commercial products, including the creation of a low-cost structured light system for monitoring bedridden people who need assistance to change their posture.

The existing literature has not found many papers related to bedridden posture classification project and those that exist use Artificial Intelligence solutions, which require large sets of labeled images to be trained, along with relatively expensive hardware (thermal and RGB-D cameras) to be massively deployed. Examples of these works are: *"A Blanket Accommodative Sleep Posture Classification System Using an Infrared Depth Camera: A Deep Learning Approach with Synthetic Augmentation of Blanket Conditions"* [3] and *"In-bed human pose estimation from unseen and privacy-preserving image domains"* [4].

The objective of the work is to develop a low-cost hardware based on structured light that, together with a proprietary algorithm, allows to scan a bedridden person and determine on which side he/she is lying, always maintaining the privacy of the scanned person. Preliminary results from the testing of the prototypes are presented at the end of this paper.

2 Hardware

To select which structured light technique was the most suitable for scanning the surface of a bedridden person, the works of J. Salví et al. *"Pattern codification strategies in structure light systems"* [5] and *"A state of the art in structured light patterns for surface profilometry"* [6] were used. Given the limitations of the use case (it cannot disturb the bedridden person and must be as low cost as possible) it was decided to use a structured light system based on infrared laser light. In the same way that in *"Laser multiple line triangulation system for real-time 3-D monitoring of chest wall during breathing"* [7] they use a laser plus camera configuration to measure the height of a patient's chest during breathing, we use the same configuration placed over patient's bed to obtain the profile formed by the legs under the sheets.

The camera being used is a Dahua IPC-HDW2231T-AS-S2, which was chosen because it is one of the main models that Raiola has deployed in its facilities. The characteristics of this camera are presented in **Table *1***.

Table 1. Dahua SD22204UE-GN characteristics

Sensor	1/2.8" CMOS
Resolution	1920 × 1080
Lens type	Fixed-focal
Focal Length	2.8 mm; 3.6 mm
Max. aperture	2.8 mm: F1.6; 3.6 mm: F1.6
Field of view	2.8 mm: Horizontal: 107°; Vertical: 56°; Diagonal: 127°. 3.6 mm: Horizontal: 88°; Vertical: 44°; Diagonal: 105°

The laser of the final product will be one whose light is in the infrared spectrum so as not to disturb bedridden people, but during the tests, to make it easier to position, a laser emitting red light was used, specifically a Picotronic DOE256-650-5-3. Together with the laser, two diffractive optical elements were used, one with a parallel line pattern (Picotronic DOE-DE-R284) and the other with a grid pattern (DOE-DE-R354).

To test the measurement capability of the laser plus camera assembly, a series of tests were performed in the laboratory before testing the setup in a real scenario. These tests consisted of, after calibrating the camera (**Fig. 1**), projecting the line pattern onto a flat wall to obtain a reference image (**Fig. 2**). Then a cylinder was placed under the projection, with and without a sheet of paper as a blanket (**Fig. 3** and **Fig. 4**), and by measuring the displacement with respect to the reference image, of the points of each line, the height of the object was obtained (**Fig. 5** and **Fig. 6**).

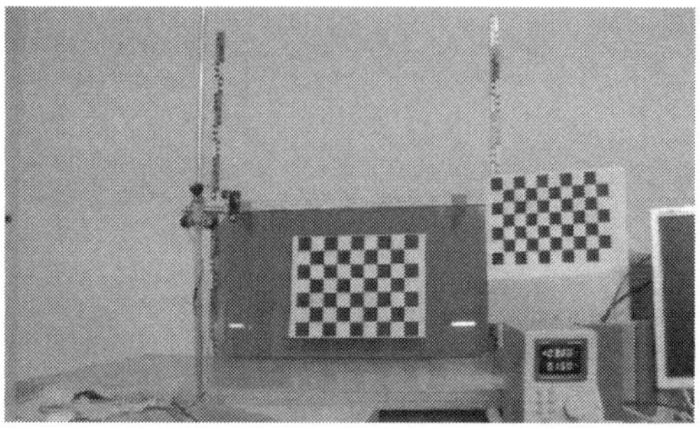

Fig. 1. System calibration.

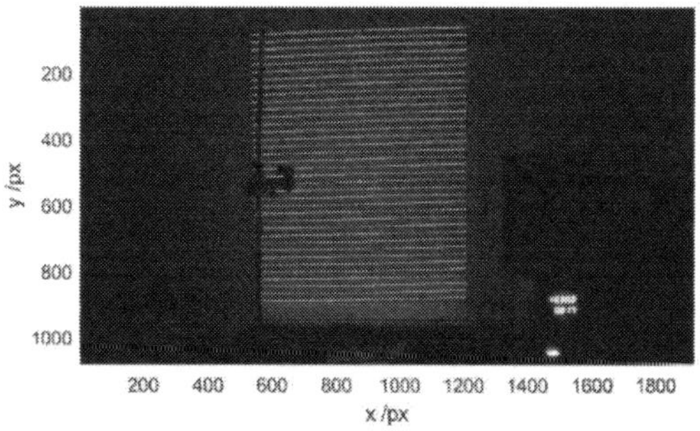

Fig. 2. Reference image.

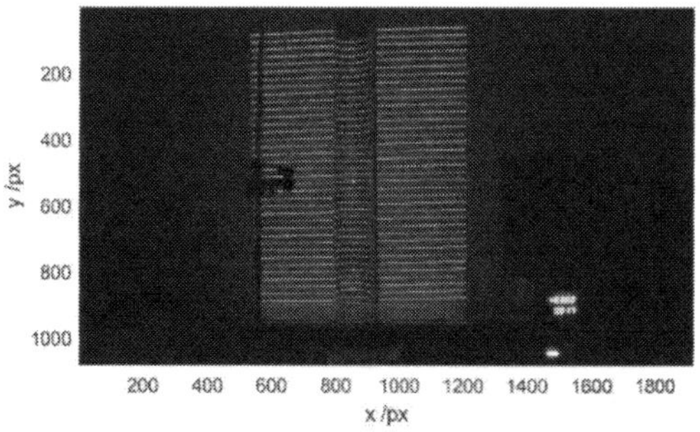

Fig. 3. Cylinder without paper sheet.

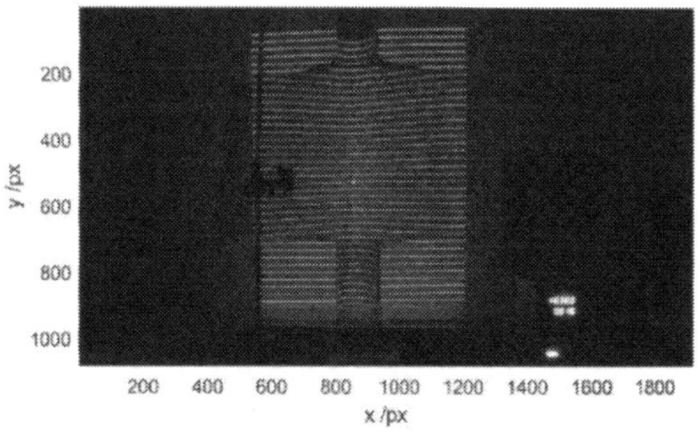

Fig. 4. Cylinder under paper sheet.

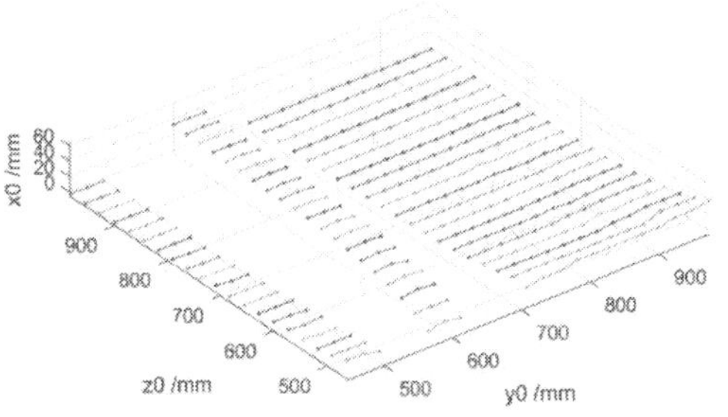

Fig. 5. Charts with the measurement of the heights of the cylinder uncovered.

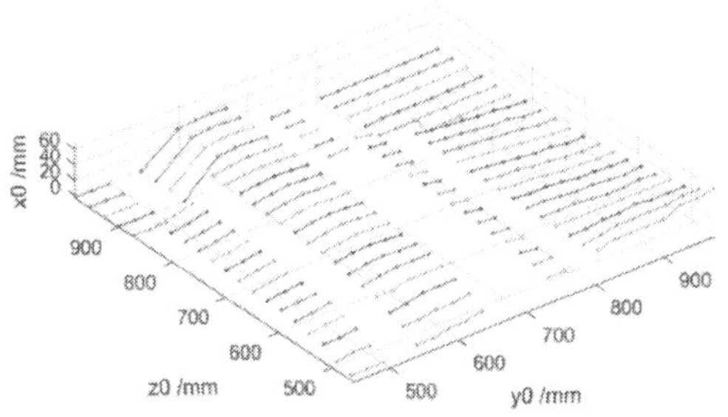

Fig. 6. Charts with the measurement of the heights of the cylinder covered with the sheet of paper.

As conclusions to these tests in the laboratory it was found that it was possible to measure with the laser plus camera set. The limitation is that it is not possible to measure the entire surface due to the shadow zones that depend on the positioning of the object with respect to the structured light system.

Once the feasibility of the system was proven in the laboratory, it was tested in a real environment. To test whether the number of points from which to obtain a height measurement could be increased, thus improving the accuracy with which the surface area of the bedridden person is obtained, we tried replacing the DOE with the parallel lines pattern by one with a grid pattern. With this new DOE, the camera plus laser assembly was first placed at a height of approximately 2m, about 1m away from the foot of the bed and pointed towards the bed at an angle of about 45°. The intention of this position was to try to get the laser pattern to cover as much of the bed surface as possible. In **Fig. 8** and **Fig. 9Fig. 9** three phenomena can be seen, the first is that the lines near the head of the bed were too close together making detection impossible, this was due to the position of the bed. Therefore, the camera was moved to a position where it was perpendicular to the bed leaving the DOE unmoved, the result of this change of position can be seen in **Fig. 10**.

Fig. 7. Assembly of the Dahua laser and camera used during the tests.

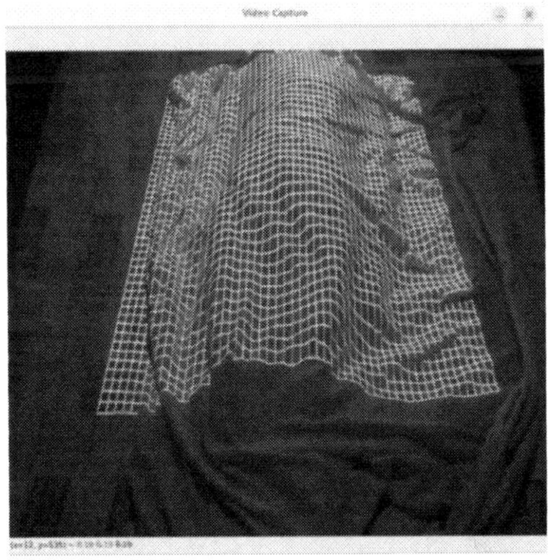

Fig. 8. Capture of an image using the DOE with grid pattern.

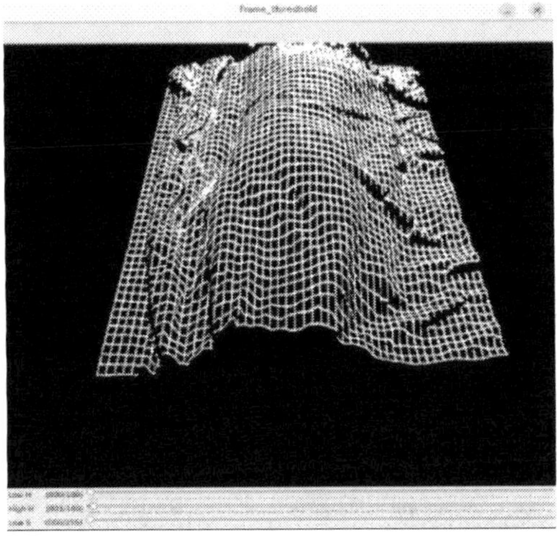

Fig. 9. Same image as **Fig. 8** after applying filtering by HSV color model values to remove the background.

Fig. 10. Image of the grid pattern after placing the camera perpendicular to the bed.

Another problematic fact is that certain colors of bed linens do not reflect laser light well (**Fig. 11** and **Fig. 12**); this is a limitation of the system that can be avoided by using white or colored bed linens that do not absorb, or are confused with, the laser light.

Fig. 11. Dummy under sheet with problematic colours.

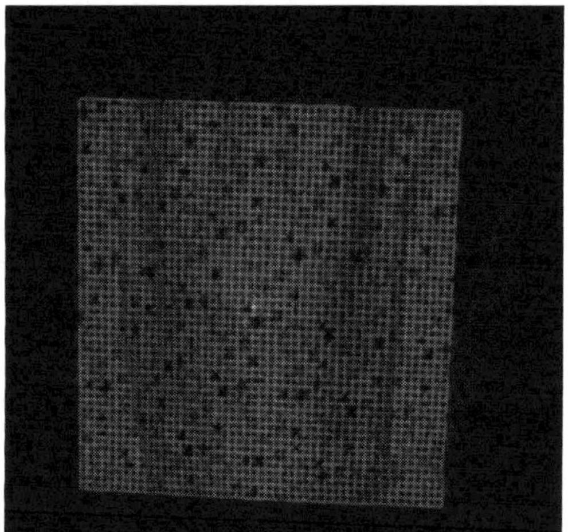

Fig. 12. Laser projection on a sheet with problematic colours.

Finally, when the dummy was placed in the bed and the grid pattern was projected onto it, it was observed that the horizontal lines provided more information about the body's height relative to the bed. The vertical lines, however, interfered with the processing of the horizontal lines. Therefore, it was decided to continue the experiments using the DOE with parallel lines.

When the DOE of parallel lines was projected onto the dummy in the bed, it was found that the issue of broken projected lines was much worse than in previous tests with the cylinder in the lab. This was due to multiple shadow zones created by the dummy under the blanket, which made it impossible to match each segment with its corresponding projected line. Finding this correspondence is crucial for the algorithm proposed in section 3 to function, as it is necessary to identify the peak of each curve reflected on the bed's surface when the pattern is projected (**Fig. 12**). Two solutions were proposed to address this problem. One involved using an array of lasers with single-line DOEs, positioned so that the lines were parallel, allowing only one laser to be turned on at a time for processing. This solution was discarded due to the high cost of acquiring multiple lasers and their corresponding DOEs.

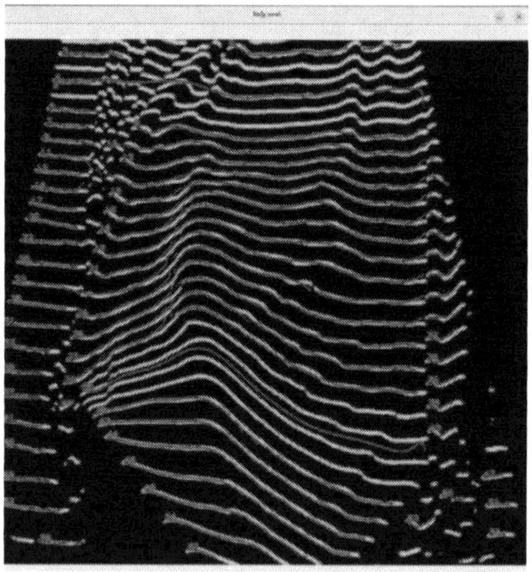

Fig. 13. Segmented curves that are drawn by projecting the pattern of lines onto a body in bed.

The other proposed solution involves placing a window in front of the DOE with parallel lines, driven by a stepper motor controlled by an ESP32 microcontroller. This window would automatically uncover each line as the peak of the curve projected onto the bedded body is calculated. This solution, with a prototype shown in **Fig. 14** and **Fig. 15**, is still under development at the time of writing.

Fig. 14. Side view of the prototype showing a single line, projected by the laser, as it is being processed.

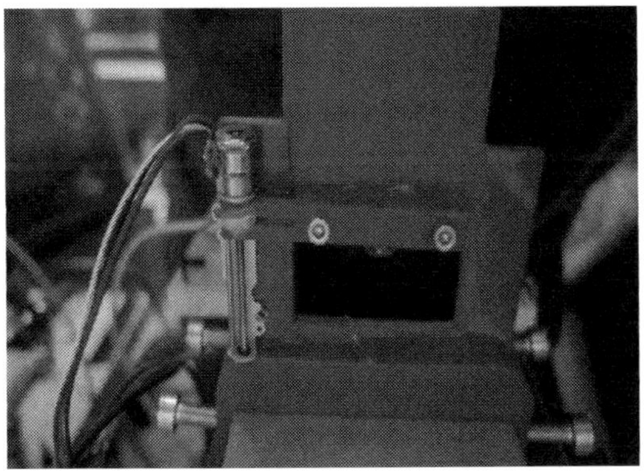

Fig. 15. Front view of the prototype shown in **Fig. 14**.

3 Proposed Algorithm

To determine on which side a person is lying, an algorithm has been proposed that calculates the concavity/convexity of the curve formed by the legs. The input to this algorithm is the maxima of each curve resulting from projecting the lines onto the bed. For those points the fitting polynomial is calculated to finally calculate the concavity/convexity of that polynomial.

To test the feasibility of this algorithm, in parallel to the development of the window to discover the lines, a version was implemented that, instead of using as input the maxima of the curves projected by the laser, uses as input the maximum points of the 3D surface obtained with a stereo camera such as the Luxonis OAK-D Pro. The results of these tests are promising, as can be seen in **Fig. 19** and **Fig. 22**, which show the change in convexity of the adjustment polynomial of the maximum points of the 3D surface, which are the ones that draw the profile of the legs, depending on the side on which the person is lying.

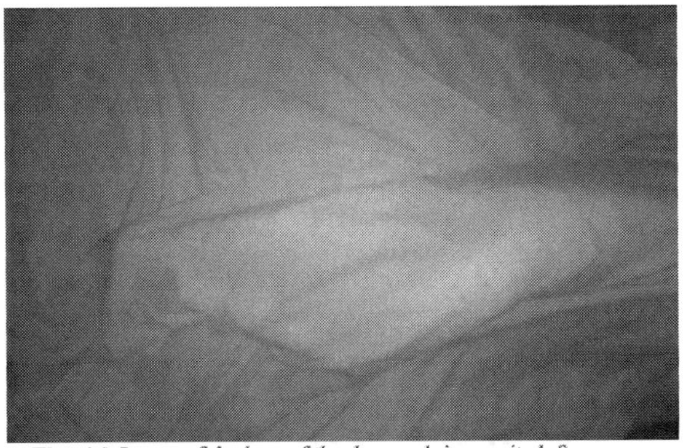

Fig. 16. Image of the legs of the dummy lying on its left side.

Fig. 17. Depth image of the legs of the dummy lying on its left side.

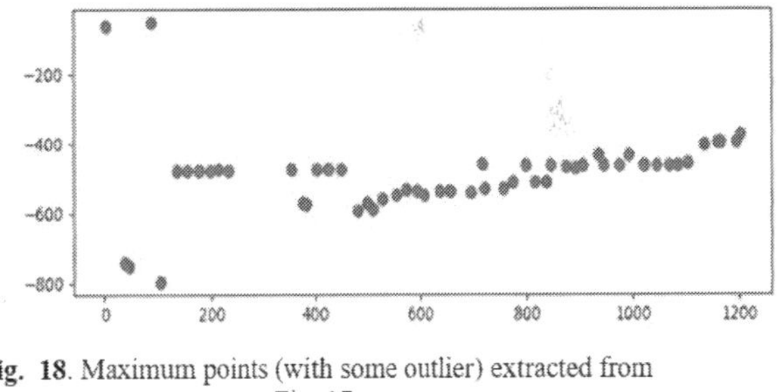

Fig. 18. Maximum points (with some outlier) extracted from Fig. 17.

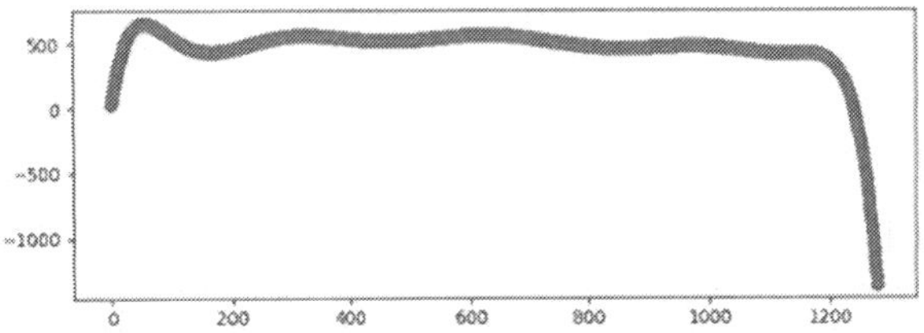

Fig. 19. Fitting polynomial, calculated from the points in Fig. 18, evaluated between 0 and the maximum width of the camera resolution

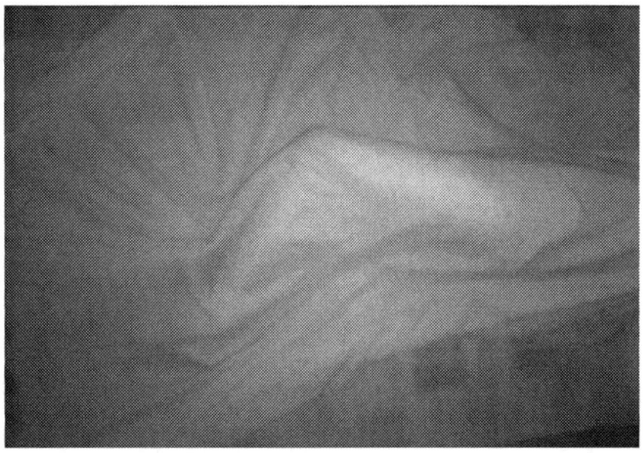

Fig. 20. Image of the legs of the dummy lying on its right side.

Fig. 21. Depth image of the legs of the dummy lying on its right side.

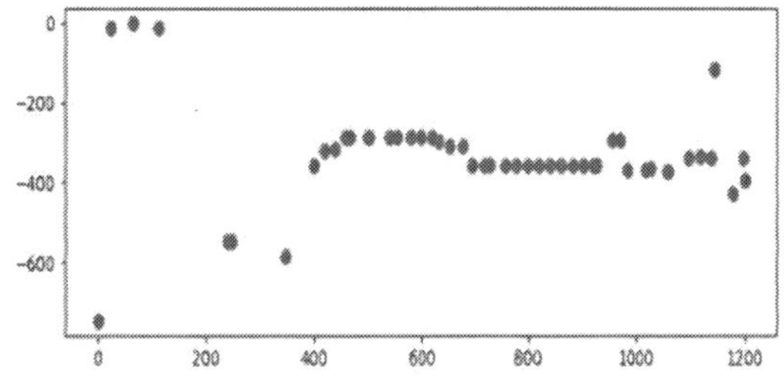

Fig. 22. Maximum points (with some outlier) extracted from Fig. 21.

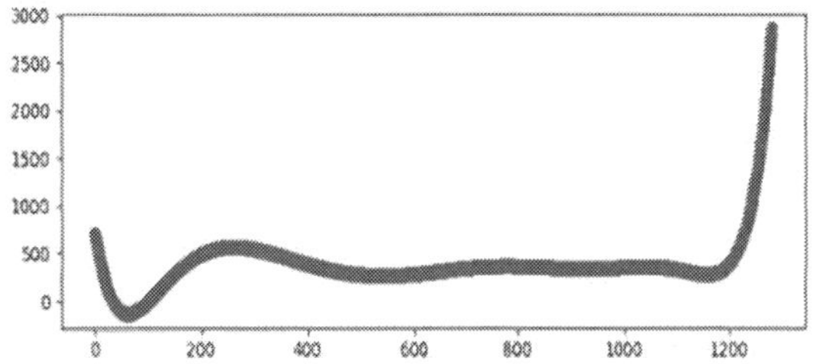

Fig. 23. Fitting polynomial, calculated from the points in Fig. 22, evaluated between 0 and the maximum width of the camera resolution.

4 Conclusions and future work

The development of a low-cost structured light system for the postural classification of bedridden individuals represents a significant step forward in enhancing patient care. The preliminary tests of our system showed promising results in the classification of lateral decubitus postures, an essential feature for the continuous monitoring of patients at risk of developing pressure ulcers. The use of infrared laser technology ensures that the system is non-intrusive, maintaining patient comfort and privacy, which is crucial in both clinical and home care environments.

However, the project has also highlighted a challenge: finding a simple and reliable way to match the segments projected onto the bed with the lines in the reference image.

Future work will focus on the following:

1. Finalizing the motorized window mechanism to enable line-by-line scanning.
2. Conducting extensive field tests in real-world settings to validate the system's effectiveness and reliability.
3. As field testing progresses, further developing and optimizing the concavity/convexity-based algorithm for posture detection.

The continuation of this project will focus on addressing the identified limitations and enhancing the system's capabilities to ensure it meets the practical demands of healthcare providers and improves the quality of life for bedridden patients.

5 References

[1] C. E. de Organizaciones Empresariales (CEOE), *El envejecimiento de la población y sus efectos en el mercado laboral, (*2016).

[2] G. Fernández-Malloralas Fernández, J. M. Labeaga Azcona, J. López Doblas, A. Madrigal Muñoz, S. Pinazo-Hernandis, V. Rodríguez Rodríguez y M. J. Vidal Domínguez, *Las Personas Mayores en España (Informe 2018), (*2021).

[3] A. Tam, B. So, T. Chan, A. Cheung, D. W.-C. Wong y C.-W. Cheung, «A Blanket Accommodative Sleep Posture Classification System Using an Infrared Depth Camera: A Deep Learning Approach with Synthetic Augmentation of Blanket Conditions» *Sensors,* **vol. 21**, p. 5553, (August 2021).

[4] T. Cao, M. A. Armin, S. Denman, L. Petersson y D. Ahmedt-Aristizabal, «In-Bed Human Pose Estimation from Unseen and Privacy-Preserving Image Domains» de *2022 IEEE 19th International Symposium on Biomedical Imaging (ISBI)*, (2022).

[5] J. Salvi, J. Pages y J. Batlle, «Pattern codification strategies in structure light systems» *Pattern Recognition,* **vol. 37**, pp. 827-849, (April 2004).

[6] J. Salvi, S. Fernandez, T. Pribanic y X. Llado, «A state of the art in structured light patterns for surface profilometry» *Pattern Recognition,* **vol. 43**, pp. 2666-2680, (August 2010).

[7] M. Jezeršek, M. Flezar y J. Mozina, «Laser multiple line triangulation system for real-time 3-D monitoring of chest wall during breathing» *Strojniski Vestnik,* **vol. 54**, pp. 503-506, (August 2008).

WDM C-band four channel demultiplexer using cascaded multimode interference on SiN strip waveguide structure

Dror Malka [1]

[1] Faculty of Engineering, Holon Institute of Technology (HIT), Holon 5810201, Israel

Abstract. Back reflection challenges significantly constrain the efficiency of optical communication networks employing dense wavelength division multiplexing (DWDM) technology, particularly those based on Silicon (Si) Multimode Interference (MMI) waveguides. To mitigate this limitation, we present a novel 1×4 optical demultiplexer design using MMI within a Silicon-Nitride (SiN) strip waveguide configuration, operating within the C-band spectrum. Our design was optimized using AI-enhanced RSoft-CAD simulations that combined the Beam Propagation Method (BPM) and Finite-Difference Time-Domain (FDTD) techniques, integrated with convolutional neural network (CNN) machine learning algorithms. The simulation results demonstrate that the proposed device efficiently transmits four channels with 10 nm spacing in the C-band, showing low power loss ranging from 1.98-2.35 dB, a broad bandwidth of 7.68-8.08 nm, and excellent crosstalk suppression between 20.8-23.4 dB. Leveraging the low refractive index of SiN, we achieve ultra-low back reflection of 40.58 dB without requiring specialized angled MMI designs, which are often necessary in Si-based MMI technology. Consequently, this SiN-based MMI demultiplexer provides an effective solution for DWDM systems, enabling high data transmission rates with minimal back reflection in optical communication networks.

1 Introduction

The rapid evolution of optical communication systems operating within the C-band spectrum demands innovative, highly efficient waveguide components that can support high-speed light transmission while minimizing power losses, broadening bandwidth, and reducing back reflection [1-3]. Among the essential components in these systems are optical demultiplexers, which can be realized through various technological approaches such as Y-branch devices [4-6], Mach-Zehnder interferometers [7], and multimode interference (MMI) couplers [8-10]. Dense Wavelength Division Multiplexing (DWDM) technology significantly increases data throughput by narrowing the spacing between wavelengths, thereby boosting the number of available communication channels within a spectral band [11]. A strip waveguide, featuring a high-index core surrounded by a low-index cladding, allows for strong light confinement and guidance via total internal reflection, making it suitable for photonic integrated circuits (PIC) [12-13]. MMI couplers, in particular, have gained prominence in PICs due to their broad optical bandwidth, low insertion losses, and compact size [14-15]. These couplers operate based on the self-imaging principle, where the input electric field replicates into multiple images as the light propagates along the waveguide [16]. One of the major challenges in optical transmitter systems is back reflection, where light reflects back toward the laser source, potentially degrading system performance. Silicon (Si)-based MMI couplers are often affected by reflections due to the refractive index mismatch between Si and the surrounding materials, such as silica (SiO2), as well as due to the self-imaging mechanism itself [17]. To overcome these limitations, Silicon Nitride (SiN) waveguides have emerged as a promising alternative. SiN, with its lower refractive index, minimizes back reflection and operates efficiently within the C-band spectrum while offering low absorption losses and high thermal stability [18-19]. In this paper, we propose a novel 1×4 wavelength demultiplexer based on an MMI coupler within a SiN strip waveguide configuration, designed to separate four channels in the C-band at wavelengths of 1530 nm, 1540 nm, 1550 nm, and 1560 nm. The device structure consists of cascaded 1×2 MMI couplers, input waveguides with tapered sections, S-bends, and output tapers. To optimize the geometrical parameters of the waveguide and MMI couplers, we employed artificial intelligence (AI)-driven techniques, utilizing convolutional neural networks (CNNs) to enhance the simulation process. Our optimization process utilized RSoft-CAD software, incorporating advanced numerical methods like the Beam Propagation Method (BPM) and Finite-Difference Time-Domain (FDTD) simulations, further boosted by machine learning algorithms. The integration of CNNs allowed us to accelerate the convergence of simulation results and refine the design parameters with greater accuracy. Python scripts were employed for post-processing and data analysis, ensuring high precision and efficiency in optimizing the device's performance. This AI-powered approach to simulation and optimization not only enhances the accuracy of the design process but also demonstrates the potential of CNN-augmented methodologies in advancing next-generation optical devices for high-performance DWDM systems.

2 Demultiplexer design

Figure 1 illustrates the cross-sectional view in the XZ plane at Y=0, displaying the three cascaded MMI couplers designed for the 1x4 demultiplexer. The width of the MMI coupler is denoted as WMMI, while their respective lengths are LMMI1, LMMI2, and LMMI3. The width of the input waveguide measures 516 nm with a height of 312 nm. The input taper lengths for the MMI1, MMI2, and MMI3 couplers are 40.6 µm, with taper widths ranging from 0.55 µm to 0.78 µm. The output taper lengths for the MMI1, MMI2, and MMI3 couplers measure 35.44 µm, with widths transitioning from 0.88 µm to 0.517 µm. The gap (Gt) between the output tapers of each MMI coupler is 0.78 µm. In the figure, the blue sections represent SiN material, whereas the white sections signify SiO2.

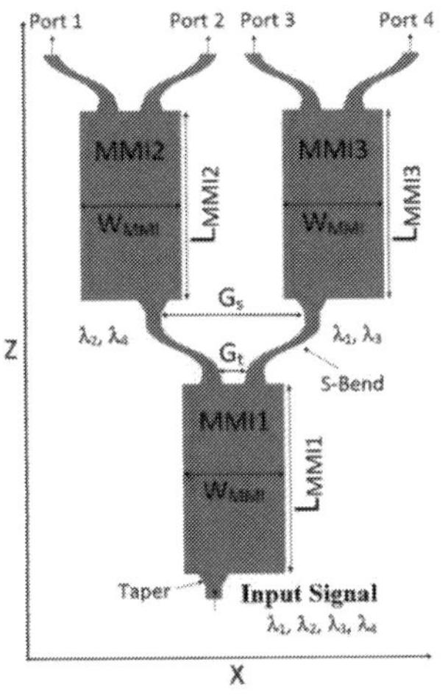

Fig. 1. Schematic diagram of the MMI demultiplexer.

The MMI1 coupler, part of the three-stage cascade, is designed to efficiently separate four targeted wavelengths: 1560 nm (λ1), 1550 nm (λ2), 1540 nm (λ3), and 1530 nm (λ4). The wavelength pair (λ2/λ4) is directed toward the MMI2 coupler, while the pair (λ1/λ3) is steered into the MMI3 coupler, as shown in Figure 1. The width of the S-Bends was adjusted to match the output taper width, set at 0.516 µm, with lengths ranging between 69 µm and 80 µm. The spacing between the S-Bend outputs (Gs) was fixed at 15.7 µm for MMI1 and 12.7 µm for both MMI2 and MMI3. The total length of the device is 6.74 mm. Based on the self-imaging effect, each wavelength entering the multimode region produces a direct or mirrored copy of itself at consistent intervals.

3 Simulation Results

The CNN model was trained using a dataset of optical simulations to predict and optimize key parameters, such as the MMI coupler lengths, width, and wavelength split efficiency. Once the dataset was generated from initial FDTD and BPM simulations, we employed the CNN model to refine these results by identifying patterns and trends within the data. The model enhanced the speed and accuracy of simulations by iterating through various design configurations and predicting outcomes more efficiently than conventional trial-and-error methods. This approach allowed us to accurately determine the optimal dimensions for the MMI couplers while accounting for fabrication tolerances, significantly improving both the design process and final performance of the device.

Figure 2 depicts the TE mode profile within the SiN strip waveguide at a wavelength of 1560 nm. The intense confinement, shown in red, reveals a mode spot size of 2.47x2.33 µm, which could be optimized by employing an adiabatic or specialized taper to enhance light coupling between the device and the input source. Similar mode profiles were observed for other operational wavelengths. The effective refractive indices for the TE mode at each wavelength were calculated using computational methods. Figure 3 shows the process of optimizing the height of the SiN layer, which was found to be 311 nm. The tolerance range around this optimal height is ±23 nm, resulting in normalized power efficiencies between 48% and 68%. The maximum power efficiency occurs at the height of 312 nm. Nonetheless, achieving this precision within current fabrication tolerances presents a significant challenge.

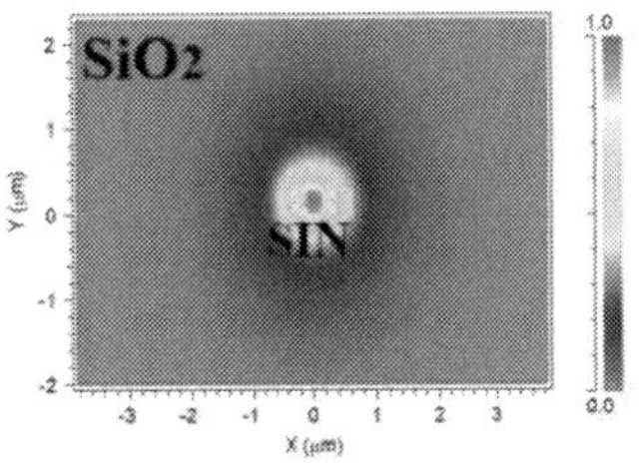

Fig. 2. TE mode solution for 1560 nm.

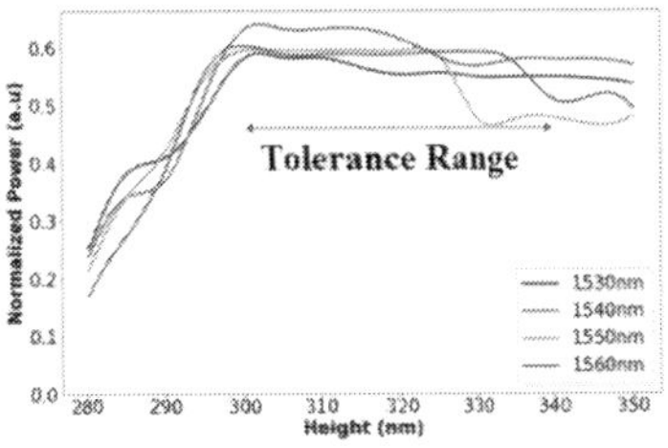

Fig. 3. Optimizations of the SiN height.

Figure 4 displays the optimal width for the MMI coupler, which is 3.53 µm, along with a tolerance range of ±23 nm around this ideal measurement. This range is chosen to ensure a normalized power output between 58% and 67%. The selected tolerance accounts for typical variations in the fabrication process, where geometric inaccuracies commonly fall within ±23 nm of the specified dimension.

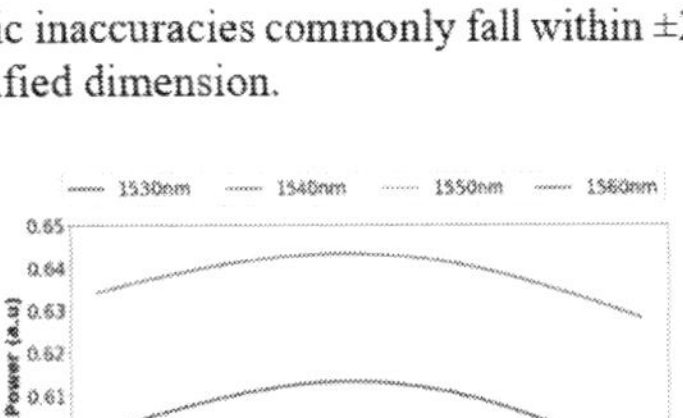

Fig. 4. Optimizations of the MMI width coupler.

By utilizing AI-driven machine learning in conjunction with BPM simulations, the optimal lengths for the MMI couplers were identified. To maximize output power while accounting for potential fabrication variations, the lengths for Lmmi1, Lmmi2, and Lmmi3 were determined to be 4168 μm, 2073 μm, and 2082 μm, respectively. Figure 5 presents the tolerance range of ±5.15 μm around these ideal lengths for each MMI coupler, corresponding to the specific wavelengths, ensuring a normalized power output ranging from 67% to 74%. This expanded tolerance range offers greater flexibility during the fabrication process.

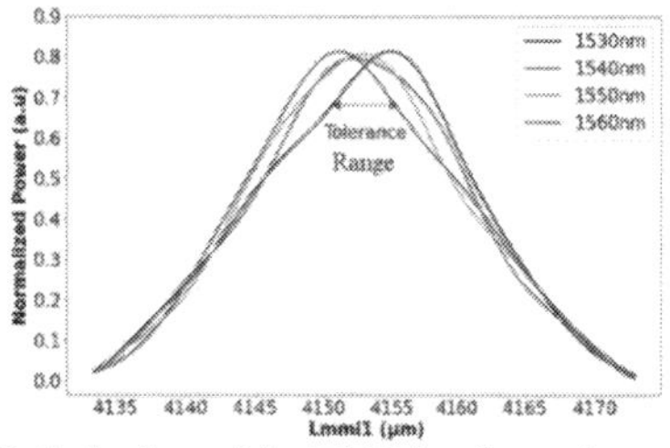

Fig. 5. Optimizations of the MMI1 length coupler.

Figure 6 shows the light intensity distribution within the SiN MMI demultiplexer for the C-band at the XZ plane. The laser light is introduced into the input waveguide at Z=0 and travels through the input taper into the MMI1 region. At Z=4255 μm, the MMI1 output splits the four wavelengths into two pairs: λ2/λ4 and λ1/λ3. The light proceeds through the S-bends to the subsequent MMI couplers, where further splitting occurs at the MMI2 output for 1540 nm (λ2) (b) and 1560 nm (λ4) (a) at Z=6515 μm, and at the MMI3 output for 1530 nm (λ1) (c) and 1550 nm (λ3) (d) at Z=6534 μm. The light finally travels through the S-bends and reaches the four output channels at Z=6632 μm.

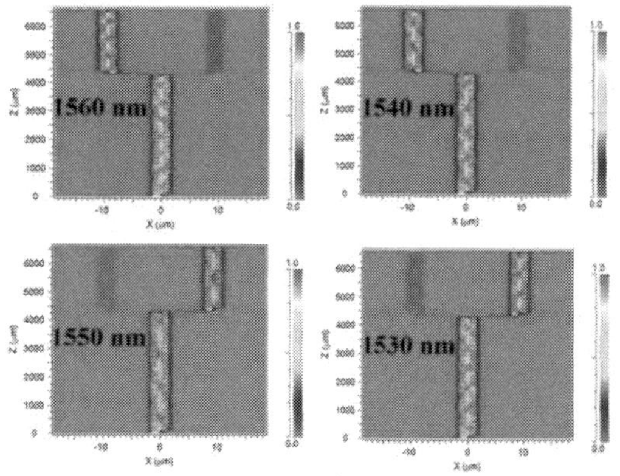

Fig. 6. Power and light intensity profile of the four C-band channels from 1530-1560 nm.

The spectral transmission for the four channels across the C-band, covering wavelengths from 1530 nm to 1570 nm, is illustrated in Figure 7. This data was obtained by solving the mode for each wavelength within this range, and then applying the results to the optimized demultiplexer design. Machine learning, specifically a CNN model, was employed to enhance the optimization process, significantly boosting the efficiency of the Python scripts used for data analysis. Parameters such as crosstalk, insertion losses, and spectral bandwidth were evaluated alongside Figure 7, with the results summarized in Table 1. A crucial design aspect is the reduction of back-reflection into the input section, which can severely impact the laser source's performance. In this study, SiN material was utilized to mitigate back-reflection caused by the self-imaging effect within the MMI coupler. To measure back-reflection, a monitor was placed in the input waveguide to detect all reflected light from the MMI coupler, as shown in Figure 8.

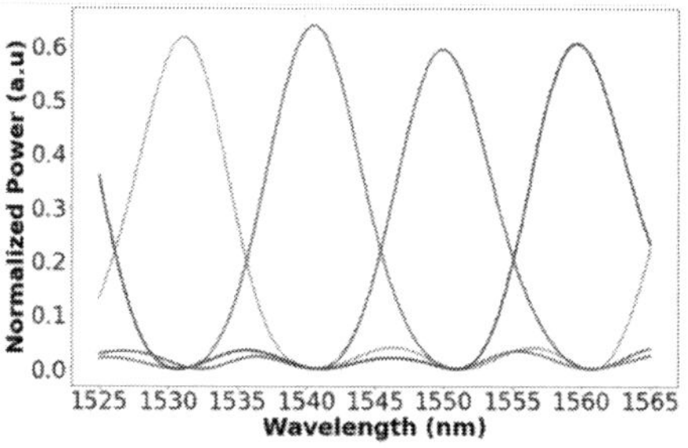

Fig. 7. C-band optical spectrum.

Table 1: Crosstalk, Losses and bandwidth values for each signal.

λ (nm)	**1530**	**1540**	**1550**	**1560**
Channel	3	2	4	1
Crosstalk (dB)	21.4	23.4	20.8	23.7
Losses (dB)	2.35	1.98	2.22	2.2
Bandwidth (nm)	7.68	7.78	8.01	8.08

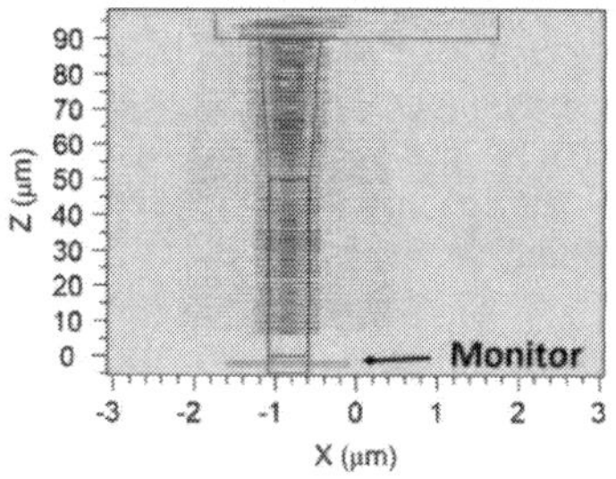

Fig. 8. Back reflection simulation and calculation.

4 Conclusions

This study presents an innovative design for a 1x4 demultiplexer in the C-band, utilizing three cascaded MMI couplers based on a SiN strip waveguide structure. By employing AI tools, specifically a CNN model, we were able to significantly enhance the optimization of BPM and FDTD simulations, allowing for more efficient analysis and design of the device. Through this advanced approach, we determined the optimal parameters required for the MMI couplers to accurately split four target wavelengths: 1530 nm, 1540 nm, 1550 nm, and 1560 nm. The total device length measures 6.63 mm, with insertion losses between 1.99 dB and 2.36 dB, and excellent crosstalk values ranging from 20.7 dB to

23.5 dB. Additionally, the optical bandwidth spans from 7.7 nm to 8.1 nm, making this demultiplexer highly suitable for long-distance optical communication networks utilizing DWDM technology in the C-band. A key achievement of this work is the demonstration of how AI-driven optimization, using the CNN model, can enhance BPM and FDTD simulations, yielding faster and more accurate results. This allowed for the fine-tuning of the device's geometry, leading to a low back-reflection loss (between 40 dB and 41 dB) without requiring a specialized angled MMI design. The use of SiN as the core material plays a pivotal role in minimizing back-reflection and maintaining signal quality, making the device highly efficient for data transmission. The results also highlight a tolerance range of 5.2 μm for the MMI coupler lengths, which provides flexibility during fabrication, ensuring compatibility with standard manufacturing processes. This robustness in the design is a significant advantage, allowing for easier production using existing semiconductor technologies. Furthermore, the simplicity of the design enables scalability, making it possible to add more MMI couplers and adjust their geometrical parameters to accommodate additional channels for broader applications. In summary, the proposed SiN-based 1x4 demultiplexer offers a highly optimized solution for advanced optical communication systems. Its low losses, strong crosstalk performance, and compatibility with DWDM applications in the C-band make it a valuable addition to photonic integration. The incorporation of AI techniques to optimize simulation and analysis has further refined its design, ensuring that it can meet the growing demands of high-speed data transmission and communication in future networks.

References

1. Isakov, O.; Frishman, A.; Malka, D. Data Center Four-Channel Multimode Interference Multiplexer Using Silicon Nitride Technology. *Nanomaterials* **2024**, *14*, 486.
2. Ohana, E.; Malka, D. O-Band Grating Couplers Using Silicon Nitride Structures. *Appl. Sci.* **2023**, *13*, 9951.
3. Zhou, X., Urata, R., & Liu, H.. Beyond 1 Tb/s Intra-Data Center Interconnect Technology: IM-DD OR Coherent? Journal of Lightwave Technology, **2020**, 38(2), 475–484.
4. Frishman, A.; Malka, D. An Optical 1×4 Power Splitter Based on Silicon–Nitride MMI Using Strip Waveguide Structures. *Nanomaterials* **2023**, *13*, 2077.
5. Brand, O.; Wolftson, B.; Malka, D. A Compact Polarization MMI Combiner Using Silicon Slot-Waveguide Structures. *Micromachines* **2023**, *14*, 1203.
6. Katash, N.; Khateeb, S.; Malka, D. Combining Four Gaussian Lasers Using Silicon Nitride MMI Slot Waveguide Structure. *Micromachines* **2022**, *13*, 1680.
7. Menahem, J.; Malka, D. A Two-Channel Silicon Nitride Multimode Interference Coupler with Low Back Reflection. *Appl. Sci.* **2022**, *12*, 11812.
8. Menahem, J.; Malka, D. 1 × 4 Wavelength Demultiplexer C-Band Using Cascaded Multimode Interference on SiN Buried Waveguide Structure. *Materials* **2022**, *15*, 5067.
9. Khateeb, S.; Katash, N.; Malka, D. O-Band Multimode Interference Coupler Power Combiner Using Slot-Waveguide Structures. *Appl. Sci.* **2022**, *12*, 6444.
10. Malka, D. A Four Green TM/Red TE Demultiplexer Based on Multi Slot-Waveguide Structures. *Materials* **2020**, *13*, 3219.
11. Guzmán-Sepúlveda, J.R.; Guzmán-Cabrera, R.; Castillo-Guzmán, A.A. Optical Sensing Using Fiber-Optic Multimode Interference Devices: A Review of Nonconventional Sensing Schemes. *Sensors* **2021**, *21*, 1862.
12. Ziman, M.; Feiler, M.; Mizera, T.; Kuzma, A.; Pudis, D.; Uherek, F. Design of a Power Splitter Based on a 3D MMI Coupler at the Fibre-Tip. *Electronics* **2022**, *11*, 2815.
13. Khalilzadeh, H., Bahrami, A., & Badri Ghavifekr, H. MMI-based all-optical four-channel wavelength division demultiplexer. Photonic Network Communications, **2018**
14. Peruzzo, A., Laing, A., Politi, A., Rudolph, T., & O'Brien, J. L. Multimode quantum interference of photons in multiport integrated devices. Nature Communications, **2011**, 2(1).
15. Krause, E.E.; Malka, D. Optimizations of Double Titanium Nitride Thermo-Optic Phase-Shifter Heaters Using SOI Technology. *Sensors* **2023**, *23*, 8587.
16. Pugachov, Y.; Gulitski, M.; Mizrahi, O.; Malka, D. Design of All-Optical Logic Half-Adder Based on Photonic Crystal Multi Ring Resonator. *Symmetry* **2023**, *15*, 1063.
17. L. B. Soldano and E. C. M. Pennings, "Optical multi-mode interference devices based on self-imaging: principles and applications," in *Journal of Lightwave Technology*, vol. 13, no. 4, pp. 615-627, April 1995, doi: 10.1109/50.372474.
18. Zhang, S., Ji, W., Yin, R., Li, X., Gong, Z., & Lv, L. "Full Bandwidth Wavelength Division Multiplexer/Demultiplexer Based on MMI, " IEEE Photonics Technology Letters, **2018** 30(1), 107–110.
19. D. J. Blumenthal, R. Heideman, D. Geuzebroek, A. Leinse and C. Roeloffzen, "Silicon Nitride in Silicon Photonics," in *Proceedings of the IEEE*, vol. 106, no. 12, pp. 2209-2231, Dec. 2018, doi: 10.1109 /JPROC.2018.2861576.
20. Pugachov, Y.; Gulitski, M.; Malka, D. Photonic Crystal Flip-Flops: Recent Developments in All Optical Memory Components. *Materials* **2023**, *16*, 6467.

High photoinduced birefringence in thermally treated layers of the azopolymer PAZO with significantly changed absorbance spectrum

Georgi Mateev[1,2*], *Deyan Dimov*[1,2], *Dimana Nazarova*[1,2], *Elena Stoykova*[1], *Keehoon Hong*[3] *and Lian Nedelchev*[1,2]
[1]Institute of Optical Materials and Technologies–Bulgarian Academy of Sciences, Acad. G. Bonchev Str, Bl. 109, Sofia 1113, Bulgaria
[2]University of Chemical Technology and Metallurgy, 8 Kliment Ohridski Blvd, Sofia 1756, Bulgaria
[3]Electronics and Telecommunications Research Institute, 218 Gajeong-ro, Yuseong-gu, Daejeon 34129, Republic of Korea

Abstract. In this experimental work we present the effect of thermal treatment of azopolymer thin films on their absorbance spectra and how that influences the photoinduced birefringence of the azopolymer. The azopolymer we use – PAZO (poly[1-[4-(3-carboxy-4-hydroxyphenylazo)benzenesulfonamido]-1,2-ethanediyl, sodium salt]) is commonly investigated material for polarization holography because it is easily accessible and has well pronounced photoanisotropic properties. In a recent study it was discovered that its absorbance spectrum gradually changes upon heating to 250°C and as a result the absorbance in the range 400-600 nm is increased. That raises the question is it possible to increase the photoinduced birefringence for wavelengths that are far away from the peak of absorbance of the azopolymer PAZO, located approximately at 360 nm. For example, there are powerful lasers at 532 nm, which could be used for optical recording instead of traditionally used lasers in the blue and UV range. To answer this question, we investigated thin film samples of PAZO deposited on quartz substrate and measured continually their spectra of absorbance while heating the samples from room temperature to 250°C. We also measured the photoinduced birefringence using pump laser at 532 nm before and after the thermal procedure. Thus, we report significant increase of the birefringence at 532 nm after the thermal treatment and mention some potential applications.

1 Introduction

Azopolymers are well-known media for many optical applications mainly in the polarization holography [1-3] but also in other areas like in advanced fabrication [4], where the authors are imprinting nanopatterns of flexible substrate or for solar-thermal energy conversion and storage [5]. With all of these optical or optically related applications it is very important to analyze properly the spectra of these materials and how you can change and influence them. Usually, the influence of the temperature is analyzed with relation of the highest laser fields illuminating the azomaterials [6-7] or other indirect ways like with thermogravimetric analysis [8-9]. There are also examples of studying the influence of temperature in other materials in polarization optics. Recently Suzuki et al. analyzed how the temperature influences the diffraction efficiency of polarization gratings recorded in liquid crystals by circularly polarized beams [10]. In that work the authors report that the increase of the temperature leads to increase of the diffraction efficiency. Earlier we studied how the temperature influences the formation of photoinduced birefringence and the investigated material was the azopolymer P_{1-2} [11]. It was shown that if you increase the temperature of the sample and then start to illuminate it, the maximal birefringence decreases its value. However, none of these shows whether heating azopolymer films could bring structural changes which can eventually improve their optical properties. So, in recent years there have been reported works when azopolymers were directly heated with the idea of checking how the chemical structure [12] or the absorbance spectra [13] will change. For example, Mateev et al. report a noticeable change of the absorbance spectra of PAZO and significant increase specifically in the range 400-600 nm [13].

That is important because of the reorientation processes of the azomolecules which are in the core of every photo induced polarization effect – converting thin films from isotropic into anisotropic, inducing chirality and creating surface relief gratings [1-3]. When a thin film of azomaterial is illuminated with linearly polarized light from the absorbance band the azochromophores start to reorient with respect to the orientation of the polarization.

So, an increase of the absorbance in the range 400-600 nm could potentially mean that the most effective pump wavelength can shift towards larger wavelength. Nedelchev et al. investigated the spectral dependence of the birefringence [14] and found that the highest values for PAZO were achieved for pump laser with

* Corresponding author: g_mateev@abv.bg

wavelength 442 nm, while around 532 nm the photoinduced birefringence (Δn) is significantly smaller which decreases the potential application using lasers at this wavelength. On the other hand, there are very powerful lasers at 532 nm like 12 W laser "Verdi" which could be used for making large optical elements if the process of photoorientation becomes more efficient. And that is the main objective of the present work. We investigate whether or not we can increase the photoinduced birefringence for pump laser at 532 nm via increasing the absorbance of PAZO at 532 nm after thermal treatment of our samples.

2 Experimental methods and results

2.1 Preparation of the samples

We have prepared our samples of PAZO with spin coating procedure using methanol as solvent. The concentration of the solution for this work was 175 mg/ml. The solution was sonicated and a magnetic stirrer was used. That provided us with samples with good optical quality and 2800-2850 nm thickness which were used for the investigation of the birefringence. Additionally, another series of samples of PAZO was made with concentration of 30 mg/ml and 200 nm thickness. The use of that samples will be discussed in the following subsection.

2.2 Spectral investigation

As pointed out in the introduction, it is reported already that heating thin films of PAZO can lead to a significant change in the absorption spectra [13]. We decided to investigate the dynamic change of the spectra while heating with thermal system THMS600 (Linkam Scientific, Salfords, UK) from room temperature to 250°C with constant rate of 1°C/min (Fig. 1). Spectral measurement was performed by a high-resolution spectrometer HR4000 (Ocean Optics, Orlando, USA).

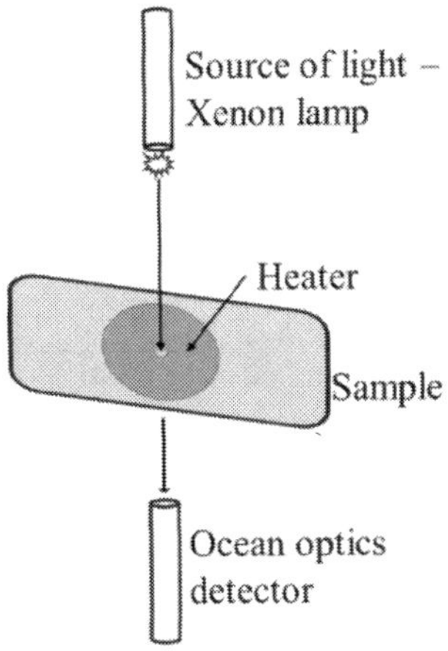

Fig. 1. Optical scheme for real-time absorbance measurement during the heating of the sample.

Heating to higher temperatures is undesirable because the polymer starts to degrade. The spectra of absorbance from 250 to 1000 nm were measured at a rate of one spectrum per minute. We are showing several spectra of absorbance corresponding to different temperatures for a sample with 200 nm thickness in Fig. 2. The sample was thinner than the samples used later for the birefringence measurement because for thicker samples there was the possibility of reaching saturation of the signal for the important range 350-600 nm where we have significant absorbance. For samples with that thickness (200 nm) we are sure that we can get reliable data.

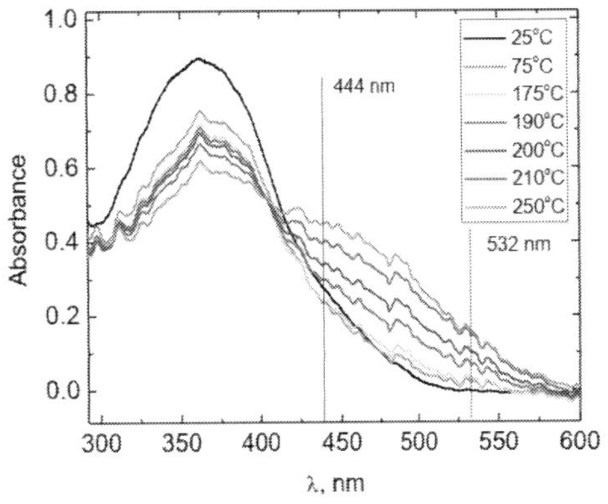

Fig. 2. Absorbance spectra of PAZO thin films with 200 nm thickness taken while heating the sample from room temperature to 250°C.

With the black curve we see the sample before the start of the heating procedure. By heating the sample, we see how the peak of absorbance at 359 nm starts decreasing while in the range between 425-600 nm the absorbance increases. Two vertical lines indicate the wavelengths of our two pump lasers (444 and 532 nm). The results correspond with previous reports [13].

2.3 Birefringence measurement

For measurement of the birefringence, we have used a well-known optical setup with polarimeter (Fig. 3). The sample is irradiated by vertically polarized pump beam.

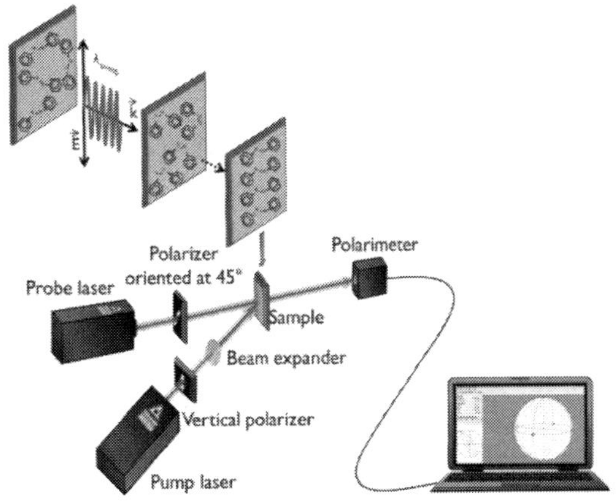

Fig. 3. Optical setup for birefringence measurement.

For our study we have applied two different pump lasers – DPSS laser at 444 nm and 12 watts DPSS Laser "Verdi" at 532 nm. As explained in the Introduction section, the second laser is very powerful but its wavelength is on the edge of the absorbance band so without the suggested thermal treatment the process of reorientation of the azomolecules is not effective. Meanwhile through the illuminated area passes also a probe beam at 635 nm (DPSS laser), linearly polarized at 45° before the sample. Its polarization after the sample is measured by polarimeter Thorlabs PAX57. Throughout the experiments we measure the background (the signal before the start of the irradiation with the pump beam) of the probe laser for several seconds. Thus, we can be sure that in the beginning of the experiment the sample is isotropic. Then we turn on the pump laser and the reorientation of the molecules begins. The process has saturation and when reached we turn off the pump laser again and record several minutes of relaxation. The intensity of the pump laser with 444 nm is approximately 125 mW/cm^2. Similarly, the intensity of the 532 nm pump laser is 300 mW/cm^2. The value of Δn in time is determined by the following formula:

$$\Delta n = \frac{\lambda}{2\pi d}\operatorname{atan}\left(\frac{S_3}{S_2}\right), \quad (1)$$

where with S_2 and S_3 we denoted the Stokes parameters and λ is the wavelength of the probe laser (635 nm).

The samples were investigated before and after thermal treatment (or annealing) which represents heating for 2h from room temperature to 190°C in oven. After reaching the desired temperature the oven holds it for another 0.5 h and then the sample is slowly cooled to room temperature again. Immediately after reaching room temperature the sample is prepared for measurement of its birefringence once at 532 nm pump laser and once at 444 nm. The results for the pump laser at 532 nm shown in Fig. 4 reveal a very significant (almost 70%) increase of the birefringence for the thermally treated sample.

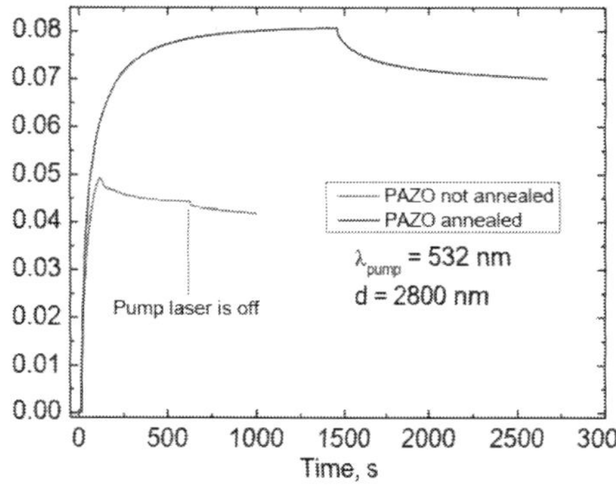

Fig. 4. Birefringence kinetics of both annealed and not-annealed sample with pump laser at 532 nm.

Even further the highest birefringence has larger value then the highest reported values of birefringence for non-treated PAZO films in any wavelength of pump laser and 635 nm for probe laser. This confirms our initial hypothesis that thermal treatment can make lasers around 532 nm more efficient for pump. And considering the existence of very powerful lasers at 532 nm that opens the possibility of creating larger polarization-selective optical elements.

It is also worth mentioning that the birefringence curve for the not-annealed sample, shown in Fig. 4, doesn't reach saturation. After a brief period of increase, the birefringence starts to decrease so the relaxation doesn't start at the highest value of the birefringence. Similar behavior was previously reported exactly for pump laser at 532 nm [14].

In Fig. 5 we are showing also the annealed and not-annealed samples when the pump laser is 444 nm.

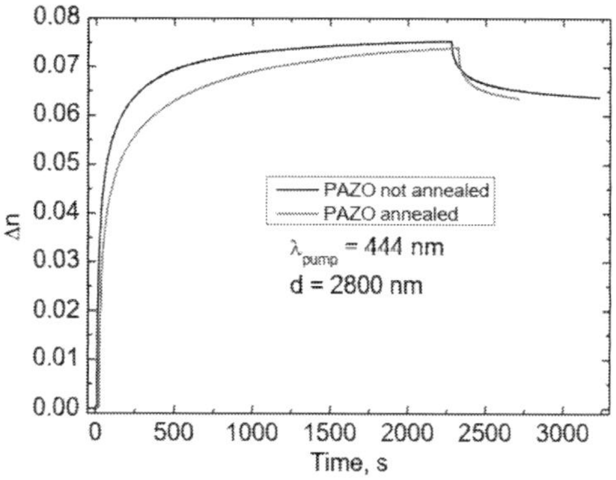

Fig. 5. Birefringence kinetics of both annealed and not-annealed sample with pump laser at 444 nm.

In this case we do not observe significant change of the birefringence curve after the annealing procedure.

3 Conclusions

We have investigated the change of the absorbance spectra of thin films deposited from the azopolymer PAZO after thermal treatment of the samples from room temperature to 250°C. As a result of the annealing, we observe higher absorbance at the range 400-550 nm which leads to significant increase of the birefringence with almost 70% when using laser with λ = 532 nm. That allows much wider use of lasers at that range for pumping PAZO for different applications as recording of polarization-selective diffractive optical elements, for digital polarization holography, digital storage of information, etc.

Acknowledgements: This work was supported by the Institute of Information & Communications Technology, Planning & Evaluation (IITP), grant No. 2019-0-00001. The authors are grateful for the funding from the European Union–NextGenerationEU, through the National Recovery and Resilience Plan of the Republic of Bulgaria, project BG-RRP-2.004-0002, "BiOrgaMCT". Research equipment of Distributed Research Infrastructure INFRAMAT, part of Bulgarian National Roadmap for Research Infrastructures, supported by Bulgarian Ministry of Education and Science, was used in this investigation.

References

1. L. Nikolova, P S. Ramanujam, Polarization holography, (Cambridge: Cambridge University Press, 2009)
2. X. Wang, Azo Polymers: Synthesis, Functions and Applications, (Springer-Verlag. Berlin Heidelberg, 2017)
3. Emoto, E. Uchida, T. Fukuda, Optical and physical applications of photocontrollable materials: Azobenzene-containing and liquid crystalline polymers, Polymers **4**, 150-186 (2012)
4. Y. Bowen, C. Feng, H. Shuai, Y. Haifeng, Athermal and Soft Multi-Nanopatterning of Azopolymers: Phototunable Mechanical Properties, Angewandte Chemie International Edition **59,** 4035-4042 (2020)
5. W. Si, B. Hans-Jürgen, Solar-Thermal Energy Conversion and Storage Using Photoresponsive Azobenzene-Containing Polymers, Macromolecular Rapid Communications **41**, 1900413 (2019)
6. A. Miniewicz, A. Sobolewska, W. Piotrowski, P. Karpinski, S. Bartkiewicz, E. Schab-Balcerzak, Thermocapillary Marangoni Flows in Azopolymers, Materials **13**, 2464 (2020)
7. A. Timóteo, J. Ribeiro, P. Ribeiro, M. Raposo, Dynamics of creation photoinduced birefringence on (PAH/PAZO) n layer-by-layer films: Analysis of consecutive cycles, Optical Materials **51**, 18-23 (2016)
8. F. Raquel, G. Junkal, E. Arantxa, T. Agnieszka, Hybrid materials based on azopolymer and sol-gel synthesized silver-containing titanium oxide nanoparticles with photoinduced birefringence, RSC Advances **5**, 15740-15748 (2015)
9. M. Ghoneim, A. El-Sonbati, M. Diab, A. El-Bindary, L.S. Serag, Supramolecular Assembly on Coordination of Azopolymer Complexes: A Review, Polymer-Plastics Technology and Engineering, **54**, 100-117 (2015)
10. M Suzuki, S. Ohara, M. Sakamoto, K. Noda, T. Sasaki, N. Kawatsuki, H. Ono, Temperature dependence of the diffraction efficiency of circular polarization gratings made by liquid crystal molecules with anisotropic absorption, Optical Materials Express, **14**, 1857-1864 (2024)
11. Nazarova, D, Mateev, G, Ivanov, D, Blagoeva, B, Kostadinova, D, Stoykova, E, Nedelchev, L. Photoinduced birefringence in thin azopolymer films recorded at different temperatures. Bulgarian Chemical Communications **48**, Special Issue G, 75-78 (2016)
12. L. Xiao, Z. Weiguang, J. Ye, H. Jianjia, C. Dongzhong, Phase Behaviors and Photoresponsive Thin Films of Syndiotactic Side-Chain Liquid Crystalline Polymers with High Densely Substituted Azobenzene Mesogens, Chemphyschem: European Journal of Chemical Physics and Physical Chemistry, e202400421 (2024)
13. G. Mateev, L. Nedelchev, A. Georgiev, D. Nazarova, Improvement of the photoinduced birefringence in azopolymer PAZO doped with TiO2 nanoparticles via thermal treatment, Open Material Sciences, **5**, 19-23 (2019)
14. L. Nedelchev, D. Ivanov, B. Blagoeva, D. Nazarova, Optical anisotropy induced at five different wavelengths in azopolymer thin films: Kinetics and spectral dependence, Journal of Photochemistry and Photobiology A: Chemistry, **376**, 1-6 (2019)

Mobile system to detect plastic particles in critical scenarios: potential approach for attachment in UAVs

Guilherme Pereira[1], Carlos Marques[1]*

[1]CICECO - Aveiro Institute of Materials, Physics department, University of Aveiro, 3810-193 Aveiro, Portugal
[*]Corresponding author: g.pereira@ua.pt

Abstract. This study aims to develop an efficient and cost-effective system for detecting small particles, with the potential for integration with unmanned aerial vehicles (UAVs). The research is divided into two main components: simulation and practical experimentation. The findings demonstrate that the system is capable of detecting particles with widths up to 400 μm. Additionally, it was established that the signal response could be modulated following a sigmoid function. These results confirm the functionality of the system; however, the simulation component indicates potential areas for improvement. Moreover, there is significant potential for the system to be adapted for UAV attachment.

1 Introduction

Year after year the problem with small sized plastic particles imposes an increasing danger to human society. First it was found in the sea, after that in the atmosphere [1]. With the passage of time, we started to find it in animals [2] implying that this hazard has entered the food chain in almost all maritime habitats. More recently studies have shown that this calamity is already present in the human body [3]. The production of plastics is expected to increase in the next years. **Fig. 1** was taken from a Nature article and shows exactly that predicted increase [4].

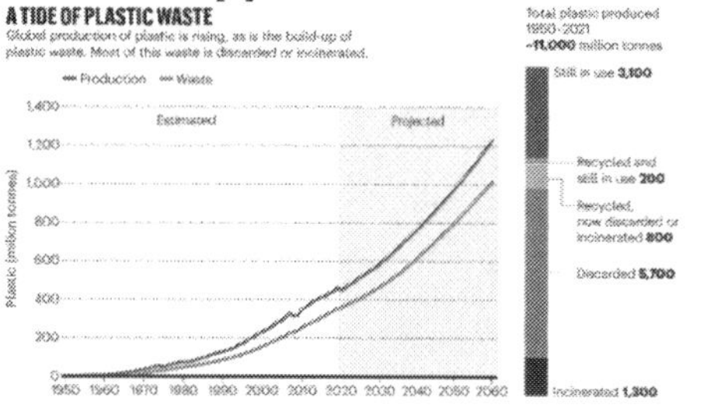

Fig. 1. Estimated and projected production and waste of plastic in the world from 1950 to 2060.

Considering this, several systems for detecting microplastics have been developed, ranging from visual identification to FTIR and Raman spectroscopy [5]. However, these methods necessitate sampling and advanced treatment processes, which are costly and time-consuming.

This study represents an initial effort to develop a cost-effective and efficient system for detecting plastic particles. The goal is to create a real-time detection system that provides immediate data, enabling on-the-spot analysis. Additionally, the system is designed with the intention of being attachable to UAVs. To achieve these objectives, the research is divided into two phases. The first phase involves simulating the projected system's performance, while the second phase consists of experimental trials to validate the hypothesis.

2 Methodology

The initial phase of this study involved simulating the system's detection capabilities and particle flow using MATLAB and SolidWorks. The subsequent phase consisted of experimental trials aimed at validating or refuting the prior simulations, with the intent to either confirm the simulations' accuracy or enhance their quality.

The system's operating principle is straightforward. It comprises a light-emitting diode (LED) and a phototransistor (PT). The phototransistor is configured inversely, such that the presence of light results in a low potential measurement, whereas the obstruction of light causes an increase in the measured value. It is crucial to avoid saturating the phototransistor during this process.

Given this configuration, the system leverages the potential spike induced by light obstruction to detect the passage of particles. A schematic representation of the system's working principle is illustrated in **Fig. 2**.

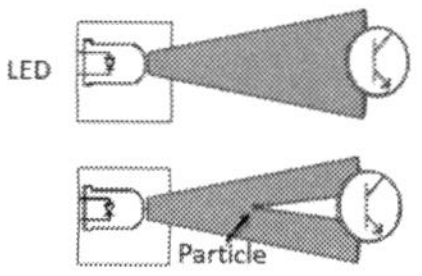

Fig. 2. Schematic of the working principle of the system.

Additionally, we conducted particle flow simulations using SolidWorks to analyze their behavior within our designed prototype (see **Fig. 3a**). MATLAB simulations were also performed to study the expected signal when an object obstructs the path between the LED and the phototransistor (PT). Based on the initial conclusions drawn from these simulations, we proceeded to experimental trials. The prototype utilized in these trials is depicted in **Fig. 3b**.

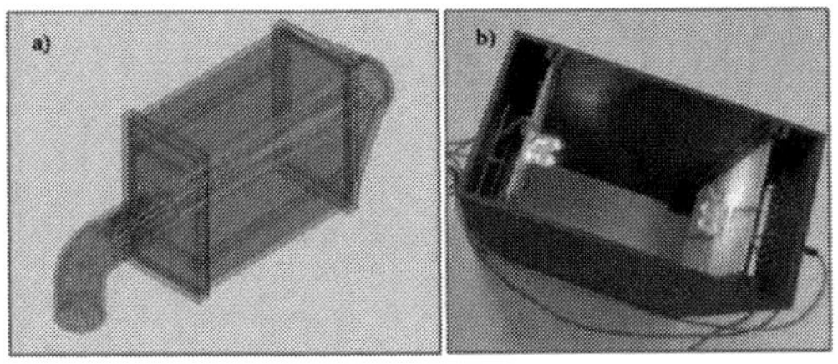

Fig. 3. a) Simulation of the flow of the particles through the prototype and **b)** prototype used to measure the signal in face of an obstruction.

The prototype was designed to facilitate obstruction experiments. Instead of particles, metal bars with a fixed width were used to create a more controlled and manageable obstruction of the PTs. This approach allows for a comparison of the two photodetectors' results and enables the study of alignment impact.

Data from the PTs was collected using an Arduino Uno R3. A MATLAB program with a graphical user interface (GUI) was employed to retrieve the data and display the system's real-time status.

3 Simulation

The first step of the simulation involved theorizing the system's operation and determining an accurate modulation method. To achieve this, we sought the most appropriate equation to describe the experiment.

Considering the system's constraints, we identified certain characteristics that the function must possess:

1. When $x \to 0 \Longrightarrow V \longrightarrow V_{min}$;
2. When $x \to \infty \Longrightarrow V \longrightarrow V_{max}$;
3. The curve must vary depending on the particle's distance to the phototransistor.

With x being the samples' width and V the signal.

Considering these properties, we concluded that the best function to describe this behavior is the sigmoid function which in this case is given by:

$$V = \frac{V_{min} - V_{max}}{1+(x/x_0)^p} + V_{max} \tag{1}$$

Where V_{min} and V_{max} are the minimum and maximum potential the PT can have, respectively, x_0 and p are constants that depend on the system.

Lacking direct knowledge of these constants' values, we decided to estimate them based on our system's initial results and predictions. We hypothesized $x_0=1400-0.5\times1100\times d$, indicating that x_0 depends only on the distance to the detector (d) and not on the particle size. The negative signal was inferred from initial results, and we also assumed p=2.

4 Results and Analysis

4.1 System response time

Firstly, we must analyze the response of our electronic devices. The initial step involves examining the Arduino to determine the optimal BaudRate for communication within the context of our experiment.

To achieve this objective, we studied the intervals between data for various BaudRates and plotted the results in **Fig. 4**.

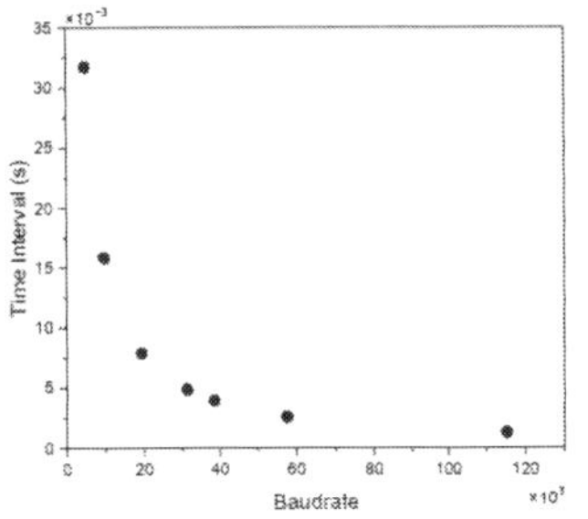

Fig. 4. Time intervals between the retrieved data in function of the BaudRate chosen in the Arduino IDE.

As observed, higher BaudRates result in smaller time intervals between data transmissions. However, this decrease reaches a limit, as depicted in the graph. We conclude that the time interval varies inversely with the BaudRate until a specific threshold, beyond which further increases in BaudRate do not affect the time interval between data.

The next step is to analyze the response time of the system. Using a signal generator and an oscilloscope, we sent a sinusoidal signal to the system and found how the system responded to the frequency applied. **Fig. 5** shows the results obtained.

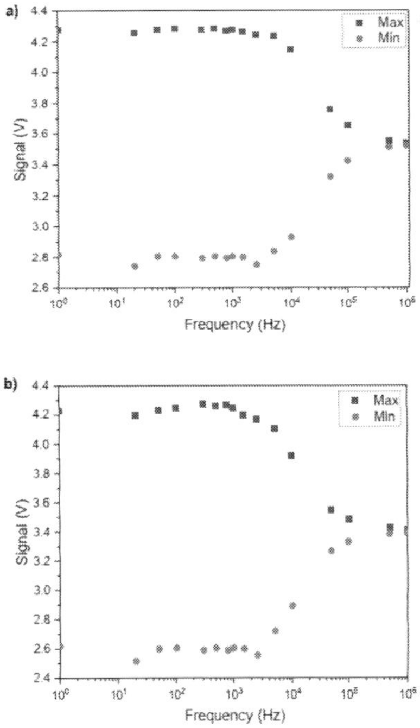

Fig. 5. Signal response for the PT used in function of the signal frequency applied to both, in **a)** PT1 and **b)** PT2.

From the graphics we can observe that the frequence of response is reasonable up to at least 10^4Hz, corresponding to time response of 0.1ms. Beyond this threshold, the system's response time fails to synchronize with the frequency of the input signal. Consequently, the system exhibits a diminished amplitude response, eventually stabilizing at zero amplitude, remaining fixed at the mean value.

It is important to consider that in this experiment, the signal is applied to the LED, so this frequency limit might not be attributed to the phototransistor (PT) but rather to the LED used. Despite this potential limitation, the value obtained is sufficient for our analysis, and thus, the possible constraint imposed by the LED does not pose a problem.

4.2 Size of the particle effect

The second part involves studying how particle size affects the value observed in the phototransistor (PT) signal. We begin by presenting the simulation results, as shown in **Fig. 6**.

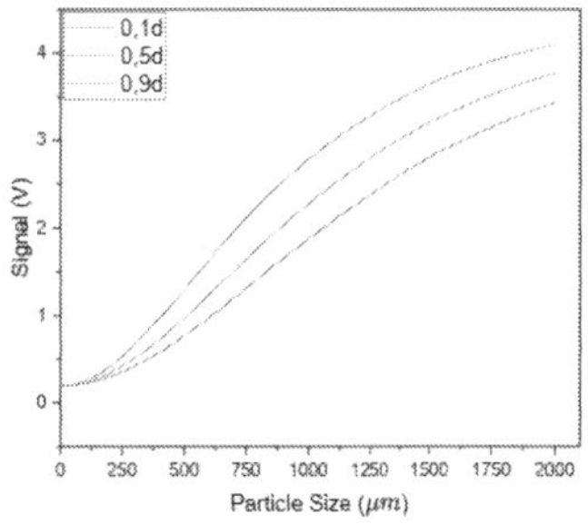

Fig. 6. Results of the simulation used to describe the signal detection in function of the sample size for some distances to the detector.

Considering the previous considerations, we can see that a 0 μm particle should not affect the system, and therefore, no alterations should be observed. Nevertheless, we can observe differences in the curves for varying distances to the detector. It is important to note that despite these differences, the rate of increase remains consistent because the p value is fixed.

Doing some trials, we get the results represented in **Fig. 7**

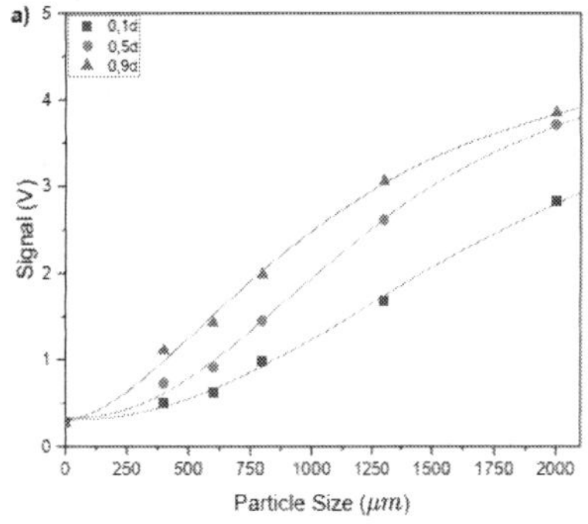

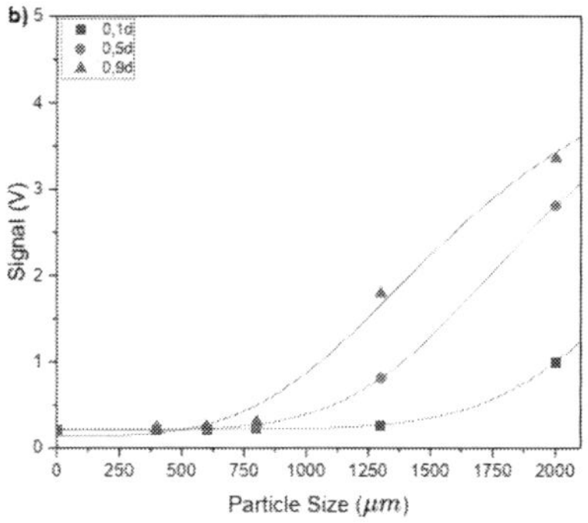

Fig. 7. Experimental results obtained for several particles along with a sigmoid fitting, in **a)** PT1 and **b)** PT2.

From our observations, the predicted evolution of the curve is noticeable confirming the simulation results. However, the differences between the curves are not as hypothesized before, leading us to conclude that the simulation could be more accurate. The oversimplification of the values of x_0 and p is inadequate to describe this system according to the experimental data.

4.3 Distance to the detector effect

To further study the distance-dependent response of the system, we conducted a MATLAB simulation using the same sigmoid function as previously described and the same assumptions for the constant values. The obtained results are represented in **Fig. 8**.

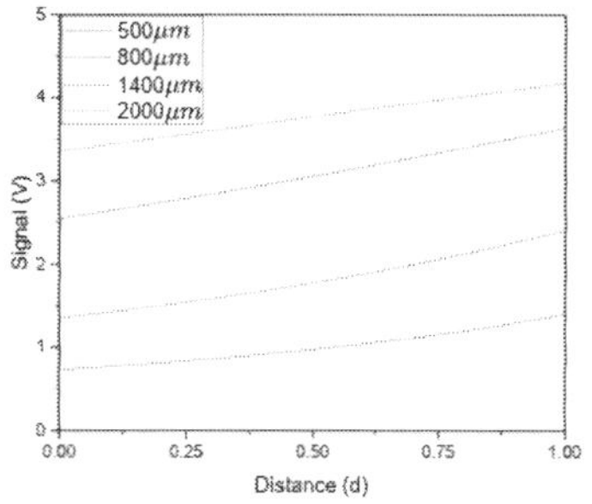

Fig. 8. Simulation of the signal in function of the distance to the detector for several sample sizes.

After some experiments of passage for particles with different width we obtained the results represented in **Fig. 9**.

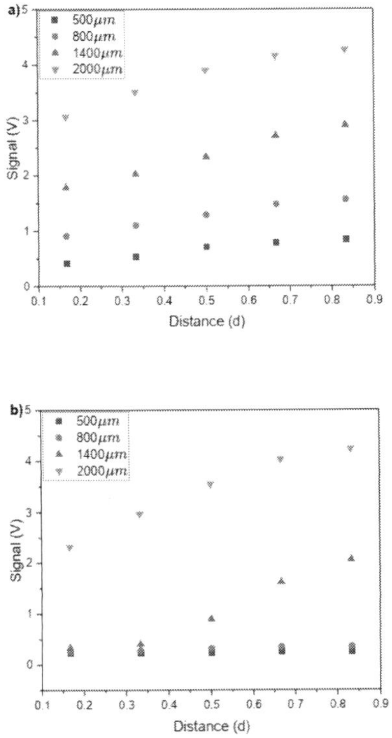

Fig. 9. Experimental results obtained for several distances to the detector, in **a)** PT1 and **b)** PT2.

These results do not appear to be neither linear nor quadratic nor any easy fit.

Considering that the initial considerations for the value of x_0 are not accurate and considering that we supposed that its value is influenced by d but also considering that the value of the constant p can also be affected by the distance we cannot determine an equation that is able to describe this behavior without getting the actual dependence of this constants on d.

5 Cost and Attachment to UAVs

Considering two of the main objectives, make this a cheap and accurate detector system and attach it to UAVs, we can consider the cost of the prototype and the cost of the material involved.

To make the prototype we invested around 30 euros. The real system would be more expensive considering the need for a smaller Arduino, a system of communication UAV-ground and a larger number of LEDs and PTs. On the other hand, we also must consider the material of the box that enclosures all those components.

To attach this system to UAVs there are some things we must consider. First, the electronics of the system must be stable to avoid any misreading and consequently misdetections. Another one is that since we have a sensible system, and the velocity of detection is limited, we must guarantee some conformity in the flow velocity.

Last, we must consider that this system is not perfect and has a limitation to the detection of plastic particles, for example smaller particles might not be detected and demand a greater investment and a greater precision that this system does not provide.

6 Conclusions

With these considerations in mind, we conclude that integrating this system with a UAV is feasible, albeit requiring significant effort to enhance its practicality.

Moreover, the system demonstrated promising results in detection. The amplitude-to-base signal ratio extended the detection range, thereby improving the system's usefulness and effectiveness.

This system proves to be simple and cost-effective. For more accurate real-time data reading, it must be complemented by software capable of self-analyzing the provided data and identifying the passage of plastic particles autonomously.

Future work should focus on accurately determining the correlations between the distance to the detector and the constants x_0 and p. Once these relationships are established, a more precise system can be simulated, facilitating the development of software and the integration of the system with UAVs.

This work was developed within the scope of the projects CICECO (LA/P/0006/2020, UIDB/50011/2020 & UIDP/50011/2020) and DigiAqua (PTDC/EEI-EEE/0415/2021), financed by national funds through the (Portuguese Science and Technology Foundation/MCTES (FCT I.P.)).

References

1. Y. Zhang, S. Kang, S. Allen, D. Allen, T. Gao, M. Sillanpää, "Atmospheric microplastics: A review on current status and perspectives," *Earth-Sci. Rev.*, **203**, 103118 (2020). doi.org/10.1016/j.earscirev.2020.103118
2. S.E. Nelms, J. Barnett, A. Brownlow, N.J. Davison, R. Deaville, T.S. Galloway, B.J. Godley, "Microplastics in marine mammals stranded around the British coast: ubiquitous but transitory?" *Sci. Rep.*, **9**, 1075 (2019). doi.org/10.1038/s41598-018-37428-3
3. C.-Y. Wong, C.-O. Leong, "Impact of Microplastics and Nanoplastics on Human Health," *Nanomaterials*, **11**(2), 496 (2021). doi.org/10.3390/nano11020496
4. Nature Editorial, "Three ways to solve the plastics pollution crisis," *Nature*, **611**, 234-237 (2023). doi.org/10.1038/d41586-023-00975-5
5. J.C. Prata, J.P. da Costa, A.C. Duarte, T. Rocha-Santos, "Methods for sampling and detection of microplastics in water and sediment: A critical review" *TrAC Trends Anal. Chem.*, **110**, 150-159 (2019). doi.org/10.1016/j.trac.2018.10.029

Emulation and Testbed Prototyping of Laser Beam Propagation Characteristics through Atmospheric Turbulence for Optical Satellite Feeder Links

Haider M. Al-Juboori [1*]

[1] South East Technological University, Faculty of Engineering, Department of Electronics Engineering and Communications, Carlow, Ireland

Abstract. Two classical scenarios in space communications can be examined to evaluate the link performance of free space optical communications (FSO). These scenarios include the downlink and uplink between earth ground stations and near earth geostationary (GEO) satellites, as well as between the earth and spacecraft that are located a significant distance from the earth (i.e., one or more astronomical units (AU) away). Deep space is often defined as being more than 0.01 AU, or around 1,500,000 km, from Earth. In the first part of this analytical research, various practical system characteristics for optical lasers, telescopes, transmitters, receivers, and atmospheric disturbances, including absorption and scintillation, are taken into consideration. The simulation outcomes were compared to significant experimental data from DLR & JAXA's OICETS/Kirari-Japan to validate the extended simulation model. Additionally, this work will propose a synthetic effects-based optical turbulence generator in order to deeply investigate atmospheric turbulence disturbances in free-space optical communication systems and support the advancement of artificial intelligence processing-based mitigation solutions. Worth mentioning, the initial outcomes and analytical comparisons demonstrate the preliminary viability of the simulation model for the link budget and scintillation estimation for FSO investigations, which can give deep insight into the development of the proposed testbed prototyping, machine learning-based laboratory measurement system, and enhancement possibility for FSO project design and execution for long- and medium-term space mission planning.

1 Introduction

Enlarging the satellite-ground data downlink capability is necessary to fulfil the high-speed transmission demands of the growing volume of satellite data and its gathering [1, 2]. The performance of realistic laser communication systems is impacted by atmospheric turbulence in free space propagation, which results in refractive index disturbances and intensity scintillation, beam drift, beam spread, and angle of arrival fluctuations in laser signals. The International Astronautical Federation (IAF) and its member organizations, including NASA and ESA, have made multiple proposals and conducted successful experiments in recent decades. Numerous space communications scenarios are considered to examine the link performance of free space optical communications (FSO) or even deep space optical communications (DSO).

Examples include those in the downlink and uplink between earth ground stations and (a) geostationary or low-earth orbit satellites, (b) spacecraft in lunar orbit, or (c) spacecraft at faraway locations from earth—roughly 1-3 AU. Several functional system characteristics for lasers, optical telescopes, transmitters, receivers, and atmospheric disturbances, including scintillation and absorption, are taken into consideration in these analytical investigations, that can help to generate deeper vision and enhance the optimization processes of optical satellite feeder links. By simulating and analysing a bidirectional free-space optical communication (FSO) link. This work attempts to address and emulate most of these challenges that might give novel chances to dive deeper into the characterization processes for optical satellite feeder links for downlink and uplink scenarios. The initial theoretical backgrounds will illustrated in section 2. The performance analyses and subsequent discussions are presented in section 3. The valuable explanation related to the advancement of artificial intelligence processing-based mitigation solutions are characterized in section 4, Concluding remarks on our work and its potential for use in GEO feeder links are presented in section 5.

2 Methodology

Generally, the S, X, and Ka bands, optical communications have benefits over traditional radio frequency communications in terms of beam divergence (up to a factor of 100), high bitrate delivery, and notably smaller flight terminal aperture sizes [3]. LLCD project (Lunar Laser Communication Demonstration) by NASA, SILEX (Semiconductor-laser Intersatellite Link

* Corresponding author: haider.aljuboori@setu.ie

EXperiment) by ESA ARTEMIS, and ARTEMIS to OGS (Optical Ground Station) are international cooperation examples of demonstration and evaluation projects that have been completed with collaboration between NASA and ESA [4-6]. Worth mentioning, some future initiatives like PSYCHE [7] and DOCS L5 [8] are also being planned. All scientific contributions and analyses, related to deep characterizations of free-space optical communication channels, might be very supportive and beneficial for these promising projects.
In principle, plane wave approximation, spherical wave propagation, and Gaussian beam propagation models are frequently employed in free space optical communications for the fundamental link budget computation [9, 10]. Gaussian beam propagation is the main topic of the current study. If one uses the same planning parameters, the methodical technique will yield the same outcomes [11].

2.1 Gaussian beam propagation

The wave front is described by the beam radius or half spot size w_0 and $w(L)$ in the Gaussian beam propagation model. The beam radius grows with distance L from the transmitter as a result of the divergence. The link budget's free space transmission efficiency, $\eta_{FS}(L)$, will essentially be substituted with the beam divergence represented by $w(L)^2/w_0^2$. Then, considering r as the lateral radius from the beam axis, L_R as the Rayleigh length, and D_R as the receiver aperture diameter, one can write the distribution of the irradiance $I(r, L)$ of the Gaussian beam with the beam radius $w(L)$ at a distance L from the transmitter as [2];

$$\frac{I(r,L)}{I(r,0)} = \frac{w_0^2}{w(L)^2} \cdot \exp\left(-\frac{2 \cdot r^2}{w(L)^2}\right) \cdot \eta_T \cdot \eta_{Atm}(L) \cdot \eta_P \quad (1)$$

$$w(L) = w_0 \cdot \sqrt{1 + \left(\frac{L}{L_R}\right)^2} \quad (2)$$

$$L_R = \frac{\pi \cdot w_0^2}{\lambda} \quad (3)$$

$$P(r,L) = I(r,L) \cdot \pi \cdot D_R^2 \cdot \eta_R \quad (4)$$

with $P(L)$ as transmitted/received power at a distance L; η_T and η_R as transmitter and receiver telescope aperture efficiencies; η_P as pointing efficiency; $\eta_{Atm}(L)$ as the atmospheric transmittance due to absorption loss, Rayleigh scattering (size of the scatterers $\ll \lambda$) and Mie scattering (size of the scatterers $\approx \lambda$).

2.2 Atmosphere absorption and scattering

Certain laser photons will be absorbed by molecules and electrons during free-space optical wave transmission through the atmosphere. Whereas electron absorption mostly occurs in the ultraviolet spectrum, molecular absorption occurs in the infrared [3]. In addition to these absorptions, water molecules exhibit characteristic resonances at various wavelengths, which result in a significant absorption loss and nearly zero transmittance "0%" (Andrews *et al.* give more related illustration in reference [9]). Typically, the atmosphere has a high transmittance at several wavelengths, ranging from 70% $\leq \tau(L) \leq$ 90%, enabling laser optical transmission. For deep space optical communications, certain common wavelengths of 800–850 nm, 1064 nm, or 1550 nm can be employed.

2.3 Turbulence, scintillation, and beam wander

The motion of the earth's atmosphere and clouds may also produce turbulences, which can range in size from a few milli-meters to eddies of several meters, in addition to Rayleigh scattering and Mie scattering at the molecular structures in the atmosphere [11].

Phase shifts and refraction results from the random fluctuation of the atmosphere's refractive index caused by all of these turbulences. Small-scale scintillation is the term for the refractive index fluctuation of eddies with sizes l0 of a few milli-meters. On the other hand, large-scale fluctuations L_0 of the refractive index cause the beams to refract in a stochastic way, changing the beam's direction of propagation randomly. The additional term for this phenomenon is beam fluctuations, or wander [12]. Scintillation on both small and large scales causes a timely stochastic change in the received power or the beam irradiance, which should be considered in any satellite feeder link analysis [13].

The Kolmogorov theory of turbulence, as exemplified by Andrews *et al.* in reference [9], can explain the turbulence. They are identified by the so-called structure parameter of the refractive index fluctuation C^2_n in [m$^{-2/3}$], which is contingent upon the height h in [m], the wind velocity v in [m/s], the temperature $\boldsymbol{T}$ in [K], and the pressure $\boldsymbol{P}$ in [mbar]. The extended mathematical descriptions were explained well by many scientific contributions such as [10,11].

3 Simulation of FSO and DSO channel at different scenarios

PILab code [11] was used in this work, which developed by DLR's Institute of Communications and Navigation, to simulate the uplink/downlink scenario, at a vertical terrestrial link length L of 40 km using Hufnagel-Vally (HUFVAL 5/7). Figure 1 shows the laser beam profile at different elevation angle, Table 1 presents further details on the definitions of other parameters that were considered.

Table 1. Values for the uplink simulation scenario's parameters.

Turbulence inner scale (l_o) [m]	$1 \cdot 10^{-5}$
Turbulence outer scale (L_o) [m]	10
Beam waist [m]	$5 \cdot 10^{-2}$
wavelength (λ) [m]	$1.55 \cdot 10^{-6}$
Pseudowind (v) [m/s]	21
Ground level turbulence [m$^{-2/3}$]	$1.7 \cdot 10^{-14}$
Turbulence Model	Kolmogorov
No. of realization	1000

By considering ground-level turbulence at daytime, it equal $1.7 \cdot 10^{-13}$ [m$^{-(2/3)}$] and using Modified HUFVAL 5/7 [14] for turbulence model, the laser beam profile at two different elevation angles shown in Figure 2, where the simulation generated results at the vertical terrestrial link length L equals 4 km, 40 km, and 40,000 km, respectively.

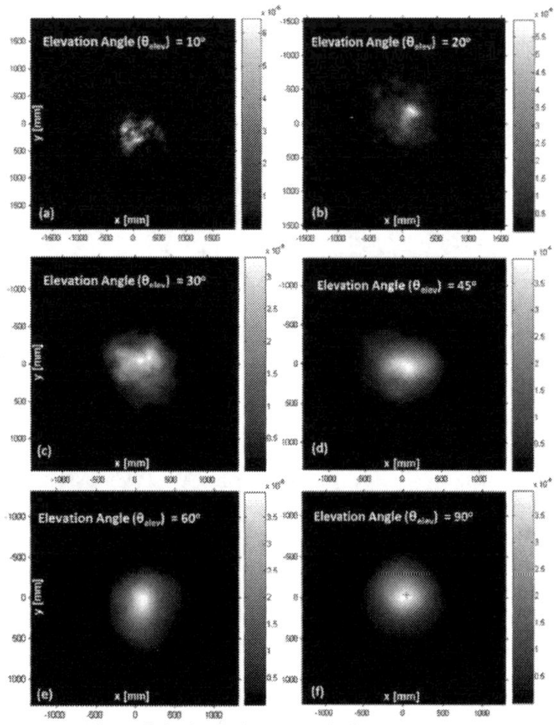

Fig. 1. The uplink simulation scenario for GEO satellite orbit beam profile at different θ_{elev}.

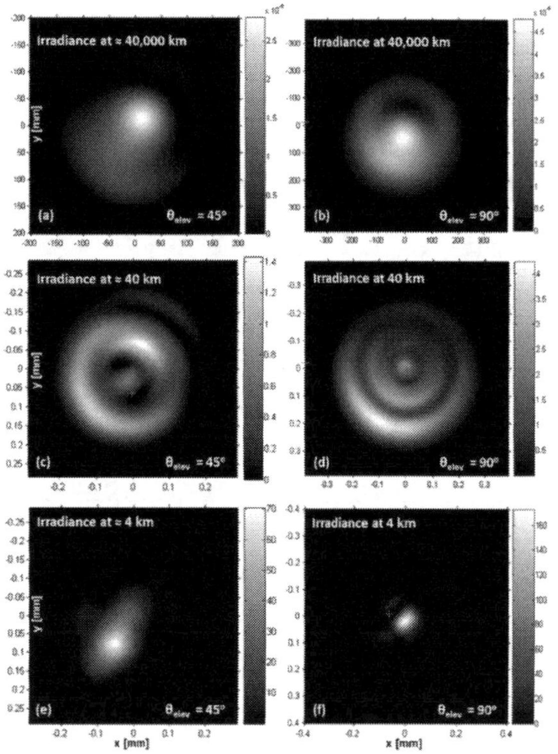

Fig. 2. The uplink simulation scenario for beam profile at different vertical terrestrial link lengths in case θ_{elev} = 45° and 90° respectively.

Under the situation of the downlink, the optical receiver arrangement for detecting the fluctuation in the laser beam is depicted in Figure 3. At a given point in time, the amplitude of the speckle pattern arising from the optical pattern aberrated by atmospheric turbulence is measured by the receiver. In the earth-to-space communications environment, the right part of Figure 3 tries to characterize and demonstrate the PILab results at three satellite heights for the analysis of the link budget at elevation angle (θ_{elev}) = 45°, λ = 847 nm and ground level turbulence (C^2_{n0}) = $1.7{\cdot}10^{-14}$ $m^{-2/3}$ at modified Hufnagel-Valley 5/7 model (night time).

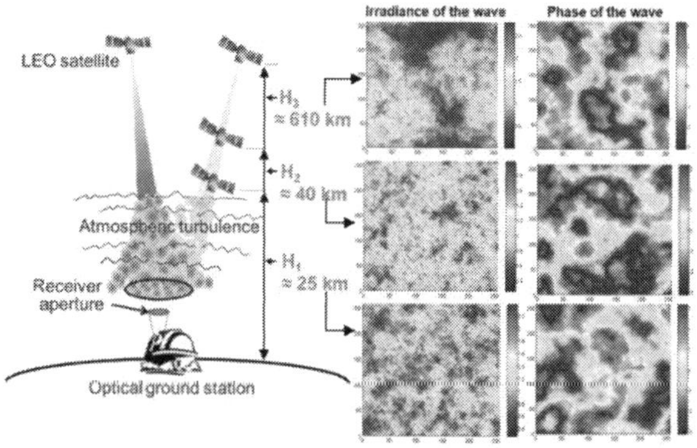

Fig. 3. Beam irradiance and phase through atmospheric turbulence at three different satellite heights (downlink scenario).

The PILab simulation results are displayed with the differential image motion monitor (DIMM) measurements in Figure 4. The focused spot motion variance was correlated with the Fried parameter (r_0) using the typical DIMM approach [3], which makes the assumption that there are no additional scintillation and Kolmogorov wavefront aberrations. It is also important to note that these assumptions could not apply to other measurements that vary significantly.

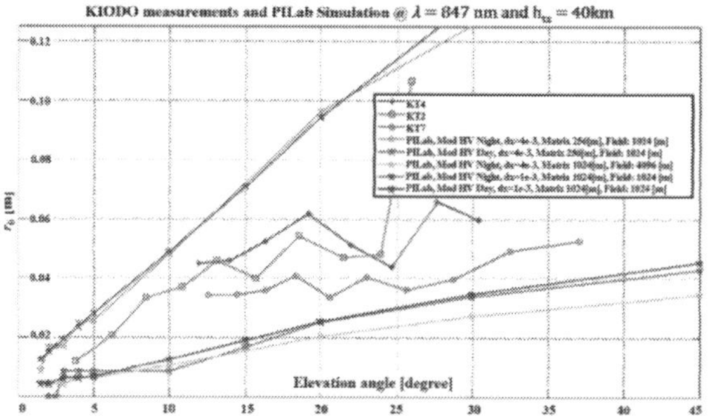

Fig. 4. Explanation and realization of simulation and experiment results for Fried's parameter at satellite height (h_{tx}) = 40km.

Figure 5 shows the experimental scintillation index of the intensity (i.e., *camera pixel intensity*). The complete scintillation dynamic could be seen as merit to the camera's 12-bit resolution. The PILab simulation results of the scintillation index in day and night scenarios, at different matrix and field sizes, were combined into the same figure for an easier comparison scheme with experiment data.

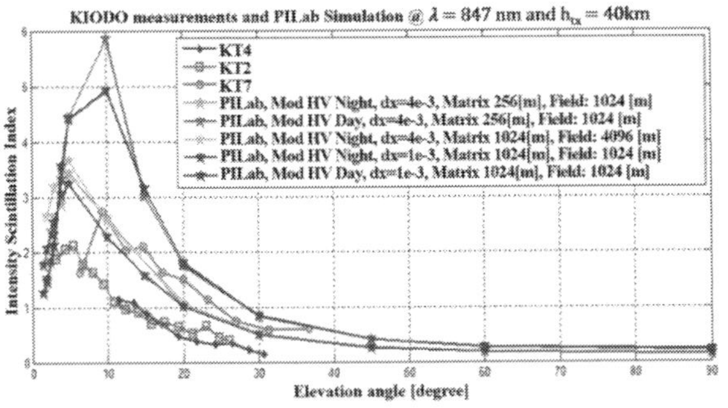

Fig. 5. Comparison of experimental and simulation results of the intensity scintillation index of the downlink scenario at satellite height (h_{tx}) = 40km.

4 Neural Networks as tool for wave-front correction

Since many categories of complex laser systems depend extensively on nonlinearity for their wide variety of optical communication applications, neural networks (NNs) techniques, and machine-learning methods in particular are highly adapted for handling nonlinear issues in the present quantum 2.0 era. This makes NNs approaches valuable in the development fields of laser technology and its applications in space specialisations. The difficulties in achieving stable operation or targeted optical transmission performances in laser communication systems originate from the multitude of effective degrees of freedom, also known as control parameters, which must be balanced.

In the context of FSO and related technology, NNs and machine learning advances might be used for:

1. Design predictions and optimizations of atmospheric turbulence regimes and mitigation;
2. Complex characterization of scintillation effects and laser beam propagation inside multilevel atmospheric disturbances;
3. Improving our understanding of the physical mechanisms underlying the operation and propagation of complex laser systems;
4. Possibility of field control and self-tuning of the lasing regime.

The integration of feedback loops and machine learning approaches holds great promise for improving the processes of FSO laser design, and control systems. This might lead to the development of lasers, that are self-starting, self-optimizing, and resilient to external atmospheric disturbances.

A route for retrieving hidden parameters and predictive control is depicted in Figure 6. In the future, even more intricate neural networks (NNs) for deep reinforcement learning (DRL) may have a chance to be embedded and developed.

Large data sets that are needed to achieve the high accuracy of NNs outcomes may be produced rapidly using prototypes of laser communication systems in a laboratory environment under artificial atmospheric turbulences. Furthermore, one may accelerate the performance of NNs even in the absence of complete parameter information by leveraging the availability of laboratory testbeds and computational models. This may be achieved, as seen in Figure 7, by either training NNs using lab-based data for link degradation in turbulence or by incorporating a priori knowledge into the architectures and loss functions of the NNs. The suggested schematics of an artificially generated data set, linked with NNs, can used to understand and predict link degradation in turbulence and develop smart algorithms for adaptive optics model for turbulence mitigation are depicted in Figure 6.

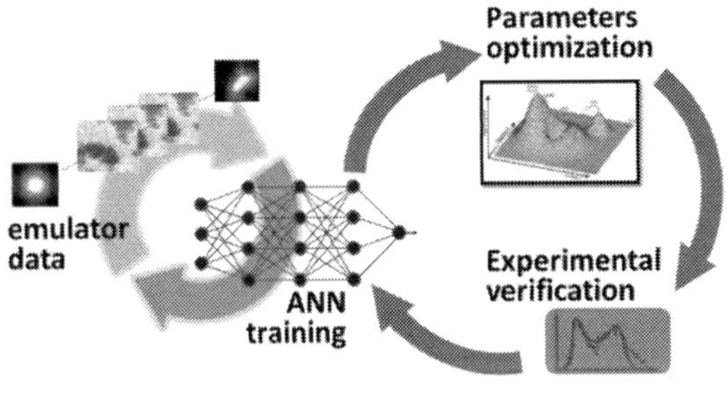

Fig. 6. Diagram of the proposed laboratory testbed-based distu-rbances for a wavefront aberration compensation techniques.

The suggested artificial atmospheric chamber design concept is shown in Figure 7, where the source and deflected laser beam, as well as the temperature, humidity, and wind speed inside the chamber, are the parameters that can indicate the degree of turbulence, which makes them the primary outcomes of the emulation. The conceptual design and laboratory-based measuring technique are depicted in Figure 7.

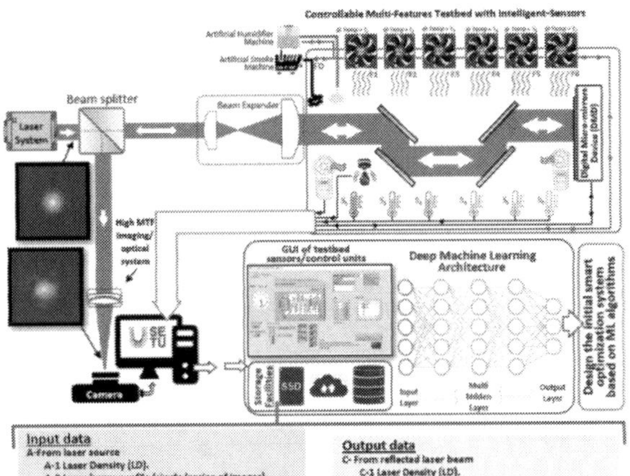

Fig. 7. Diagram of the suggested intelligent detection of labertorial turbulence based identification system in a simplified form, with insets showing beam density distribution at transmitter and receiver.

Integration of the input data from synthetic characterizations of internal testbed conditions with the resulting laser beam after propagating through artificial atmospheric turbulence can be a very good candidate for the training and validation stages of the proposed machine learning algorithm.

5 Conclusions

Certain simulation approaches were showcased for both initial deep-space optical communications (DSO) and free-space optical communications (FSO). Two possibilities for a link have been examined using the simulation algorithm: (a) between the near-earth GEO satellite (downlink) and (b) between the GEO satellite and the optical ground station 'OGS' (uplink).

The effect of the scintillation index and Fried parameter were also illustrated at different turbulence conditions, as well as compared with DLR & JAXA's OICETS/Kirari-Japan experimental data. In particular, the significance of scintillation estimation with practicable parameters and the link budget computation is verified by the use of comparable procedures and relevant measurement data. Since the structure function of the refractive index changes over a large distance and also varies due to changes in temperature, pressure, wind, azimuth, and elevation angles, one cannot find an analytical set of parameters which are valid for all complex scenarios.

The comparison demonstrates the feasibility of using the simulation models for the scintillation estimation and link budget for free space and deep space optical communications research. The initial investigation of Neural Networks (NNs) based turbulence testbed prototyping might help to find a reliable empirical set from a laboratory turbulence-based identification system for many projects with typical constellations, daytimes, and weather conditions.

Acknowledgements
We would like to thank, Dr Dirk Giggenbach and all members of DLR-IKN, for their invaluable input and support throughout the research process. Their insights and expertise were instrumental in shaping the direction of this project.

References

[1] C. Zhai, "Performance of rectangular QAM/FSO communication systems in the anisotropic non-Kolmogorov ground-to-satellite uplink," J. Opt. Commun. Netw. 14(9), 713–724 (2022). https://doi.org/10.1364/JOCN.456657

[2] R. Samy, H. C. Yang, and T. Rakia, et al., "Space-air-ground FSO networks for high-throughput satellite communications," IEEE Commun. Mag. 61(3), 82–87 (2023). https://doi.org/10.1109/MCOM.002.2200018

[3] Hemmati, H., Biswas, A., Djordjevic, I.B.: Deep Space Optical Communications: Future Perspectives and Applications. Proceedings of IEEE 99(11),(November 2011). https://doi.org/10.1109/JPROC.2011.2160609

[4] Sans, M., Sodnik, Z., Zayer, I., Daddato, R.: ' 'Design of the ESA Optical Ground Station for Participation in LLCD", Proc. International Conference on Space Optical Systems and Applications (ICSOS) 2012, 3-1, Ajaccio, Corsica, France, October 9–12, 2012. https://doi.org/10.1117/12.2045510

[5] Sodnik, Z., Czichy, R.: "Design Data Summary of the ESA Optical Ground Station for In-Orbit Check-Out of Laser Communication Payloads and for the Observation and Registration of Space Debris". ESA Specifications for the Artemis OGS OPALE project. Ref.: TEC-MMO/267/ZS. May 2, 2005. https://doi.org/10.1117/12.464106

[6] Internal ESA Artemis OGS OPALE Measurement Results. 2003-2004

[7] Sodnik, Z., Smit, H., Sans, M., Giggenbach, D., Becker, P., Mata-Calvo, R., Fuchs, C., Zayer, I., Anucara, M.L., Schulz, K., Widmer, J., Arnold, F., Mosberger, M., Alonso, A., Montilla, I.: "Results from a Lunar Laser Communication Experiment between NASA's LADEE Satellite and ESA's Optical Ground Station", Proc. International Conference on Space Optical Systems and Applications (ICSOS) 2014, S2-1, Kobe, Japan, May 7-9, 2014. https://doi.org/10.2514/6.2014-1919

[8] Sodnik, Z., Heese, C., Arapoglou, P.-D., Schulz, K.-J., Zayer, I., Daddato, R., Kraft, S.: "Deep-Space Optical Communication System (DOCS) for ESA's Space Weather Mission to Lagrange Orbit L5". IEEE International Conference on Space Optical Systems and Applications (ICSOS), November 14-16,2017. https://doi.org/10.1109/ICSOS.2017.8357207

[9] Andrews, L.C., Phillips, R.L., Hopen, C.Y.: Laser Beam Propagation through Random Media. SPIE Press, Bellingham, Washington USA (2005). https://doi.org/10.1117/3.626196

[10] Andrews, L.C., Phillips, R.L., Hopen, C.Y.: Laser Beam Scintillation with Applications. SPIE Press, Bellingham, Washington USA (2001). https://doi.org/10.1117/3.412858

[11] Al-Juboori, Haider M., and Dirk Giggenbach. "Design of Computer Identification System for GEO Satellite Link Characterization.", Journal of Al-Ma'moon College, 25, 2015. https://doi.org/10.6595.224/621.

[12] Moll, Florian, and Haider M. Al-Juboori. "Investigation and Improvement of Adaptive System Model for Free-Space Optical Communication." Journal of Al-Ma'moon College 27 (2016). https://doi.org/10. 7293.215/823

[13] Al-Juboori, Haider M., and James Garland. "Laser Beam Propagation Features via Atmospheric Turbulence-Induced Beam Wander: Testbed Conceptual Design." In COAT-2023-workshop (Comm. & Observations through Atmospheric Turbulence), 2023. https://hal.science/hal-04246206/

[14] Giggenbach, D. "Optimierung der optischen Freiraumkommunikation durch die Atmosphäre-Focal array receiver." Universität der Bundeswehr, München (2004).

Modelling Investigation of Atmospheric Turbulence-Induced Beam Deviation for LEO/GEO FSO Communication Link

Haider Al-Juboori[1*]

[1] South East Technological University (SETU), Faculty of Engineering, Department of Electronics Engineering and Communications, Carlow, Ireland

Abstract. The uplink analytical simulation scenario of atmospheric turbulence-induced beam deviation can give the insight needed to understand and create the mitigation techniques that could support novel applications in the fields of free-space optical communications (FSOC) and the generation of laser guide stars. One of the significant challenges that causes wandering and fading of the beam from the receiver aperture is beam deviation. In this study, two uplink numerical modelling scenarios for low-earth orbit (LEO) and geostationary (GEO) satellites will be investigated and discussed. The work seeks to visualize the effective parameters and analytically simulate atmospheric optical propagation, especially for the wandering effect, which can give concrete milestones for developing a new scientific investigation era and deeper insight into the system's theoretical demonstration of free-space optical communications.

1 Introduction

In order to innovate the design of future optical ground stations and to demonstrate and quantify the possible upper boundaries of intelligent solution performance in free-space optical communications, continued science efforts are capable of investigating trends in performance by studying a broad parameter space.

Primarily, free-space optical communications (FSOC) uses lasers between the Earth and space to transmit data safely and quickly. Instead of more conventional radio frequencies, information is sent between the satellite and the ground station by laser beams. Because of its line-of-sight characteristics and minimal divergence, this compact and low-power technology is also seen as being more secure and having the potential for larger data bandwidths [1, 2]. Moreover, quantum key distribution (QKD) techniques are compatible with FSOC and can open new horizons in the near future [3, 4]. Up to now, the turbulence in the earth's atmosphere has the potential to significantly restrict the stability and effectiveness of FSOC. The turbulence causes "jitter," or beam wander, and "scintillation," where the beam divides into multiple speckles, both of which contribute a variance in the received signal. These effects diminish the intensity received at the target and cause beam spread and "jitter" to occur. In FSOC applications, the variations in received strength brought on by the atmospheric channel result in surge and fade events, considerably reducing the bandwidth that may be achieved [5].

The characterization of correcting beam deviations caused by atmospheric turbulence is crucial for several aspects of free-space optical communications (FSOC), which might help:

- Maintaining signal integrity and correcting beam deviations ensures that the beam remains aligned with the receiver, maintaining the integrity and strength of the received signal.
- preventing signal loss by minimizing the related risks, and ensuring consistent communication.
- Enhancing data transmission reliability and supporting FSOC systems, which can stabilize the signal, reduce the impact of scintillation, and ensure more reliable data transmission.
- Improving communication efficiency, which is particularly important in scenarios with high data rate requirements for all categories of satellite communications.
- Enabling long-distance communication and reducing the need for larger apertures can increase the possibility of using smaller, more cost-effective receiving apertures while still achieving reliable communication by correcting beam deviations.
- Supporting high-precision space techniques where correcting these deviations is vital to ensuring the precision and accuracy of these high-rigour applications, such as satellite-to-satellite communication, laser ranging, or ground-to-space communications, in a wide range of scenarios.

The main influence of atmospheric turbulence is in an altitude range of 20 km height, between the tropopause atmospheric layer and ground level, as Fig. 1 explains. The presented work in this paper will explore the effect of atmospheric turbulence-induced beam deviation on laser beams propagating to space in

* Corresponding author: haider.aljuboori@setu.ie

both low-earth orbit (LEO) and geostationary orbit (GEO) scenarios. The analytical simulation in this work might play a crucial role in understanding beam wandering in free-space optical communications by allowing specialised researchers to model and initially analyse the effects of atmospheric turbulence on laser beams without the need for costly or logistically challenging experiments.

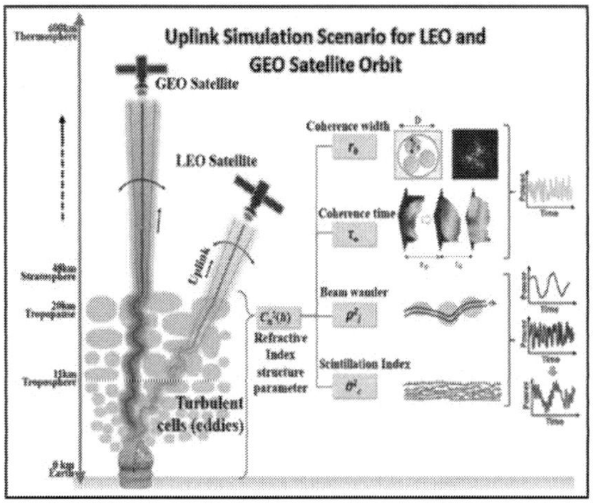

Fig. 1. General scheme of the most optical uplink/downlink and well-known causes of atmospheric turbulences.

2 Methodology

The non-uniform refractive index between transmitter and receiver is a major source of turbulence, and its result is beam deviation and phase deviation. The beam deviation from the receiver aperture causes a decrease in power density, and the phase deviation causes incoherent light intensity on the receiver side. The turbulence efficacy of a free-space optical (FSO) link is like fading in radio communication; therefore, the analysis method and the mitigation approaches of turbulence could be pretty much the same as fading.

In present beam wander (BW) models, BW is spatially represented by a Gaussian distribution, which is characterized by the size of the received beam without any averaging (the short-term beam), designated by ω_{ST}, and the long-term average radial variance, $\langle r_c^2 \rangle$ [6], as shown in Fig. 2.

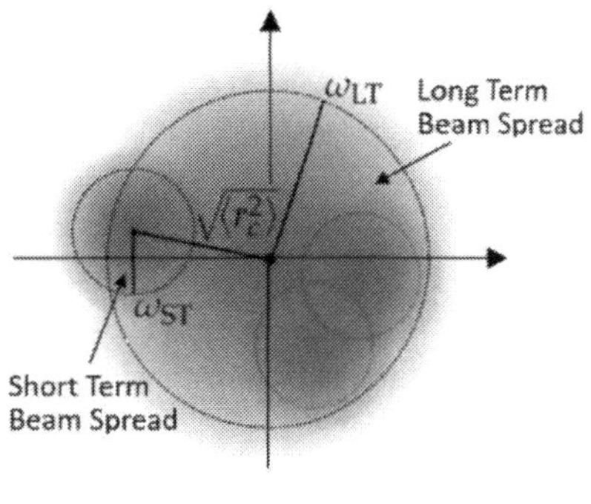

Fig. 2. A graphic depicting the link between short-term and long-term beam wander characteristics [7].

According to the central limit theorem [7], it makes sense to assume that a shorter, wandering short-term beam will result in a longer, Gaussian-shaped beam. The long-term beam's ω_{LT} size is specified by

$$\omega^2_{LT} = \omega^2_{ST} + \langle r_c^2 \rangle \tag{1}$$

There are several analytical approaches to the beam wandering radial variance. The beam wander radial variance for infinite outer scale Kolmogorov turbulence is given by

$$\langle r_c^2 \rangle = 2.42\, C_n^2\, L^3\, \omega_o^{-1/3}\, {}_2F_1(1/3,1{:}4;\, 1-|\Theta_o|), \tag{2}$$

where L is the propagation distance, ω_o is the starting beam waist, ${}_2F_1$ is the hyper-geometric function, and Θ_o is the beam parameter [8]. C_n^2 is the refractive index structure constant that specifies the intensity of the turbulence. With $\Theta_o = 1$, a collimated beam, the equation becomes

$$\langle r_c^2 \rangle = 2.42\, C_n^2\, L^3\, \omega_o^{-1/3} \tag{3}$$

and equation 3 will be more accurate for a collimated laser beam and a limited outer scale turbulence, as follow;

$$\langle r_c^2 \rangle = 2.42\, C_n^2\, L^3\, \omega_o^{-1/3} \left[1 - \left(\frac{K_o^2 \omega_o^2}{1+K_o^2 \omega_o^2}\right)^{1/6}\right] \tag{4}$$

The refractive index structural parameter C_n^2 ($m^{-2/3}$) values determine how extensively the turbulence may be categorized into three regimes. The three regimes can be used to classify turbulence [9]:

$$\begin{aligned} C_n^2 &\geq 10^{-13} \rightarrow \text{Strong turbulence} \\ 10^{-16} < C_n^2 &> 10^{-13} \rightarrow \text{Moderate turbulence} \\ C_n^2 &\geq 10^{-16} \rightarrow \text{Weak turbulence} \end{aligned}$$

Theoretically, the degree of freedom parameter $1/3\times((C_n^2)^{-2/3}\times\lambda)^{1/2}$ for the eddy size (R) is considered to be Chi-Square distributed [10, 8]. Additionally, the Gamma distribution is usually used for moderate and high turbulence regimes. Fig. 3 displays the eddy size (R) in Chi-Square and Gamma distributions for a given turbulence value. Large-scale eddies are primarily responsible for the beam wander phenomenon. Compared to the Chi-Square distribution, the mean magnitude of the eddies produced by the Gamma distribution is significant. In comparison to Chi-Square distribution, Gamma distribution exhibits very few instances of small-scale eddies in conditions of strong turbulence, making Gamma appropriate for the distribution of eddy size.

For a random variable x, the probability density functions (PDF) for Gamma distribution is given by Mukherjee *et al.* [11]:

$$f_x(x; s, \theta) = \frac{x^{(s-1)} e^{-\frac{x}{\theta}}}{\theta^s\, \Gamma(s)}, 0 < x < \infty \text{ and } s, \theta > 0 \tag{5}$$

where s and θ are the shape and scale parameters of the distribution, respectively, and $\Gamma(.)$ denotes the Gamma function.

Assumption, the centers of the turbulent eddies follow a uniform distribution within a circle of radius R centered at the propagation axis. Then, simulation of the effect on how the beam wanders under various scenarios. A beam encounters the maximum number of turbulent eddies in these scenarios for a given propagation distance and turbulence condition.

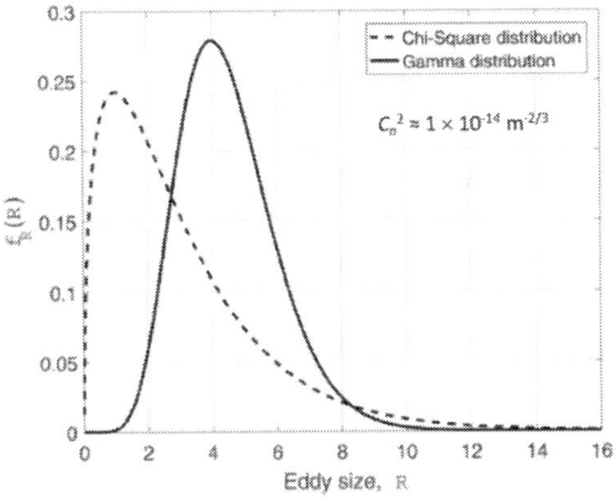

Fig. 3. Chi-Square and Gamma distributions for eddy size (R) [11].

A laser beam's divergence in free space is roughly represented by the ratio λ/D, where D is the transmit aperture diameter and λ is the laser beam's wavelength. Eddies greater than the beam diameter tend to deflect the beam when there is turbulence. In contrast, smaller eddies tend to spread the beam [12].

The beam spot wanders in the aperture plane of the receiver if the exposure time is much less than the time, δt, where $\delta t = D/|v|$ and v is the transverse wind velocity. The beam form is almost identical to that of free space. If the exposure time is much larger than δt, we would see a single beam spot with a larger diameter, as shown in Fig. 2. There are, on average, $L/2R$ eddies encountered by the propagating beam for a horizontal terrestrial link of length L. The mean square radius of beam wander is given by Ishimaru [13]:

For $L < L_t$

$$\langle \rho_l^2 \rangle = \frac{\omega_0^2}{2}[(\alpha_r L)^2 + (1 + \alpha_i L)^2] + \frac{4L^2}{k^2 \rho_0^2} \quad (6a)$$

and, for $L \geq L_t$

$$\langle \rho_l^2 \rangle = \frac{\omega_0^2}{2}[(\alpha_r L)^2 + (1 + \alpha_i L)^2] + 2.2\, C_n^2\, l_o^{-1/3} L^3 \quad (6b)$$

where ω_0 = beam radius, $\alpha_r = \lambda/\pi\, \omega_0^2$, $\alpha_i = 1/F_0$, $k = 2\pi/\lambda$, $L_t = (0.39k^2 C_n^2\, l0^{5/3})^{-1}$, F_0 = focal length of the Gaussian beam, and coherence length, $\rho_0 = (0.55k^2 L C_n^2)^{-3/5}$.

The last two terms in (6a) and (6b) represent the impacts of air turbulence, whereas the first two terms in (6a) and (6b) represent beam spread in free space. Most of the time, condition (6a) is met. However, condition (6b) is applicable under extreme turbulence. Experimental data [7] validate (6a) and (6b) requirements. Similar analytical conclusions are also mentioned in reference [14] for beam wander analysis in a ground station-satellite uplink and reference [15] for a focused Gaussian beam in a horizontal trajectory.

3 RESULTS

The created digital modelling system called - PILab (short for Propagation and Imaging Lab) is a collection of MATLAB toolboxes which facilitate the construction of optical system simulations, developed by DLR's Institute of Commun-ications and Navigation. PILab is organized in packages as shown in Fig. 4. PILab Core (PC) is the main package. It contains the core stock of system components, facilities for optical system construction, and essential visualisation functions. PILab Modelling Atmospheres (PMA) is an extension package to support simulations of atmospheric optical propagation. PILab Reporting (PREP) provides advanced facilities for simulation monitoring.

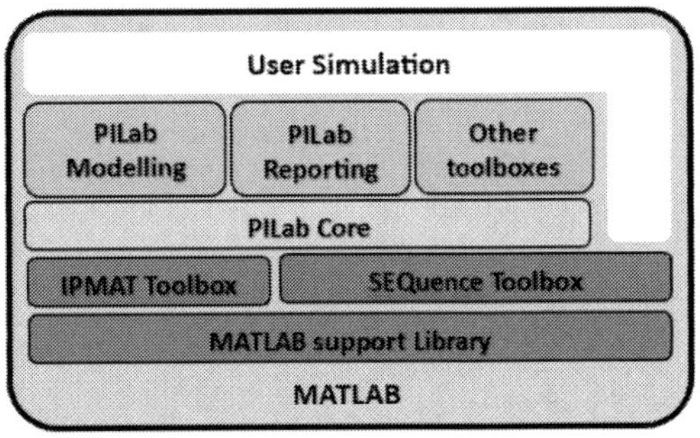

Fig. 4. Toolboxes, packages and inter-dependencies of PILab [17].

The subject of a PILab numerical simulation is the so-called slowly varying envelope $u(\rho; z)$ of the transverse scalar electromagnetic field. The paraxial approximation is assumed to be valid and the paraxial approximation's exact meaning, prerequisites and implications are thoroughly discussed in many optics textbooks such as [16,18]. In Fig. 4 a level above the integral system description and an operator description can be introduced.

The simulation results of the beam wander effect were considered at λ = 1550 nm for beam waist = 50 mm and for LEO and GEO scenarios. In a moderate turbulence regime, Fig. 5 shows the distribution of beam peaks throughout the aperture area using the suggested model at elevation angles (θ_{elev}) = 90°.

The radial displacement was simulated at three different realization numbers (i.e., 500, 1000, 2000). The distribution of beam peaks over the receiver aperture (in case of uplink LEO simulation scenario) at height of the receiver (h_{sat} = 300 km) is illustrated in Fig. 5a, b and c, respectively. The radial displacement of the beam centre (R_C) can show too in Fig. 5(d), (e) and (f), correspondingly.

The probability density function (PDF) in Fig. 6a describes the probability of beam peaks throughout the aperture area at three different realization numbers, and Fig. 6b shows the cumulative distribution function

(CDF) which describes the probability of beam peaks throughout the aperture area that will take on a value less than or equal to a centre value for LEO simulation scenario.

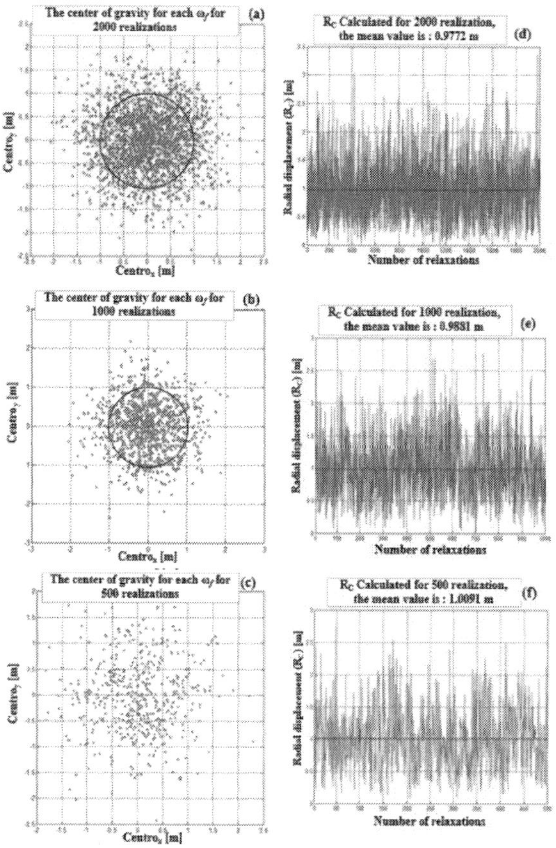

Fig. 5. Random beam wandering and distribution of beam peaks over the receiver aperture (uplink LEO simulation scenario) for a vertical terrestrial link length of 300 km.

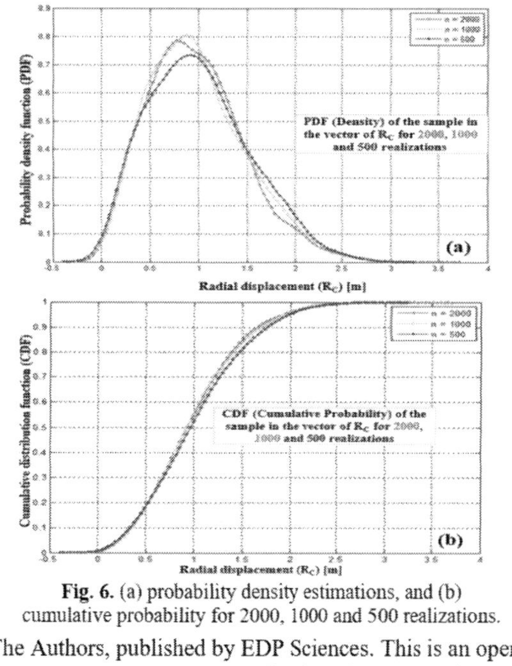

Fig. 6. (a) probability density estimations, and (b) cumulative probability for 2000, 1000 and 500 realizations.

Due to turbulence-induced beam wander, beam peaks are unpredictably displaced away from the aperture's center. The effect of beam wander caused by air turbulence is more noticeable the further away the beam peaks are from the center of the receiver aperture. Fig. 7 illustrates the dataset and mean radial displacement (R_C) values for 1000 realizations at three different elevation angles (θ_{elev}).

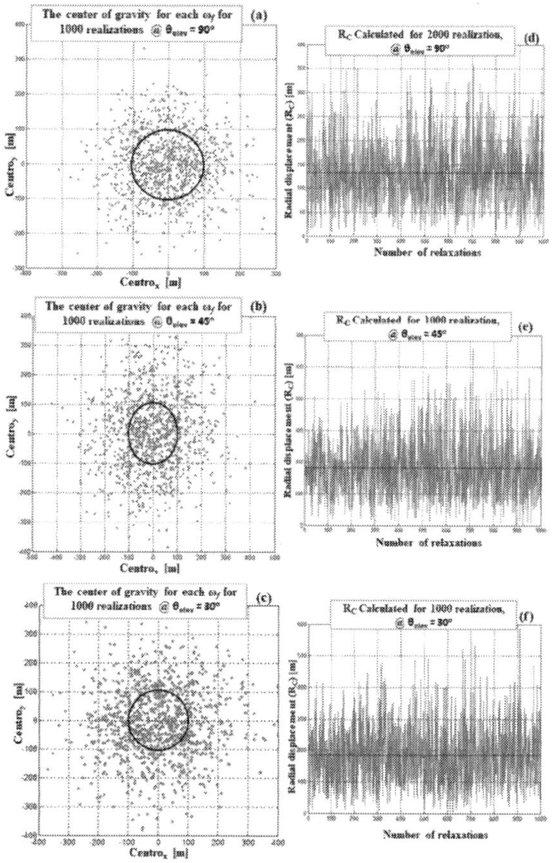

Fig. 7. Random beam wandering and distribution of beam peaks over the receiver aperture (uplink GEO simulation scenario) for a vertical terrestrial link length of 40,000 km.

Particularly, Fig. 7 shows the beam wander effect simulation results at $\lambda = 1550$ nm for a vertical terrestrial link length of 40,000 km and beam waist = 50 mm. In a moderate turbulence regime, sub-figures 7(a), (b) and (c) illustrate the distribution of beam peaks throughout the aperture area using the suggested model at three different elevation angles (θ_{elev}).

The study highlights significant differences in beam wander effects between LEO and GEO scenarios. In LEO situations, the beam experiences rapid but less severe deviations due to the shorter propagation path and the varying atmospheric layers it traverses. The turbulence encountered is more dynamic, leading to quicker but less intense beam wander. In contrast, GEO scenarios exhibit more pronounced beam wander due to the longer propagation path through the atmosphere. The cumulative effect of atmospheric turbulence over this extended distance results in larger deviations and more significant signal degradation. The study's simulations show the beam wander in GEO FSOC links is more challenging to reduce without advanced mitigation techniques.

The effect of beam wander caused by atmospheric turbulence is more noticeable the further away the beam peaks are from the center of the receiver aperture. Obviously, Fig. 7(a), (b) and (c) illustrate the dataset and mean radial displacement (R_C) values for the same realizations at three different elevation angles.

For each simulation at a specific elevation angle, the intensity variations of the received signal that are caused by atmospheric turbulence and the beam wander effect have been computed using PDF as well as CDF, as shown in Fig. 8. The results tend to match the distribution of eddies based on the Gamma model explained in Fig. 3. The deviation of the laser beam for a long distance (i.e., uplink GEO scenario) is ~100 times greater than if compare with uplink LEO scenario. Therefore fading in this turbulence situation is hard and the bit error rate (BER) of the system will be high because the maximum receiver aperture is 50 cm and laser wandering and scintillation on the aperture are high too.

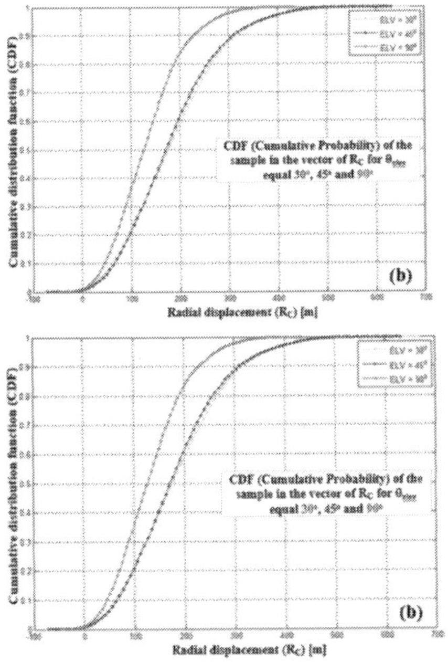

Fig. 8. Uplink GEO simulation scenario, (a) probability density estimations, and (b) cumulative probabilities at elevation angle (θ_{elev}) = 30°, 45° and 90°.

4 CONCLUSIONS

The study explains and conceptualizes the mathematical modelling of turbulent eddies in moderate and high turbulence regimes to address the beam wander effect. The results regarding the impact of GEO beam wander are consistent with the outcomes from references [12,13]. Additionally, the paper investigates two primary numerical modelling scenarios for FSOC links involving LEO and GEO satellites. These scenarios consider the unique characteristics of each orbital regime. For LEO satellites, which operate at altitudes ranging from 700 to 3,000 kilometres, the beam path encounters varying degrees of atmospheric turbulence as it passes through different layers of the atmosphere. The short propagation distance in LEO scenarios means that the beam experiences less overall turbulence but is more susceptible to rapid changes in different atmospheric conditions. In contrast, GEO satellites, situated at an altitude of approximately 40,000 kilometres, experience a longer propagation path through the atmosphere. This extended distance leads to a higher cumulative effect of atmospheric turbulence, resulting in greater beam wander and signal degradation. Furthermore, the study uses the PILab simulation framework to model these scenarios, focusing on the beam wander effects at different elevation angles and turbulence conditions.

The beam wander simulation findings in a mild turbulence domain and the suggested Gamma model agree well. Under this regime, the connection availability rapidly decreases. In high and moderate regimes, the link availability substantially decreases. Using an aperture averaging effect, or having a larger aperture area at the receiver, is one method of making up for it. Nevertheless, this results in a trade-off by adding extra background noise to the system. Even so, the impact of background noise can be reduced by utilizing an expensive adaptive optics approach [19], a high-order pulse position modulation (PPM), or an optical heterodyne receiver [20]. With some additional system complications, the optical regenerate-and-forward (ORF) approach in a multi-hop system can significantly reduce the influence of background noise [21].

In the current case study, the beam wander is especially important; therefore, direct and comparable measurements are crucial. Therefore, Fig. 9 suggests the concept of an atmospheric measurement system that can help with the proposed mathematical model validation. On the ground and satellite/UAV sides, a weather station with an anemometer, hygrometer, and temperature is employed. The multi-point scintillometer, the PSF camera, and the power meters can be set up on the ground and satellite sides, respectively. A typical time stamp is produced by a GPS clock. The data might be kept on an accessible raid system where an anemometer, hygrometer, and thermometer provide meteorological measurements on both sides in addition to the turbulence classification. Additionally, the data acquisition device (DAQ) gathers the raw data from the raid along with a time stamp. Both the uplink and the downlink have a pointing, acquisition, and tracking beam, but only the uplink has a communication beam.

Worthy mention: the unsupervised or supervised machine learning or artificial intelligence (AI) algorithms might be used to integrate the real data that was extracted from the proposed experimental testbed (the atmospheric measurement system), which provides convenient support for the development and testing of procedures related to adaptive nonlinear characterization for free-space optical communication.

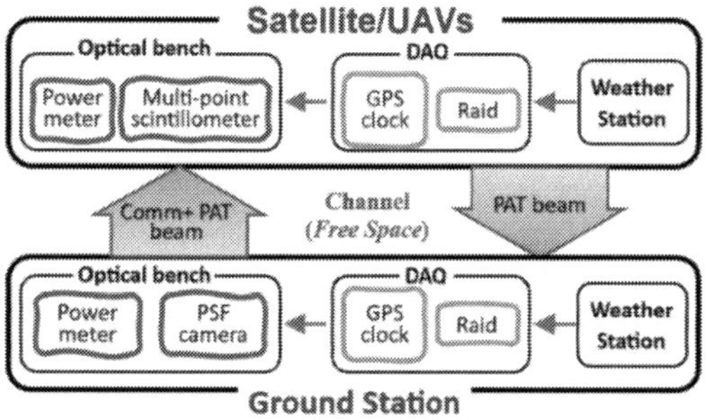

Fig. 9. An experimentally proposed model for the characterization of an atmospheric channel.

Generally, correcting beam deviations caused by atmospheric turbulence is essential for maintaining the performance, reliability, and efficiency of FSOC systems, principally in scenarios where stable and high-quality communication links are critical.

Moreover, simulations are useful tools for gaining insights into the dynamics of beam wandering, leading to better design and optimization of free-space optical communication systems. Additionally, digital implementations can help predict beam behaviour, test various scenarios, develop mitigation strategies, validate theoretical models, and conduct cost-effective and safe analyses.

Acknowledgment

We would like to thank, Dr. Dirk Giggenbach and all members of DLR-Institute of Communication and Navigation, for their invaluable input and support throughout the research process. Their insights and expertise were instrumental in shaping the direction of this project.

References

[1] Chan, Vincent WS. "Free-space optical communications." Journal of Lightwave technology 24, no. 12, 4750-4762, 2006. http://doi.org/10.1109/JLT.2006.885252

[2] Calvo, Ramon Mata, Juraj Poliak, Janis Surof, Andrew Reeves, Mathias Richerzhagen, Helawae Friew Kelemu, Ricardo Barrios et al. "Optical technologies for very high throughput satellite communications." In Free-Space Laser Communications XXXI, vol. 10910, pp. 189-204. SPIE, 2019. https://doi.org/10.1117/12.2513819

[3] Erven, Christopher, C. Couteau, R. Laflamme, and G. Weihs. "Entangled quantum key distribution over two free-space optical links." Optics express 16, no. 21, 16840-16853, 2008. https://doi.org/10.48550/arXiv.0807.2289

[4] Polnik, Mateusz, Luca Mazzarella, Marilena Di Carlo, Daniel KL Oi, Annalisa Riccardi, and Ashwin Arulselvan. "Scheduling of space to ground quantum key distribution." EPJ Quantum Technology 7, no. 1(3), 2020. https://doi.org/10.1140/epjqt/s40507-020-0079-6

[5] Motlagh, A. Chaman, V. Ahmadi, Z. Ghassemlooy, and K. Abedi. "The effect of atmospheric turbulence on the performance of the free space optical communications." In 2008 6th International Symposium on Communication Systems, Networks and Digital Signal Processing, pp. 540-543. IEEE, 2008. https://doi.org/10.1109/CSNDSP.2008.4610725

[6] Rodriguez-Gomez, Alejandro, Federico Dios, Juan Antonio Rubio, and Adolfo Comeron. "Temporal statistics of the beam-wander contribution to scintillation in ground-to-satellite optical links: an analytical approach." Applied optics 44, no. 21:4574-4581, 2005. https://doi.org/10.1364/ao.44.004574

[7] Fante, Ronald L. "Electromagnetic beam propagation in turbulent media." Proceedings of the IEEE 63, no. 12:1669-1692, 1975. 1975IEEEP.63.1669F

[8] Ghasemi, Abdollah, Ali Abedi, Farshid Ghasemi, Abdollah Ghasemi, Ali Abedi, and Farshid Ghasemi. "Selected topics in radiowave propagation." Propagation engineering in wireless communications 357-396, 2012. https://doi.org/10.1007/978-1-4614-1077-5_8

[9] Nor, Norhanis Aida Mohd, Zabih Ghassemlooy, Jan Bohata, Prakriti Saxena, Matej Komanec, Stanislav Zvanovec, Manav R. Bhatnagar, and Mohammad-Ali Khalighi. "Experimental investigation of all-optical relay-assisted 10 Gb/s FSO link over the atmospheric turbulence channel." Journal of Lightwave Technology 35, no. 1:45-53, 2016. https://doi.org/10.1109/JLT.2016.2629081

[10] Al Juboori, Haider M., and James Garland. "Laser Beam Propagation Features via Atmospheric Turbulence-Induced Beam Wander: Testbed Conceptual Design." In COAT-2023-workshop (Communications and Observations through Atmospheric Turbulence). 2023. https://doi.org/hal-04246206

[11] Mukherjee, Arka, Subrat Kar, and Virander Kumar Jain. "Analysis of FSO link under atmospheric turbulence from first principle." In Optics and Photonics Japan, p. 30aOD6. Optica Publishing Group, 2017. https://doi.org/10.5772/intechopen.80275

[12] Firoozmand, Mohsen, and M. N. Moghadasi. "Modeling and simulation of fading due of atmospheric turbulence by chi-squared and exponential pdf for a FSO link." NNGT Int. Journal Networking and Computing 2:1-9, 2015. https://doi.org/10.15308/Sinteza-2020-193-198

[13] Ishimaru, Akira. Wave propagation and scattering in random media. Vol. 2. New York: Academic

press, 1978.

[14] Dios, Federico, Juan Antonio Rubio, Alejandro Rodríguez, and Adolfo Comeron. "Scintillation and beam-wander analysis in an optical ground station-satellite uplink." Applied optics 43, no. 19:3866-3873, 2004. https://doi.org/10.1364/AO.43.003866

[15] Recolons, Jaume, Larry C. Andrews, and Ronald L. Phillips. "Analysis of beam wander effects for a horizontal-path propagating Gaussian-beam wave: focused beam case." Optical engineering 46, no. 8:086002-086002,2007. https://doi.org/10.1117/1.2772263

[16] Goodman, Joseph W. Introduction to Fourier optics. Roberts and Company publishers, 2005.

[17] Aljuboori, Haider M., and Dirk Giggenbach. "Design of Computer Identification System for GEO Satellite Link Characterization." Journal of Al-Ma'moon College 25, 2015. https://doi.org/10.2019-1650

[18] Schmidt, Jason D. "Numerical simulation of optical wave propagation with examples in MATLAB.", 2010.

[19] Kaushal, Hemani, and Georges Kaddoum. "Optical communication in space: Challenges and mitigation techniques." IEEE communications surveys & tutorials 19, no. 1:57-96, 2016. https://doi.org/10.1109/COMST.2016.2603518

[20] Kopeika, Norman S., and Joseph Bordogna. "Background noise in optical communication systems." Proceedings of the IEEE 58, no. 10 :1571-1577,1970. https://doi.org/10.1109/PROC.1970.7982

[21] Kazemlou, Shabnam, Steve Hranilovic, and Shiva Kumar. "All-optical multihop free-space optical communication systems." Journal of Lightwave Technology 29, no. 18:2663-2669, 2011. https://doi.org/10.1109/JLT.2011.2160615

Nanoparticles Enhanced Laser-Induced Breakdown Spectroscopy for Characterisation and Discrimination of Gemstones

Haider Al-Juboori[1*], *Zhaib* Haider[2], *Wei* Hang[2], *Sufi* Roslan[3]

[1]Dept. of Electronics Engineering and Communications, South-East Technological University, Carlow, Ireland.
[2]Department of Science and Mathematics, Center for Diploma Studies, Universiti Tun Hussein Onn Malaysia (UTHM), Edu Hub Pagoh, 84600 Pagoh, Johor, Malaysia
[3]Department of Physics and Chemistry, Faculty of Applied Sciences and Technology (FAST), Universiti Tun Hussein Onn Malaysia (UTHM), Edu Hub Pagoh, 84600 Pagoh, Johor, Malaysia

Abstract. Laser-induced breakdown spectroscopy (LIBS) is an excellent technique for rapid on-site investigations that attracts interest from diverse research areas. Gemmology is no exception. The application of LIBS for gemstone characterisation is limited due to ineffective ablation and crack formation, even more so with cost-effective, non-gated LIBS systems. Nanoparticle-enhanced LIBS (NELIBS) has proven to improve the effectiveness of LIBS by minimizing sample damage and enhancing the spectral features. Therefore, this study is dedicated to exploring the advantages of NELIBS, for characterising Sapphire and Opal and discrimination based on spectral differences. Our objective is to explore enhancing spectral features and performing discriminant analysis using the PLS-DA algorithm. Nanoparticles (NPs) were deposited in two layers by sequentially drying two drops (2µL) of a colloidal solution of 20nm gold nanoparticles (AuNPs) on the sample surface. Targeted areas were shot with 3 pulses of Nd:YAG laser (~350mJ, 10ns, 1064nm, 1Hz) for collecting NELIBS spectra with the OceanOptics HR4000 spectrometer. The procedure was repeated without NPs for comparative analysis with conventional LIBS (CLIBS). Results have shown a significant enhancement in spectral features, i.e., the emergence of several new spectral lines of major gemstone elements in the UV-Vis regions of the NELIBS spectra, while the CLIBS spectra were devoid of any meaningful spectral information. The PLS-DA model was trained and validated using a 4-fold cross-validation approach. The model discriminated gemstones with 99.48% accuracy at the 4th fold and exhibited a mean cross-validation accuracy of 98.97%. This preliminary investigation demonstrates the effectiveness of NELIBS for characterization and the potential for onsite identification of gemstones.

1 Introduction

Laser-Induced Breakdown Spectroscopy (LIBS) has proven to be useful for a wide variety of applications [1–6]. Herman and Senesi provide their in-depth analysis of LIBS as a geochemical tool for this century [7]. This is because of its simplicity, versatility, and immense potential, which is yet to be realized, especially for geochemical investigations. LIBS offers an attractive set of features for application in gemmology. It is a rapid and non-destructive method for studying gemstones [8]. It can do simultaneous multi-elemental analysis in a single laser shot. This makes LIBS an invaluable tool for identifying different types of natural gemstones but also for authentication, i.e., discriminating between original and artificial replicas [9]. Its ability to detect light elements, require minimal sample preparation, and offer portability for in-field analysis makes it particularly suitable for studying geological materials, including gemstones. However, when it comes to transparent or semi-transparent samples, the ineffective laser-sample coupling renders the performance rather inefficient. The high refractive index causes laser energy to absorb beyond the superficial layer, resulting in cracks and irreparable damage to the sample.

It is proven that a layer, of nanoparticles, NP, on the target surface can enhance the signal intensity by several folds and minimize sample damage [10], particularly for semi-transparent and transparent samples [11]. Moreover, Khan et al. [12] have reported an improvement in the sensitivity and repeatability of measurements with Cu nanoparticles.

The method of enhancing laser-induced breakdown spectroscopy with nanoparticles is referred to as nanoparticle-enhanced laser-induced breakdown spectroscopy and is represented by the acronym NELIBS. In this technique, a layer of nanoparticles, usually Au or Ag, is deposited on the sample surface simply by pouring a drop of colloidal solution. Due to faster evaporation at the boundary, the drop of colloidal solution leaves a ring of NPs on the sample surface, which leads to inhomogeneous NP distribution and is commonly referred to as the "coffee ring" effect. Sànchez-Aké et al. [11] have attempted to address this

* Corresponding author: haider.aljuboori@setu.ie

issue of non-uniform distribution by depositing a layer of Au on a glass substrate, but it adds complexity to the procedure. Therefore, deposition drying with a drop of NP colloidal solution is usually preferred, and the central region of the ring is used where the NP distribution is relatively homogeneous [13].

When the sample is irradiated with a laser in the presence of a deposited layer of metallic NPs, the laser primarily interacts with the NPs through surface plasmon resonance. It explodes the NPs layers and couples laser energy to the superficial layers of the sample evaporating them into plasma [14]. It minimizes the damage and effectively avoids the internal cracking of transparent and semi-transparent samples. NELIBS has received considerable interest from scientists in recent years for enhancing quantitative investigations for a variety of samples [15–18].

The enhancement of spectral features [12] is attractive for classification and authentication applications. It motivated us to study the effectiveness of NELIBS with AuNPs for the characterization and discrimination of sapphire and opal gemstones by enhancing the spectral features that were hard with the conventional LIBS technique. We employed a cost-effective LIBS system in its simplest configuration and reported a significant outcome in spectral feature enhancement and discrimination using PLS-DA.

2 Methodology

2.1 Sample Preparation

Gemstone samples were acquired from a reputed gemstone dealer in the local market. Before being exposed to laser pulses, the sample surface was wipe-cleaned using isopropyl alcohol. Distinct sites on the sample surface were cleaned with NELIBS and CLIBS. With the help of a micropipette, two layers of gold nanoparticles (20 nm, ~6.54E+11 particles/ml, Sigma Aldrich) (AuNPs) were sequentially deposited on the sample surface to have a sufficient number of nanoparticles at the ablation site. Each layer is a drop of 2µL volume; after pouring the first drop onto the sample surface, it was left for 30 minutes to air dry in the laboratory environment. The second layer was deposited similarly and dried before subjecting it to NELIBS.

2.2 Experimentation

The LIBS system utilised in this project consisted of an Nd:YAG (350 mJ, 1064 nm, 1 Hz, ~10 ns) laser aligned orthogonally to the sample surface. With the help of a converging lens (f = 10 cm), the laser was loosely focused onto the sample surface to a 2 mm spot size to irradiate a sufficient number of nanoparticles for significant signal enhancement [13]. The emission from the resulting laser-induced plasma was collected by a 600µm fibre and delivered to an HR4000 spectrometer to resolve this radiation into its constituent wavelengths and display the spectrum on a laptop screen. The parameters for spectra collection were controlled using Spectrasuite software (OceanOptics, Inc.).

The spectrometer covered the wavelength range from 197 nm to 1100 nm, however, our region of interest was within the UV-Vis (220 nm–530 nm) region. The spectrometer showed reasonable sensitivity in this region, and it was sufficient to observe distinctive spectral signatures of the major elements of gemstones i.e., Al and Si. The schematic of the experimental setup is illustrated in Fig. 1.

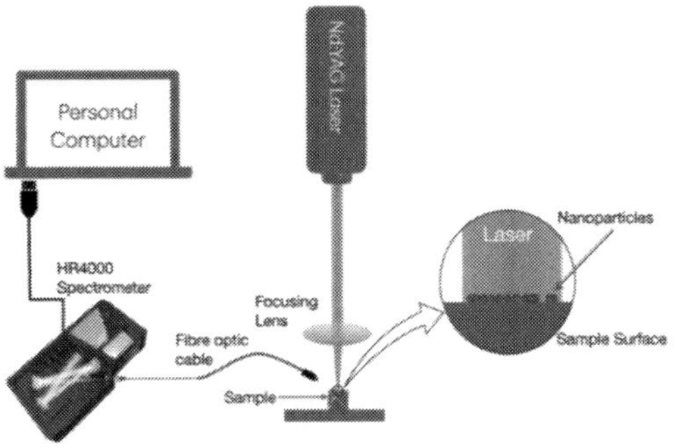

Fig. 1. Schematic illustration of the experimental setup.

3 Results and Discussion

Nanoparticles play a crucial role in enhancing LIBS performance by the surface plasmon resonance (SPR) phenomenon. The collective oscillations of electrons on the surface of NPs are called surface plasmons. The electromagnetic field of the laser governs these oscillations, causing surface plasmon resonance (SPR) with the laser field. This results in efficient laser energy absorption by NPs, rapidly heating them [13]. This energy is effectively coupled to the sample surface underneath NPs [10, 19]. This causes material to evaporate from highly localized regions on the sample surface, minimizing sample damage and preventing internal cracking. The ablated material forms an excited plasma plume, that emits significantly strong spectral lines upon de-excitation.

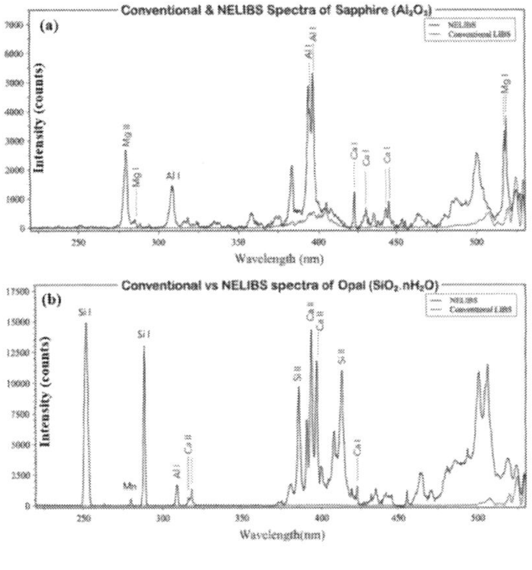

Fig. 2. Comparison between conventional and NELIBS spectra of two gemstones (a) Sapphire, and (b) Opal.

Table 1. Elements identified in Gemstones using NELIBS and wavelengths of corresponding spectral lines

	Element	Wavelength (nm)
Sapphire	Al I	308.22, 309.27, 394.40, 396.15
	Ca I	422.68, 430.25, 443.51, 445.49
	Mg I	285.21
	Mg II	279.55
Opal	Al I	308.215, 309.27
	Ca I	422.68
	Ca II	315.88, 318.93, 393.37, 396.85
	Mn I	279.48
	Si I	251.6, 288.16
	Si II	385.6, 413.07

The resonance of the electromagnetic field of the laser with the surface plasmons of NPs causes a strong coupling between laser energy and NPs. Most of the laser energy is absorbed by NPs, instantly heating them and their surroundings. NPs transfer this energy to the sample surface, causing localized evaporations, confined to regions equivalent to the size of NPs [13]. The evaporated target material appears in the form of excited plasma and emits characteristic spectral lines, as shown in Fig. 2. The comparison between the CLIBS and NELIBS spectra of sapphire and opal shows clear differences.

Several prominent atomic and ionic lines of major elements of gemstones, i.e., Al for Sapphire and Si for Opal, in addition to impurities or minor elements, Ca, Mg, Al, and Mn, cause specific physical properties and give different colours to the gemstones. The emission lines are identified using the Atomic-Spectral Database compiled by the National Institute for Science and Technology (ASD-NIST) [20] and verified using published literature [14, 21].

CLIBS spectra are almost void of any significant spectral features. It is evidence of ineffective laser-sample coupling that has led to inefficient ablation and hence near-to-no significant spectral emissions.

The spectral lines were found and counted by using the "find_peaks" method in the "SciPy" library for Python [22]. To find peaks, we used "prominence" as the key parameter. It identifies a data point as a peak if it has a certain local prominence as defined by the user. In the case of the Sapphire spectra, the value of prominence was set to 140 counts and 205 counts for Opal, to accommodate the maximum prominent peaks and avoid the identification of noise points as peaks at the same time.

The algorithm identified 40 and 43 prominent spectral lines in the NELIBS spectra of Sapphire and Opal, respectively, as indicated by a red spot in the four subplots in Fig. 3. The majority of these spectral lines can be easily attributed to Al and Si elements, which are the major elements for sapphire and opal gemstones, respectively, in addition to the minority elements Mg, Ca, and Mn in these gemstones.

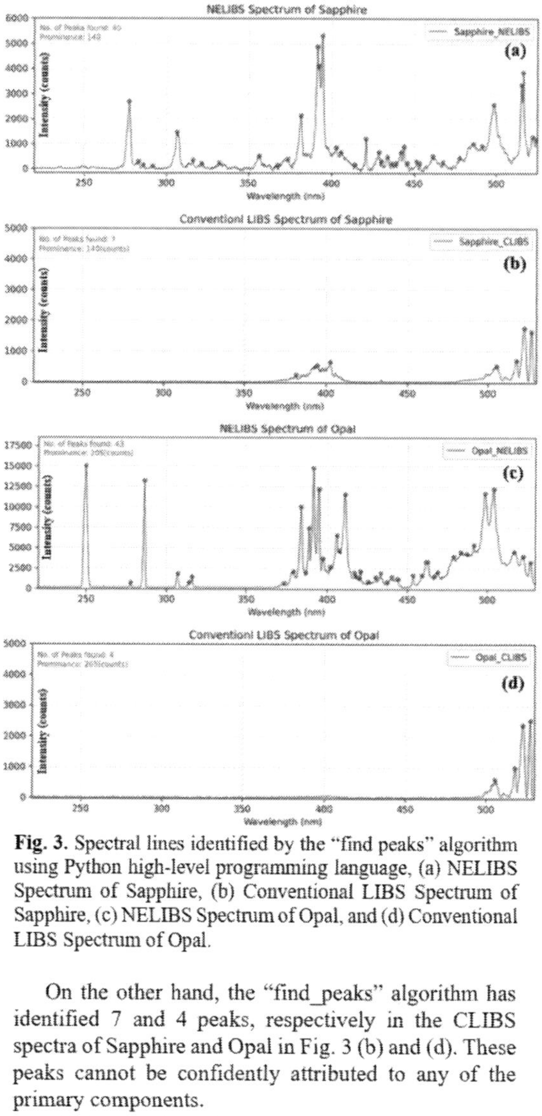

Fig. 3. Spectral lines identified by the "find peaks" algorithm using Python high-level programming language, (a) NELIBS Spectrum of Sapphire, (b) Conventional LIBS Spectrum of Sapphire, (c) NELIBS Spectrum of Opal, and (d) Conventional LIBS Spectrum of Opal.

On the other hand, the "find_peaks" algorithm has identified 7 and 4 peaks, respectively in the CLIBS spectra of Sapphire and Opal in Fig. 3 (b) and (d). These peaks cannot be confidently attributed to any of the primary components.

The bar graph in Fig. 4 gives a comparison between the number of spectral lines found in the CLIBS and NELLIS spectra of sapphire and opal gemstones. The differences are well observed, especially since the number of spectral lines based on the NELIBS technique was more than five times higher if compared with the diagnosed number of spectral lines based on the CLIBS technique.

3.1 Partial Least Square - Discriminant Analysis (PLS-DA)

Gemstone discrimination is an important application. The present work attempted it using partial least squares discriminant analysis (PLS-DA). The model was trained on the labelled dataset consisting of 10 instances of NELIBS spectra of Sapphire and Opal samples, each instance having 1184 features. The model was trained and validated using a 4-fold cross-validation approach.

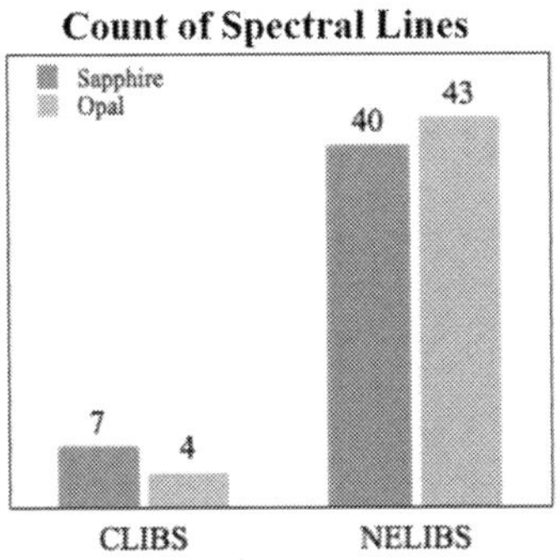

Fig. 4. Number of spectral lines in CLIBS and NELIBS spectra of Sapphire and Opal Gemstones.

The original data had only 3 spectra per sample because NPs ablate away after the first two laser shots. Therefore, the analytical efforts in this paper only consider the spectra obtained with the first three laser shots. New instances of the data were generated by scaling and adding Gaussian noise to the existing data. It is a rather small dataset that lacks diversity and robustness for training a discrimination and prediction model. It may lead to overfitting and less than optimal performance [23, 24]. An efficient characterization plan was considered to overcome this issue in future extensive investigations.

The PLS-DA algorithm provided by scikit-learn machine learning library for Python was used in this project. Since the model was trained and validated using a four-fold Cross-Validation approach which means, the data is divided into four subsets, three of which were used as training sets and one as the test set. The process is repeated until the model's performance has been tested on all subsets and the overall performance is averaged over all iterations. The performance of the PLS-DA model improves over each iteration of validation, at the fourth fold of cross-validation, the accuracy is 99.49%, while the mean cross-validation accuracy is 98.97%. As shown in Fig. 5, in the PLS-DA classification plot, there is a clear separation between the components of opal and sapphire with no overlapping.

PLS-DA was not tested on CLIBS data since there were no significant spectral signatures of major gemstone elements in the spectral region of interest.

4 Conclusion

In this paper, the characterisation of sapphire and opal gemstones was presented using nanoparticle-enhanced laser-induced breakdown spectroscopy (NELIBS) and compared with conventional laser-induced breakdown spectroscopy (CLIBS). Several key observations have emerged. NELIBS revealed significant spectral features, surpassing those found in the CLIBS spectra of sapphire and opal gemstones. CLIBS spectra lacked distinct spectral signatures, while NELIBS spectra exhibited prominent spectral lines corresponding to major and minor/impurity elements in sapphire and opal gemstones.

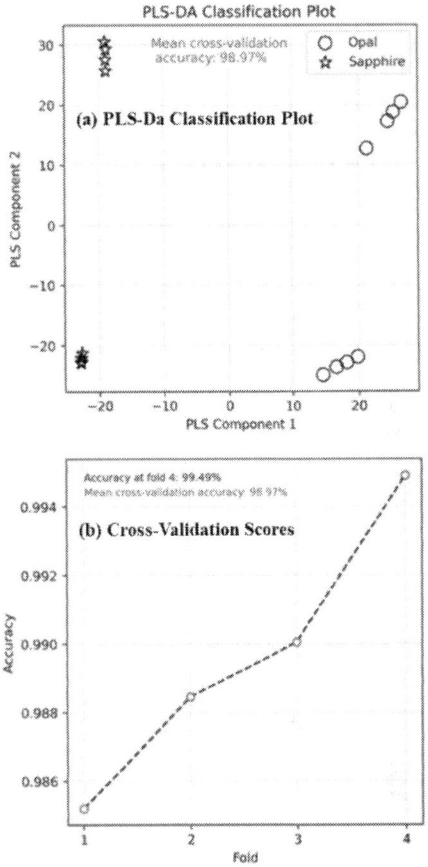

Fig. 5. PLS-DA classification of gemstones based on dataset from NELIBS spectra.

Additionally, the Partial Least Squares-Discriminant Analysis (PLS-DA) classification plot demonstrates a clear separation between the two gemstone classes. Remarkably, the model achieved near-perfect accuracy (99.49%) in four-fold cross-validation, with an average cross-validation accuracy of 98.97%. These findings underscore the efficacy of NELIBS for the characterization and classification of gemstones.

ACKNOWLEDGMENTS

This research was supported by Universiti Tun Hussein Onn Malaysia (UTHM) through Tier 1 (vot H848).

References

1. P.R. Villas-Boas, M.A. Franco, L. Martin-Neto, H.T. Gollany, D.M.B.P. Milori, Applications of laser-induced breakdown spectroscopy for soil characterization, part II:

Review of elemental analysis and soil classification., European Journal of Soil Science. 71 (2020) 805–818.

2. O.A. Al-Najjar, Y.S. Wudil, U.F. Ahmad, O.S. Baghabra Al-Amoudi, M.A. Al-Osta, M.A. Gondal, Applications of laser induced breakdown spectroscopy in geotechnical engineering: a critical review of recent developments, perspectives and challenges, Applied Spectroscopy Reviews. 58 (2023) 687–723. https://doi.org/10.1080/05704928.2022.2136192.

3. L. Sun, H. Yu, Z. Cong, H. Lu, B. Cao, P. Zeng, W. Dong, Y. Li, Applications of laser-induced breakdown spectroscopy in the aluminum electrolysis industry, Spectrochimica Acta Part B: Atomic Spectroscopy. 142 (2018) 29–36. https://doi.org/10.1016/j.sab.2018.02.005.

4. S.W. Hudson, J. Craparo, R. De Saro, D. Apelian, Applications of Laser-Induced Breakdown Spectroscopy (LIBS) in Molten Metal Processing, Metallurgical and Materials Transactions B. 48 (2017) 2731–2742. https://doi.org/10.1007/s11663-017-1032-7.

5. A. p. m. Michel, M. j. Lawrence-Snyder, S. m. Angel, A. d. Chave, Oceanic applications of laser induced breakdown spectroscopy: laboratory validation, Proceedings of MTS/IEEE, OCEANS 2005. (2005) 741. https://doi.org/10.1109/OCEANS.2005.1639841.

6. H.B. Andrews, A.M. Wymore, E.E. Wetter, E.M. Herndon, H. Li, S.A. Martin, N.A. Griffiths, X. Yang, W. Muchero, D.J. Weston, M.Z. Martin, Rapid screening of wood and leaf tissues: investigating silicon-based phytoliths in Populus trichocarpa for carbon storage applications using laser-induced breakdown spectroscopy and scanning electron microscopy–energy dispersive X-ray spectroscopy, Journal of Analytical Atomic Spectrometry 38 (2023) 2353–2364. https://doi.org/10.1039/d3ja00186e.

7. R.S. Harmon, G.S. Senesi, Laser-Induced Breakdown Spectroscopy – A geochemical tool for the 21st century, Applied Geochemistry. 128 (2021) 104929. https://doi.org/10.1016/j.apgeochem.2021.104929.

8. N.J. McMillan, B.L. Dutrow, Laser-induced breakdown spectroscopy analysis of tourmaline: protocols, procedures, and predicaments, Eur. J. Mineral. 36 (2024) 369–379. https://doi.org/10.5194/ejm-36-369-2024.

9. N.M. Saufi, N.N. Adnan, S.Z.H. Rizvi, M.S. Roslan, S.N. Razak, N. Hasim, N.I. Mansor, Element Identification of Different Gemstones by Using LIBS-PCA Method, Proceedings of the 7th International Conference on the Applications of Science and Mathematics 2021, 155–163 Springer Nature Singapore, Singapore, (2022). https://doi.org/10.1007/978-981-16-8903-1_16.

10. A. De Giacomo, R. Alrifai, V. Gardette, Z. Salajková, M. Dell'Aglio, Nanoparticle enhanced laser ablation and consequent effects on laser induced plasma optical emission, Spectrochimica Acta Part B: Atomic Spectroscopy. 166 (2020) 105794. https://doi.org/10.1016/j.sab.2020.105794.

11. C. Sánchez-Aké, T. García-Fernández, J.L. Benítez, M.B. de la Mora, M. Villagrán-Muniz, Intensity enhancement of LIBS of glass by using Au thin films and nanoparticles, Spectrochimica Acta Part B: Atomic Spectroscopy. 146 (2018) 77–83. https://doi.org/10.1016/j.sab.2018.05.007.

12. M.R. Khan, S.U. Haq, Q. Abbas, A. Nadeem, Improvement in signal sensitivity and repeatability using copper nanoparticle-enhanced laser-induced breakdown spectroscopy, Spectrochimica Acta Part B: Atomic Spectroscopy. 195 (2022) 106507. https://doi.org/10.1016/j.sab.2022.106507.

13. M. Dell'Aglio, R. Alrifai, A. De Giacomo, Nanoparticle Enhanced Laser Induced Breakdown Spectroscopy (NELIBS), a first review, Spectrochimica Acta Part B: Atomic Spectroscopy. 148 (2018) 105–112. https://doi.org/10.1016/j.sab.2018.06.008.

14. C. Koral, M. Dell'Aglio, R. Gaudiuso, R. Alrifai, M. Torelli, A. De Giacomo, Nanoparticle-Enhanced Laser Induced Breakdown Spectroscopy for the noninvasive analysis of transparent samples and gemstones, Talanta. 182 (2018) 253–258. https://doi.org/10.1016/j.talanta.2018.02.001.

15. S. Xiao, Y. Liu, Y. Luo, Y. Zhu, W. Wang, J. Nie, W. Huang, C. Niu, A. Gong, L. Guo, Sensitivity and stability improvement on slippery surface-aggregated substrate for trace heavy metals detection using NELIBS, Talanta. 275 (2024) 126001. https://doi.org/10.1016/j.talanta.2024.126001.

16. L. Yuan, Q. Wang, H. Yu, P. Lang, H. Li, X. Gao, J. Lin, High-sensitivity determination of heavy metal elements in water with circular grooves and nanoparticle-enhanced LIBS, J. Anal. At. Spectrom. 39 (2024) 2097–2105. https://doi.org/10.1039/D3JA00397C.

17. Q. Wang, F. Zhang, X. Han, X. Gao, W. Chen, K. Huan, N. Ying Cui, Y. Liu, W. Liang, A. Chen, Emission and stability improvement of AlO molecular bands in NELIBS using laser pretreatment, Journal of Analytical Atomic Spectrometry. (2024). https://doi.org/10.1039/d4ja00102h.

18. A. Safi, J.E. Landis, H.G. Adler, H. Khadem, K.E. Eseller, Y. Markushin, S. Honarparvaran, A. De Giacomo, N. Melikechi, Enhancing biomarker detection sensitivity through tag-laser induced breakdown spectroscopy with NELIBS, Talanta. 271 (2024) 125723. https://doi.org/10.1016/j.talanta.2024.125723.

19. A. De Giacomo, M. Dell'Aglio, R. Gaudiuso, C. Koral, G. Valenza, Perspective on the use of nanoparticles to improve LIBS analytical performance: nanoparticle enhanced laser induced breakdown spectroscopy (NELIBS), J. Anal. At. Spectrom. 31 (2016) 1566–1573. https://doi.org/10.1039/C6JA00189K.

20. Atomic Spectra Database, NIST. (2009). https://www.nist.gov/pml/atomic-spectra-database (accessed July 30, 2024).

21. A. Zamira, S.Z.H. Rizvi, M.S. Roslan, N.N. Adnan, S.N.A. Razak, M.A. Jalil, Discrimination of Precious and Semi-Precious Gemstones Using Laser-Induced Breakdown Spectroscopy and Machine Learning Approaches, Proceedings of the 7th International Conference on the Applications of Science and Mathematics 2021, 165–174 Springer Nature Singapore, Singapore, (2022). https://doi.org/10.1007/978-981-16-8903-1_17.

22. find_peaks — SciPy v1.14.0 Manual, (n.d.). https://docs.scipy.org/doc/scipy/reference/generated/scipy.signal.find_peaks.html#scipy.signal.find_peaks (accessed July 30, 2024).

23. K. Liu, D. Tian, H. Wang, G. Yang, Rapid classification of plastics by laser-induced breakdown spectroscopy (LIBS) coupled with partial least squares discrimination analysis based on variable importance (VI-PLS-DA), Anal. Methods. 11 (2019) 1174–1179. https://doi.org/10.1039/C8AY02755B.

24. M.D. Dyar, C.R. Ytsma, Effect of data set size on geochemical quantification accuracy with laser-induced breakdown spectroscopy, Spectrochimica Acta Part B: Atomic Spectroscopy. 177 (2021) 106073. https://doi.org/10.1016/j.sab.2021.106073.

Comparative Analysis of Ethanol Gas Sensors Based on Bloch Surface Wave and Surface Plasmon Resonance

João P. M. Carvalho [1,2,*], *Miguel A. S.* Almeida[1,2], *João P.* Mendes[1], *Luís C. C.* Coelho[1] and *José M. M. M. de* Almeida[1,3]

[1] INESC TEC – Institute of Systems and Computer Engineering, Technology and Science, and Department of Physics, Faculty of Sciences, University of Porto, 4169-007 Porto, Portugal
[2] Department of Engineering Physics, Faculty of Engineering, University of Porto, 4200-465 Porto, Portugal
[3] Department of Physics, School of Sciences and Technology, University of Trás-os-Montes e Alto Douro, 5001-801 Vila Real, Portugal

Abstract. Ethanol plays a crucial role in modern industrial processes and consumer products. Despite its presence in human activity, short and long-term exposure to gaseous ethanol poses risks to health conditions and material damage, making the control of its concentration in the atmosphere of high importance. Ethanol optical sensors based on electromagnetic surface waves (ESWs) are presented, with sensitivity to ethanol vapours being achieved by the inclusion of ethanol-adsorptive zinc oxide (ZnO) layers. The changes in optical properties modulate the resonant conditions of ESWs, enabling the tracking of ethanol concentration in the atmosphere. A comprehensive comparative study of sensor performance is carried out between surface plasmon resonance (SPR) and Bloch surface wave (BSW) based sensors. Sensor efficiency is simulated by transfer matrix method towards optimized figures of merit (FoM). Preliminary results validate ethanol sensitivity of BSW based sensor, showcasing a possible alternative to electromagnetic and plasmonic sensors.

1 Introduction

1.1 Volatile Organic Compounds

Volatile organic compounds (VOCs) are a group of organic compounds of high vapour pressure, typically exhibiting low boiling points at ambient pressure [1]. Several parameters have been selected for VOC classification, namely their boiling point, molecular structure and molecular polarity. According to their boiling points, VOCs can be further dissected into four categories: very volatile organic compounds, volatile organic compounds, semi-volatile organic compounds and particulate organic compounds. Regarding molecular structure, VOCs can be characterized with one of seven different designations: alkanes, alkenes, aromatic hydrocarbons, halogenated hydrocarbons, alcohols, aldehydes or ketones. As for molecular polarity, they can either be polar or apolar [2].

VOCs are ever present in human activity, commonly as products or byproducts of industrial processes. Their volatility, together with thermal processes in industrial environments, can lead to significant concentrations of VOCs in the atmosphere, posing health risks both to short-term (airways irritation, headaches, dizziness, etc.) and long-term exposure (central nervous system damage, disruption of endocrine system, foetal development problems, etc.) [3-4].

Ethanol is one of the most present VOCs, with important applications in fuel processing, alcoholic beverage distilleries and chemical plants for disinfectants and antiseptic products [2]. In such backgrounds, the monitorization of gaseous ethanol is of extreme importance, and has been primarily achieved with resistive electromagnetic sensors.

Optical sensors, however, are immune to electromagnetic interference, lightweight, low power consuming and enable easy remote detection capabilities [5]. Optical VOC sensors have been implemented through control of optical phenomena such as fluorescence, interferometry, plasmonic resonances or in photonic crystals (PhCs) [6].

The implementation of plasmonic or PhC based sensing platforms has been pursued for their high sensitivity and potential for sensor miniaturization. Sudheer et al. [7] developed a SPR based fibre optic sensor incorporating a metallic layer with graphene and MXene layers for the detection of both acetone and ethanol, attaining sensitivities exceeding 5000 nm/RIU. He et al. [8] grew a metal-organic framework (MOF) on top of a layer of Au nanoparticles, previously coated on top of a fibre optics tip. The MOF increased adsorption specificity, attaining sensitivities of 13.7 and 15.5 nm/%

* Corresponding author: joao.p.miranda@inesctec.pt

for acetone and ethanol concentration measurements, respectively.

Gryga et al. [9] introduced a one-dimensional PhC of alternating layers of TiO_2 and SiO_2 showing sensitivity to ethanol vapours up to 3272 nm/RIU. Veeraghattam et al. [10] grew a VOC reactive layer of PVA on top of a similar PhC, showing sensitivity to propanol and isopropanol interaction [10]. More complex structures have been studied to enhance VOC sensing. Gao et al. [11] developed porous MOFs of large surface area for visual detection of benzene concentration upon adsorption into de MOF matrix down to a limit of detection of 0.8 g/m^3.

Despite exhibiting VOC sensitivity, some of the introduced devices rely on complex fabrication techniques, such as layer-by-layer assembly of MOFs or nanoparticles growth from precursor solutions. In this work, simple and compact planar thin films structures are studied for enhanced figure of merit in VOC sensing, resorting to metal-oxide ZnO for VOC adsorption.

VOC adsorption in Metal-Oxides

The application of metal-oxides in gas sensors has been researched previously due to their distinct physicochemical properties, namely their high surface area and reactivity due to the d^0 or d^{10} band arrangement that facilitates formation of electron-hole pairs, that grant selectivity towards gas detection via adsorption. Moreover, they exhibit stability, tuneable band edges, and easy and low-cost fabrication [12].

Zinc oxide (ZnO) is one of the most resourced metal-oxides for gas sensing applications. Being a thermal oxide, its fabrication is simple, and its thermal stability and high surface area grant it great adsorption capacities for gas adsorption sensing applications [13-15].

Antoine Equation

The vapour pressure of an ideal gas in the atmosphere can be estimated by the Antoine equation, given by

$$p = A - \frac{B}{C+T} \quad (1)$$

with p being the vapour pressure, T being the temperature, and A, B and C being constant parameters of the gas [16].

Knowing the vapour pressure and the volume of the system, the concentration of an ideal gas in a closed system of can be estimated [17].

1.2 Electromagnetic Surface Waves

Electromagnetic Surface Waves (ESW) are planar electromagnetic modes propagating along an interface normal to the plane of incidence. These resonant modes are characterized by a strong electric field enhancement on the interface that exponentially decays into the materials at the interface.

The confinement of the electric field on the interface renders sensitivity of the resonant frequency to the optical properties of the media.

Surface Plasmon Resonance

Different kinds of ESWs are known, depending on the type of materials used and their topological configuration. The surface plasmon resonance (SPR) is arguably the most widely reported ESW, much due to its simple configuration and high sensitivity to media perturbations. These modes are excited at a metal-dielectric interface due to the resonance of free electrons at the interface, resulting in an absorption band in a well-defined region of the electromagnetic spectrum, sensitive to external conditions [18].

The presence of a metallic layer at the interface results in optical losses that cause absorption bands to damp. For this reason, noble metals of reduced optical absorption such as Au and Ag are usually applied.

Bloch Surface Wave

Bloch Surface Waves (BSW) are electromagnetic modes generated in a truncated PhC [19].

A PhC can be defined as a periodic modulation of refractive index (RI), composed of alternating layers of different materials, each with a fixed thickness. These structures select guided modes, reflecting internally in the high index layers, according to layer thickness, layer material and number of unitary cells of the PhC. Their all-dielectric composition results in few optical losses, allowing sharp, narrow bands of reduced full width at half maximum (FWHM). These guided modes occur along a range of frequencies of allowed modes. These allowed bands are separated by a range of frequencies where guided mode propagation is prohibited – the photonic bandgap.

The addition of a capping layer with symmetry-breaking thickness on top of a PhC will allow the excitation of a different mode, present in the photonic bandgap of the PhC, called BSW. The local field enhancement of this electromagnetic mode, and its decay into de capping layer and the external media, also grants it refractometric sensitivity to media properties, although usually of lower expression than in the case of SPR.

2 Numerical Simulation

An optimized configuration was simulated for highest figure of merit (FoM), defined as in Eq. (2),

$$FoM = \frac{\partial\lambda/\partial c}{FWHM} \rightarrow \frac{\Delta\lambda}{FWHM} \quad (2)$$

both for SPR and BSW-carrying structures. A measure of FoM usually considers a spectral sensitivity to the analyte, $\partial\lambda/\partial c$, and the FWHM of the resonant mode. The simulation of the spectral sensitivity to gas adsorption, however, is challenging without very specific data according to layer material and gas. In a first approach, this sensitivity was simply measured by the wavelength shift of the sensing optical bands, $\Delta\lambda$. Although this approach will not likely provide an accurate absolute value of FoM, it should still provide information on what structure would enable an optimized FoM, regardless of its actual magnitude.

The simulation work was carried out with the transfer matrix method (TMM), which allows a quick simulation of reflectance spectra and dispersion relations for planar slabs of arbitrary materials and thicknesses [20]. The optical constants of all materials were extracted from literature [21-25], and all layers were swept for varying thicknesses. For all resulting spectra, the FoM of the sensing band was determined, and an optimized configuration was selected for fabrication based on highest FoM within experimental restrictions, namely the available spectral interrogation range.

All simulation parameters were defined as to match the experimental conditions. Light is incident from a semi-infinite SiO_2 coupling prism, at 74°, with an estimated 2° beam divergence.

2.1 Surface Plasmon Resonance

For the plasmonic structure, a metallic layer of Au was selected due to its low optical losses and physical and chemical stability. The ethanol-adsorptive ZnO layer is placed on top of the Au layer. Due to the poor adhesion of Au thin films, a 3 nm layer of Cr metal is considered in the glass substrate before the gold layer and another before the ZnO capping layer. These Cr layers should be considered in the TMM simulation, as they can significantly damp the resultant optical bands.

Spectra for this configuration were simulated for a spanning range of Au and ZnO thicknesses, and the respective FoMs were determined. Fig. 1 maps the determined FoM according to the simulated thicknesses for both layers.

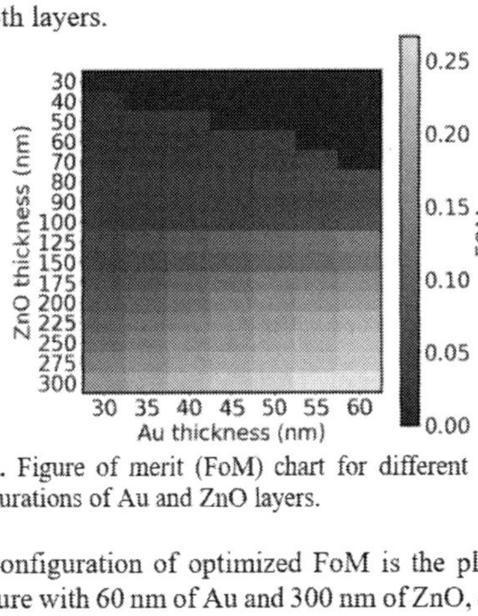

Fig. 1. Figure of merit (FoM) chart for different thickness configurations of Au and ZnO layers.

The configuration of optimized FoM is the plasmonic structure with 60 nm of Au and 300 nm of ZnO, depicted in Fig. 2. The respective simulated spectrum is plotted in Fig. 3, with a plasmonic band deep into the NIR range. Given the generally higher refractometric sensitivity of plasmonic sensors in the NIR, it does not come as a surprise for the optimized configuration to fall in such region of the spectrum, requiring a larger thickness of ZnO.

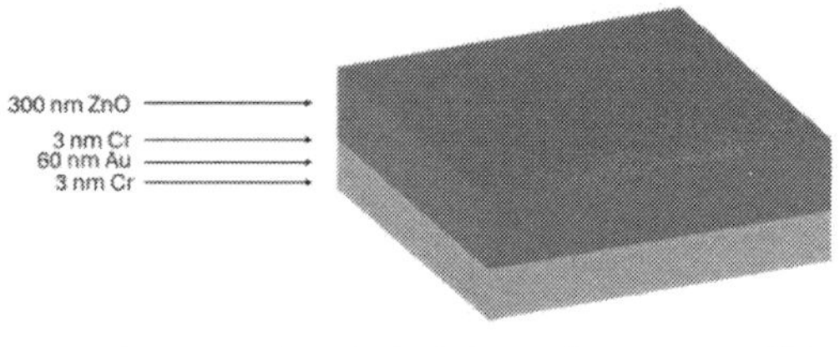

Fig. 2. Diagram of the optimized plasmonic structure, with 60 nm of Au and 300 nm of ZnO. Notice de Cr adhesion layers, considered in the simulation work.

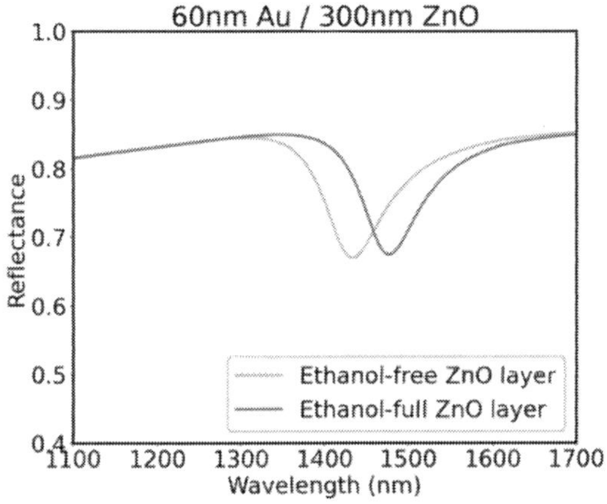

Fig. 3. Reflectance spectrum of the optimized plasmonic configuration. High Au and ZnO thicknesses result in a sensitive structure while limiting band widening.

This drift into the NIR range, however, tends to lead to a widening of the plasmonic band. In this configuration, this effect is being limited as the thick 60 nm layer of Au keeps the FWHM low by reflecting a significant share of incident light, limiting the range of resonant wavelengths, and resulting in a FoM = 0.267.

2.2 Bloch Surface Wave

For the BSW structure, TiO_2 and SiO_2 were used as dielectric materials for the truncated PhC. The high RI contrast between these two materials ($n_{TiO_2} = 2.52$ and $n_{SiO_2} = 1.45$, at $\lambda = 800nm$) enables strong bound states with well-defined absorption bands. Over the capping layer, a 3 nm Cr layer was considered, followed by the ethanol-adsorptive ZnO layer.

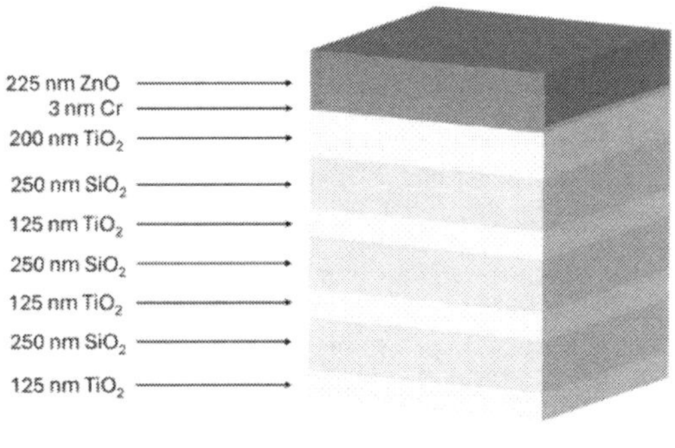

Fig. 4. Diagram of the optimized plasmonic structure, with 3 cells of 125 nm TiO_2 and 250 nm SiO_2, a 200 nm TiO_2 capping layer, and a 225 nm layer of ethanol-adsorptive ZnO.

Both the number of unitary cells of the PhC, the thickness of each material and their topological placement were simulated towards the highest FoM.

The optimized configuration consists of 3 unitary cells PhC of 125 nm TiO_2/250 nm SiO_2 bilayers, capped by a 200 nm TiO_2 layer, and followed by a 3 nm Cr adhesion layer with the 225 nm ZnO layer on top, returning FoM = 0.256. A schematic of the optimized structure is shown in Fig. 4, and the respective simulated spectrum is plotted in Fig. 5. Guided modes and BSW modes are visible, and readily identified by inspection of the dispersion relation in Fig. 6, obtained by the TMM by imposing bound-state conditions on the final matrix.

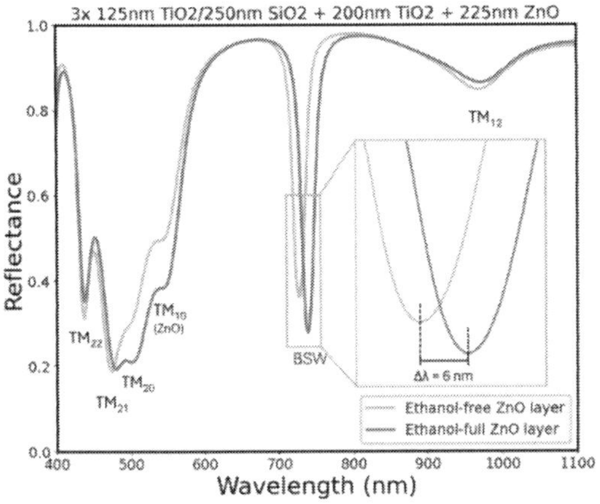

Fig. 5. Reflectance spectrum of the optimized BSW configuration. The symmetry-breaking capping layer near the ZnO layer enables ethanol adsorption sensitivity.

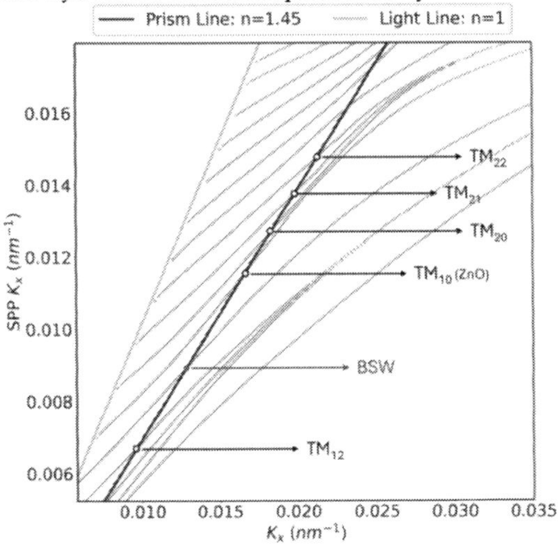

Fig. 6. Dispersion relation of the simulated optimized structure. Guided modes can be seen diverging towards the NIR, and BSW propagating in the photonic bandgap. The prism dispersion line determines mode wavelength in the Kretschmann configuration.

3 Experimental Validation

A preliminary characterization of the BSW carrying structure was carried out. The sample was fabricated and characterized for ethanol adsorption sensitivity.

3.1 Thin Film Fabrication

The fabricated sample was deposited on a commercial soda-lime glass substrate. The substrate was cleaned in successive ultrasonic baths of acetone, ethanol and de-ionized water, and later rinsed with an air blowgun.

The SiO2 and TiO2 layers of the truncated PhC, as well as the Cr adhesion layer, were deposited in a RF magnetron sputtering system. The top ZnO layer started as a metallic Zn thin film deposited on the sample in a e-beam evaporator system. The structure then oxidated in a thermal oxidation process at 400°C for one hour, at ambient atmosphere, forming the porous ZnO layer [26].

3.2 Sensor Characterization

The thin film structure was characterized in an acrylic gas cell of known volume. This gas cell, containing two valve inlets, for ethanol gas and a vacuum pump, was placed inside an oven for very stable temperature monitorization. The ethanol gas comes from a glass flask containing liquid ethanol sitting inside the oven, ensuring thermal equilibrium with the gas cell.

Inside the gas cell, a Kretschmann setup was installed, with a polarizer and a SiO_2 coupling prism (35 mm × 25 mm × 17 mm). Light from a tungsten-halogen light source, covering the VIS-NIR spectral range, was coupled into an optical fibre patch cord that goes into the gas cell to couple light to the prism. A similar patch cord collects light out of the prism, guiding it out of the gas cell, and to a spectrometer. A diagram of the experimental setup is depicted in Fig. 7.

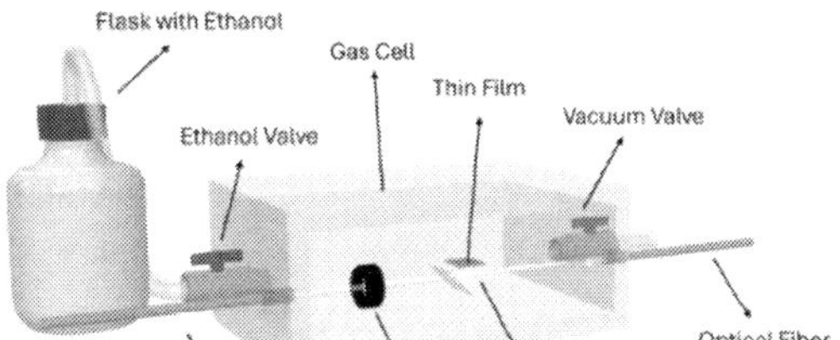

Fig. 7. Experimental setup for characterization of ethanol sensitivity in Kretschmann configuration. The oven ensures controlled temperature monitorization, while the gas cell of known volume allows ethanol concentration to be estimated.

The substrate with the deposited thin film structure was placed on top of the coupling prism, and the chamber closed. A rough vacuum is initially set on the chamber by opening the vacuum valve. This vacuum ensures a more controlled atmosphere, free of other potential VOC contaminants. The vacuum valve is then closed, and the ethanol gas valve is opened. The temperature of the entire system was slowly increased, resulting in an increased concentration of ethanol in the gas cell. As the concentration increases, optical spectra of the thin film were recorded.

The measurements were taken until the BSW mode appeared to stop moving, indicating a saturation point for ethanol adsorption in the sensing structure.

3.3 Preliminary Results and Discussion

Before the ethanol adsorption characterization and Zn deposition, an experimental spectrum of the truncated PhC was recorded and is plotted in Fig. 8 together with the simulated spectrum of the same structure. Despite a slight blueshift, a fair match between simulated and experimental spectra is obtained. All guided modes from the PhC are visible and in fair agreement with the respective dispersion relation (Fig. 9).

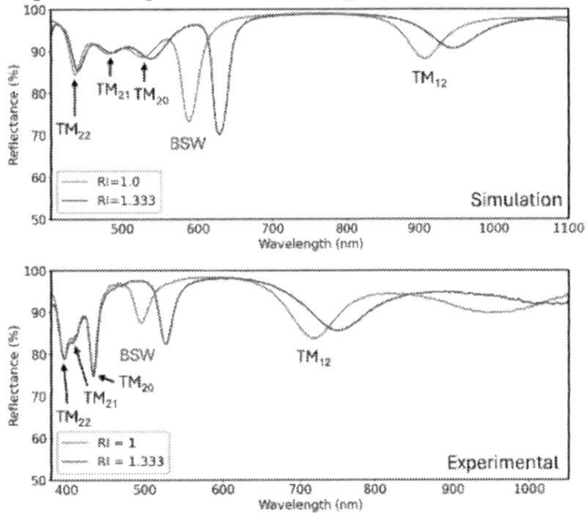

Fig. 8. Simulated and experimental spectra of the PhC generating the BSW, prior to Zn deposition, for the external medium of air and water. A blueshift in experimental data may indicate errors in deposited thicknesses.

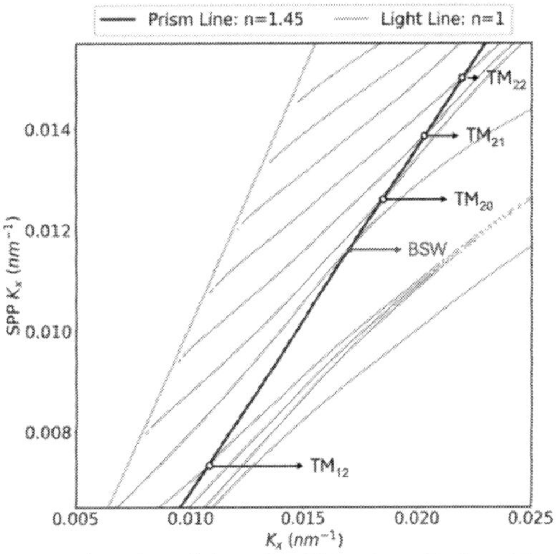

Fig. 9. Dispersion relation of the PhC prior to Zn deposition. The respective spectrum is plotted in Fig. 8.

The spectra for the ethanol sensitive structure prior to ethanol exposure, and for different ethanol concentrations are plotted in Fig. 10(a) and 10(b). Despite the guided modes being in fair agreement with the simulated spectra, the same does not occur for the BSW mode, exhibiting a strongly damped band.

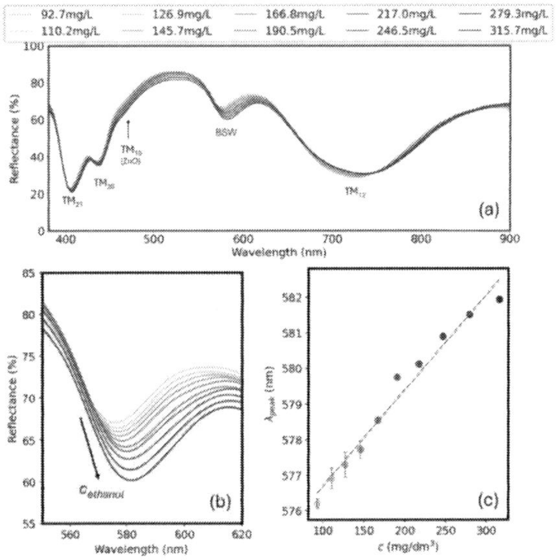

Fig. 10. (a) Experimental reflectance spectrum of the deposited thin film. Despite all visible guided modes, the BSW appears damp. (b) Close-up reflectance spectra of the BSW. A redshift is clear for increased ethanol concentration. (c) Linear sensitivity is roughly observed for the studied range of ethanol concentrations, with a sensitivity of (0.027±0.002) nm·L/mg.

Given the reasonable match for the truncated PhC exhibited in Fig. 8, it is likely that the error relies on the later fabricated ZnO layer, either in the deposition of metallic Zn or in the oxidation process into ZnO.

The shift of the BSW mode could be attributed to an error in the deposited thickness of the Zn layer, later oxidized and expanded into ZnO, however, it would not entirely explain the wide damping of the optical bands seen in Fig. 10(a). An explanation for this phenomenon could be attributed to the remaining of some non-oxidated metallic Zn in the ZnO film, as metals are known to greatly absorb light.

Most likely, both effects may be occurring to some point, and without proper characterization techniques to measure film thickness (ellipsometry, SEM, etc.) or material composition (XRD, XPS, etc), no answer is definite. Regardless, a sensitivity to ethanol of (0.027±0.002) nm·L/mg is exhibited from the 100 to 300 mg/L operational range (Fig. 10(c)). For higher ethanol concentrations, the sensitivity decreases non-linearly until saturation.

3.4 Conclusions

A comparative simulation study of SPR and BSW-based sensors for VOC detection was carried out by TMM. The simulation of ethanol adsorption in ZnO capping layers of both SPR and BSW bearing structures was achieved considering a perturbation of the permittivity of the ZnO layer. FOMs were determined for both ESWs, with the SPR sensor FOM being governed by the spectral sensitivity to the ZnO perturbation, whereas the BSW FOM was governed by the reduced FWHM of the resulting bands.

The preliminary results for the fabricated BSW structure unveiled a fair match of optical bands, despite a strong damping of the BSW mode. Despite this, the redshift of the BSW mode with loss of reflectance for increasing ethanol concentrations matches that of the simulated spectrum, attaining a sensitivity of (0.027±0.002) nm·L/mg and evidencing the potential application of BSW all-dielectric structures for VOCs monitorization.

References

[1] Khatib, M., & Haick, H. (2022). Sensors for volatile organic compounds. *ACS nano*, *16*(5), 7080-7115. 10.1021/acsnano.1c10827

[2] Xie, Y., Lyu, S., Zhang, Y., & Cai, C. (2022). Adsorption and degradation of volatile organic compounds by metal–organic frameworks (MOFs): a review. *Materials*, *15*(21), 7727. 10.3390/ma15217727

[3] Khan, A., Kanwal, H., Bibi, S., Mushtaq, S., Khan, A., Khan, Y. H., & Mallhi, T. H. (2021). Volatile organic compounds and neurological disorders: from exposure to preventive interventions. In *Environmental contaminants and neurological disorders* (pp. 201-230). Cham: Springer International Publishing. 10.1007/978-3-030-66376-6_10

[4] Wei, C., Pan, Y., Zhang, W., He, Q., Chen, Z., & Zhang, Y. (2023). Comprehensive analysis between volatile organic compound (VOC) exposure and female sex hormones: A cross-sectional study from NHANES 2013–2016. *Environmental Science and Pollution Research*, *30*(42), 95828-95839. 10.1007/s11356-023-29125-0

[5] Pathak, A. K., & Viphavakit, C. (2022). A review on all-optical fiber-based VOC sensors: Heading towards the development of promising technology. *Sensors and Actuators A: Physical*, *338*, 113455. 10.1016/j.sna.2022.113455

[6] Qu, X., Hu, Y., Xu, C., Li, Y., Zhang, L., Huang, Q., ... & Fu, Y. (2024). Optical sensors of volatile organic compounds for non-invasive diagnosis of diseases. *Chemical Engineering Journal*, 149804. 10.1016/j.cej.2024.149804

[7] Sudheer, V. R., Kumar, S. S., & Sankararaman, S. (2020). Ultrahigh sensitivity surface plasmon resonance–based fiber-optic sensors using metal-graphene layers with Ti 3 C 2 T x MXene overlayers. *Plasmonics*, *15*, 457-466. 10.1007/s11468-019-01035-3

[8] He, C., Liu, L., Korposh, S., Correia, R., & Morgan, S. P. (2021). Volatile organic compound vapour measurements using a localised surface plasmon resonance optical fibre sensor decorated with a metal-organic framework. *Sensors*, *21*(4), 1420. 10.3390/s21041420

[9] Gryga, M., Ciprian, D., & Hlubina, P. (2020). Sensing concept based on Bloch surface waves and wavelength interrogation. *Optics Letters*, *45*(5), 1096-1099. 10.1364/OL.387593

[10] Veeraghattam, B., Mukherjee, S., Guha, P. K., & Bhaktha, B. S. (2024). Sensing of volatile organic compounds using one-dimensional photonic crystal Bloch surface waves and internal optical modes. *Optics & Laser Technology*, *175*, 110818. 10.1016/j.optlastec.2024.110818

[11] Gao, L., Kou, D., Ma, W., & Zhang, S. (2023). Biomimetic Metal–Organic Framework-Based Photonic Crystal Sensor for Highly Sensitive Visual Detection and Effective Discrimination of Benzene Vapor. *ACS Applied Materials & Interfaces*, *15*(25), 30606-30618. 10.1021/acsami.3c03673

[12] Saruhan, B., Lontio Fomekong, R., & Nahirniak, S. (2021). Influences of semiconductor metal oxide properties on gas sensing characteristics. *Frontiers in Sensors*, *2*, 657931. 10.3389/fsens.2021.657931

[13] Franco, M. A., Conti, P. P., Andre, R. S., & Correa, D. S. (2022). A review on chemiresistive ZnO gas sensors. *Sensors and Actuators Reports*, *4*, 100100. 10.1016/j.snr.2022.100100

[14] Wang, C. N., Li, Y. L., Gong, F. L., Zhang, Y. H., Fang, S. M., & Zhang, H. L. (2020). Advances in doped ZnO nanostructures for gas sensor. *The Chemical Record*, *20*(12), 1553-1567. 10.1002/tcr.202000088

[15] Kang, Y., Yu, F., Zhang, L., Wang, W., Chen, L., & Li, Y. (2021). Review of ZnO-based nanomaterials in gas sensors. *Solid State Ionics*, *360*, 115544. 10.1016/j.ssi.2020.115544

[16] NIST 2013, *Ethanol*, NIST Chemistry WebBook, SRD 69, accessed 31 July 2024, <https://webbook.nist.gov/cgi/cbook.cgi?ID=C64175&Mask=4&Type=ANTOINE&Plot=on#Top>.

[17] Mougkogiannis, P., Turner, M., & Persaud, K. (2020). Amine detection using organic field effect transistor gas sensors. *Sensors*, *21*(1), 13. 10.3390/s21010013

[18] Yesudasu, V., Pradhan, H. S., & Pandya, R. J. (2021). Recent progress in surface plasmon resonance based sensors: A comprehensive review. *Heliyon*, *7*(3). 10.1016/j.heliyon.2021.e06321

[19] Gryga, M., Ciprian, D., & Hlubina, P. (2020). Bloch surface wave resonance based sensors as an alternative to surface plasmon resonance sensors. *Sensors*, *20*(18), 5119. 10.3390/s20185119

[20] Markos, P., & Soukoulis, C. M. (2008). *Wave propagation: from electrons to photonic crystals and left-handed materials*. Princeton University Press. 10.1515/9781400835676

[21] Johnson, P. B., & Christy, R. W. (1974). Optical constants of transition metals: Ti, v, cr, mn, fe, co, ni, and pd. *Physical review B*, *9*(12), 5056. 10.1103/PhysRevB.9.5056

[22] Johnson, P. B., & Christy, R. W. (1972). Optical constants of the noble metals. *Physical review B*, *6*(12), 4370. 10.1103/PhysRevB.6.4370

[23] Malitson, I. H. (1965). Interspecimen comparison of the refractive index of fused silica. *Josa*, *55*(10), 1205-1209. 10.1364/JOSA.55.001205

[24] DeVore, J. R. (1951). Refractive indices of rutile and sphalerite. *JOSA*, *41*(6), 416-419. 10.1364/JOSA.41.000416

[25] Bond, W. L. (1965). Measurement of the refractive indices of several crystals. Journal of Applied Physics, 36(5), 1674-1677. 10.1063/1.1703106
[26] Coelho, L., Viegas, D., Santos, J. L., & De Almeida, J. M. M. M. (2016). Characterization of zinc oxide coated optical fiber long period gratings with improved refractive index sensing properties. *Sensors and Actuators B: Chemical, 223*, 45-51. 10.1016/j.snb.2015.09.061

Efficient generation of two coherent spectral lines using two mutually injection locked DFB lasers

Leonardo Rama [1,*], *Manuel* Violas[1] and *Miguel* Drummond [1]

[1]Instituto de Telecomunicações and Universidade de Aveiro, Aveiro, Portugal

Abstract. This work proposes a new energy efficient method for generating two mutually coherent laser lines. This method combines mutual optical injection locking with external modulation, which can be implemented with an electro-absorption modulator (EAM). Numerical simulations demonstrate that the proposed method is feasible within a narrow locking range. In addition, it is demonstrated that suppressing the optical carrier of the modulated signal extends the locking range by a factor of 3.

1 Introduction

Microwave photonics (MP) intertwines radio-frequency technology with photonics to generate and process RF signals in the optical domain. This field takes advantage of the fact that converting RF signals to the optical domain reduces the wavelength by three orders of magnitude, thereby enabling a decrease in the size, weight, and power consumption (SWaP) [1]. One of the most important basic functionality that a MP circuit may provide is an RF signal generator [2]. In fact, there are many different techniques for building such a circuit.

Generating a single tone RF signal can be accomplished by beating two laser lines in a photodiode. However, beating two incoherent spectral lines generates an RF signal with wandering RF frequency and strong phase noise. External modulation [3] enables producing two mutually frequency- and phase-locked spectral lines. However, the power of the generated spectral lines is low due to low modulation efficiency. As a result, optical frequency combs [4] and mode locking lasers [5] have been proposed since they are able to generate several spectral lines with high efficiency. However, these methods end up being inefficient given that only two spectral lines are selected, resulting in a significant amount of energy being wasted on unused lines. Optical injection locking (OIL) can be combined with external modulation to overcome the aforementioned shortcomings [3]. However, OIL requires optical isolation of the master laser, which cannot be achieved at least directly in photonic integrated circuits (PICs). In short, a method to efficiently generate two and only two coherent spectral lines with high power efficiency is still missing.

This work proposes a new energy-efficient methodology for the generation of two mutually coherent spectral lines, which combines mutual optical injection locking (MOIL) with external modulation produced by an electro-absorption modulator (EAM).

This paper is structured to first introduce the theoretical background of MOIL, followed by the presentation of the proposed concept, numerical results, and finally, the conclusion and future work.

2 Theoretical background

2.1 Optical Injection locking

In OIL, the electric field of a master laser is injected in the cavity of a slave laser, as illustrated in Figure 1. As result, the electric field of the slave laser will operate at the frequency of the master laser and with a constant phase difference. Therefore, in the end the two lasers are coherent.

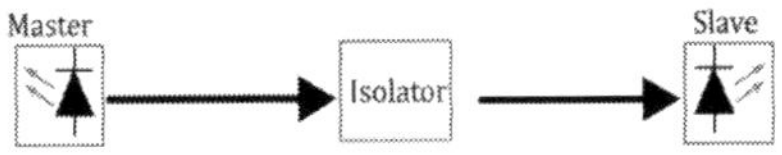

Figure 1. Optical Injection Locking scheme.

The influence of the injected field can be expressed by a differential equation [6,7]:

$$\frac{dE_s(t)}{dt} = \frac{1}{2}(1+i\alpha)g(N(t)-N_{th})E_s(t) + \kappa E_m(t)e^{i\Delta\omega t}, \quad (1)$$

where $E_s(t)$ is the electric field of the slave laser and $E_m(t)$ is the electric field of the master laser. α is the linewidth enhancement factor, g is the coefficient gain and N_{th} is the threshold carrier. κ is proportional to the injection ratio, r_{in}, which it is defined as the power that is injected into the slave laser: $\kappa = 1/\tau_{in}(1-r_0^2)r_{in}/r_0$. The other terms are constant and detailed in [7]. The first term of the previous equation is related to the normal functioning of the laser without injection locking. Nonetheless, the second term governs the participation of the master laser in the slave laser cavity.

* Corresponding author: leonardorama@av.it.pt

The authors of [6,7] derived the expression for the locking range in OIL. The locking range can be defined as the frequency difference between the lasers within which injection locking is achievable. The locking range is usually determined as function of the injection ratio.

2.2 Mutual optical injection locking

In MOIL both lasers are slave and masters. In others words the electric field of both lasers have influence in each other as it is represented in Figure 2.

Figure **2**. Mutual optical injection locking scheme.

The main advantage of MOIL is the possibility to synchronize the lasers without the use of optical isolators. However, the set of equations that describe MOIL are more complex. They can be expressed as (based on [7,8]):

$$\frac{dE_1(t)}{dt} = \frac{1}{2}(1 + i\alpha)g(N_1(t) - N_{th})E_1(t) + \kappa E_2(t - \tau)e^{+i(\omega_2\tau+\Delta\omega t)}, \quad \textbf{(2)}$$

$$\frac{dE_2(t)}{dt} = \frac{1}{2}(1 + i\alpha)g(N_2(t) - N_{th})E_2(t) + \kappa E_1(t - \tau)e^{+i(\omega_1\tau-\Delta\omega t)}, \quad \textbf{(3)}$$

$$\frac{dN_1(t)}{dt} = J_1 - \gamma_N N_1(t) - [\gamma_p + g(N_1(t) - N_{th})]A_1(t)^2, \quad \textbf{(4)}$$

$$\frac{dN_2(t)}{dt} = J_2 - \gamma_N N_2(t) - [\gamma_p + g(N_2(t) - N_{th})]A_2(t)^2 \quad \textbf{(5)}$$

$E_1(t)$ and $E_2(t)$ are the electric field of the laser 1 and 2. $N_1(t)$ and $N_2(t)$ denote the carriers of the laser 1 and 2, respectively. $J_1(t)$ and $J_2(t)$ are the current for both lasers. τ is the time of flight between the two lasers. The other constants are described in [7]: $\alpha = 3.5$ is the linewidth-enhancement factor, $g = 7.0 \times 10^{-6}\,ns^{-1}$ is the linear gain coefficient, $N_{th} = 2.214 \times 10^8$ is the threshold carrier number, $\gamma_p = 500\,ns^{-1}$ is the photon decay rate, $\gamma_N = 0.5\,ns^{-1}$ is the carrier decay rate and $\tau_{in} = 7\,ps$ is the round-trip time in the cavity.

The interaction between both lasers creates the potential for several phenomena to occur: free run, chaos, or MOIL. Consequently, it is essential to identify the conditions under which MOIL occurs. Employing a similar approach to that used by the authors of [6,7] for OIL, we derived an expression for the locking range for MOIL:

$$-2\kappa\sqrt{1 + \alpha^2}\, cos(\omega_R\tau + \psi) < \Delta\omega < 2\kappa\sqrt{1 + \alpha^2}\, cos(\omega_R\tau + \psi)\,, \quad \textbf{(6)}$$

where ω_R is the resonance frequency and $\psi = \arctan(\alpha)$.

The numerical results will be compared to this expression, to check that the simulations are produced correctly. It should be noted, that a similar expression was obtained by the authors of [9], who used slightly different set of equations to describe MOIL.

3 Proposed concept

The main goal of this paper is to demonstrate a method for energy efficient photonic generation of mm wave signals within a PIC. The setup includes an electro-absorption modulator (EAM) between two lasers, as it is illustrated in Figure 3.

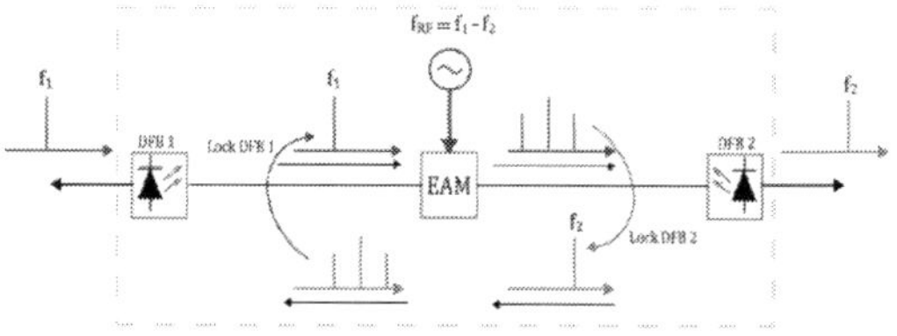

Figure **3**. Proposed 2-tone generator.

Given that the RF frequency is identical to the frequency spacing between distributed feedback (DFB) lasers, i.e., $f_{RF} = f_1 - f_2$, DFB1 and DFB2 can be mutually optical injection locked through the generated RF sidetones. Consequently, two mutually coherent tones are produced and the phase noise is reduced. The main advantage of the proposed approach is that it does not require optical isolators, thus being implementable in a PIC.

In this paper, we investigate the viability of the proposed concept using analytical and numerical assessment. To simulate the proposed concept, a new set of equations which include the modulation of the EAM are considered:

$$\frac{dE_1(t)}{dt} = \frac{1}{2}(1 + i\alpha)g(N_1(t) - N_{th})E_1(t) + \kappa\mathbf{EAM}(t - \frac{\tau}{2})E_2(t - \tau)e^{+i(\omega_2\tau+\Delta\omega t)}, \quad \textbf{(7)}$$

$$\frac{dE_2(t)}{dt} = \frac{1}{2}(1 + i\alpha)g(N_2(t) - N_{th})E_2(t) + \kappa\mathbf{EAM}(t - \frac{\tau}{2})E_1(t - \tau)e^{+i(\omega_1\tau-\Delta\omega t)}, \quad \textbf{(8)}$$

$$\frac{dN_1(t)}{dt} = J_1 - \gamma_N N_1(t) - [\gamma_p + g(N_1(t) - N_{th})]A_1(t)^2, \quad \textbf{(9)}$$

$$\frac{dN_2(t)}{dt} = J_2 - \gamma_N N_2(t) - [\gamma_p + g(N_2(t) - N_{th})]A_2(t)^2, \quad \textbf{(10)}$$

However, the simulations can only be realized, if an analytical model for the EAM is considered. For now, let us assume that the EAM operates as an amplitude modulator. Hence, the EAM can be described as:

$$\mathrm{EAM}(t) = a_0 + a_1 \sin\omega_{RF}t. \quad \textbf{(11)}$$

The modulation produced by the EAM generates three laser lines. Consequently, several types of interactions can occur between the laser lines of both lasers. It can occur interaction between the main tones of laser 1 and 2, the side tones of laser 1 and 2 and finally, between the main tone of laser 1 and side tone of laser 2, and vice-versa.

Thus, MOIL should only occur between the main tone of laser 1 and the side tone of laser 2, and vice versa. This implies that interactions between the main tones of both lasers should result in free-running lasers, as should interactions between the side tones. Nevertheless, the interaction between the side tones of both lasers can be disregard, since the modulation efficiency of EAM is low.

In short, two conditions must be fulfilled:

1. The interaction between the main tones of laser 1 and 2 should result in free-run lasers. On other words, the main tones should not interact between them.
2. MOIL should only occur between the main tone of laser 1 and side tone of laser 2, and vice-versa.

4 Methodology

In the methodology of this paper, the requirements in terms of injection ratio and detuning needed to meet the previously established conditions are assessed. The work flow realized is the following:

1. MOIL, where the lasers remain unmodulated, $a_1 = 0$. The goal of this stage is to validate simulations with the derived curve and determine the locking range diagram. In particular, this stage is important for understanding the conditions under which interference between the main tones leads to the generation of free-running lasers.
2. MOIL with two tones modulation. Each laser is modulated by single sinusoidal tone: $a_0 = 0$, $\text{EAM}(t) = a_1 \sin \omega_{RF} t$. This type of modulation can be assumed as a double side band modulation with suppression carrier (DSB-SC). The goal of this stage is to determine by simulation the locking range between the main tone and side tone.
3. MOIL with three tones. Each laser is amplitude modulated. The equation that describes the EAM is (11). The goal of this stage is to determine the feasibility of our concept.

Stages 1 and 2 are implemented with the same injection ratio for comparison purposes. For now, the delay between lasers were set to 0. The threshold current for both lasers was 35 mA, and they were both supplied with 50 mA. The RF frequency was constant for the three stages, $f_{RF} = 20\ GHz$. The set of equations were solved using the dde23 routine from Matlab.

5 Results

5.1 MOIL with unmodulated lasers

Figure **4a)** represent the locking range diagram for unmodulated lasers. It is illustrated three different regions: free run, chaos and MOIL. Free-running lasers are achieved when two incoherent main laser lines are present. Chaos occurs when the output becomes unstable, resulting in several unstable laser lines in the spectrum. In MOIL, however, only a single laser line is produced.

As expected, MOIL occurs within the region defined in (6). Moreover, with the increase in injection ratio, the locking range also increases. Free-running lasing is achieved with lower injection ratios but higher detunings. For the proposed modulation frequency, free-run lasers are obtained for injection ratios lower than 0.2%.

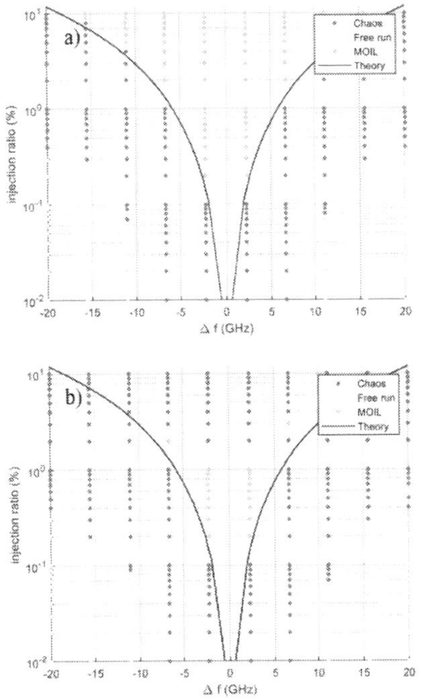

Figure **4.** Locking range diagram for a) unmodulated lasers b) MOIL with DSB-SC modulation.

5.2 MOIL with two tones

The goal of this section is to measure the locking range between the main tone and side tone. In this way, the detuning between the two lasers was defined as $\Delta f' = 20GHz + \Delta f$. Therefore, Δf is the difference between the main tone and the side tone, allowing the utilization of the derived curves for the locking range.

Figure **4b)** represents the locking range diagram for the double side modulation with suppression carrier. It is illustrated that the locking range occurs between derived curves.

Furthermore, it can be observed that for effective injection ratios exceeding 1%, only chaos is observed. Thus, the optimal injection ratio is 1%, as it corresponds to the widest locking range (~ 5.8 GHz). In addition, it is possible to conclude that higher the injection ratio greater is the locking range.

5.3 MOIL with three ones

Through the section 5.1 and 5.2 is possible to achieve some remarks:

1. From section 5.1, it is possible to obtain free-running lasers for lower injection ratio and higher detuning between the carriers, which means that a_0 should be lower. For the simulations conditions previously mentioned, free-running lasers is obtained for injection ratios lower than 0.2 %.

2. From section 5.2, it was shown that to achieve larger locking range, the effective injection ratio for the main tone-side tone should be large, which means that a_1 should be high. The maximum locking range possible is 1 % and the corresponding locking range is 5.8 GHz.

From these results, some remarks can be drawn for the three tones modulation. Realistically, in the modulation produced by the EAM, $a_0 > a_1$, indicating that, injection ratios higher than 0.2 % can be excluded, as interactions between the main tones lead to chaos (Figure **4a**). Consequently, the maximum locking range at 1 % of injection ratio (Figure **4b**) is unattainable, as demonstrated in Figure **5**.

However, the strength of both interferences depends on the amplitudes of the modulation produced by the EAM:

$$\mathrm{EAM}(t) = a_0 + a_1 \sin \omega_{RF} t = a_0 + \frac{1}{2i} a_1 e^{j\omega_{RF} t} - \frac{1}{2i} a_1 e^{-j\omega_{RF} t}. \quad \textbf{(12)}$$

As a result, the power of the generated spectral lines is related to parameters a_0 and a_1. Therefore, the injection ratio can be defined as a function of these parameters for the analyzed use cases. In use case 5.1 (MOIL unmodulated), the injection ratios can be expressed as $r_{m-m} = r_{in} a_0^2$. For use case 5.2 (MOIL with two tones), the injection ratio can be described as $r_{m-s} = \frac{r_{in} a_1^2}{4}$.

In this way, if ideal modulation is assumed, $a_0 = a_1 = 1$, the power of the side tones is one quarter of the main tone. In other words, $r_{m-s} = \frac{1}{4} r_{m-m}$. This implies that, $r_{m-m} = 0.2$ % and $r_{m-s} = 0.05$ %. Consequently, the maximum locking possible, assuming ideal modulation, is 1.7 GHz.

Figure **5** represent the scheme behind the derivation of the maximum locking range possible.

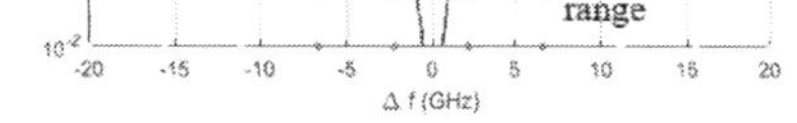

Figure **5**. Derivation of the maximum locking range for three tones modulation.

6 Conclusion

This paper proposes a novel method for generating two mutually coherent laser lines, based on MOIL. Some remarks should be highlighted.

The derived curves are good base to understand the locking range mechanisms, even assuming modulation between the lasers. The locking range determined for MOIL and MOIL with two tones modulation was similar until 1% of injection ratio.

MOIL with two-tone modulation (DSB-SC) offers the widest locking range of 5.8 GHz. However, the practical implementation of this modulation is complex and power-inefficient, as a setup with just one EAM is not viable.

MOIL with three tones modulation (DSB) produces two coherent laser lines, in a narrow locking range (1.7 GHz). However, such a narrow locking range requires a precise monitoring and control system. Nevertheless, this limitation can be reduced by raising the RF frequency at which the EAM is modulated, as a greater spacing between the lasers, free-run lasers is easily achieved.

European Space Agency (ESA) under contract with reference 4000137619/22/NL/MM/cna.

References

1. Marpaung, D., Roeloffzen, C., Heideman, R., Leinse, A., Sales, S., & Capmany, J. (2013). Integrated microwave photonics. Laser & Photonics Reviews, 7(4), 506-538.
2. Seeds, A. J., & Williams, K. J. (2006). Microwave photonics. Journal of lightwave technology, 24(12), 4628-4641.
3. Yao, J. (2009). Microwave photonics. Journal of lightwave technology, 27(3), 314-335.
4. Wang, W. T., Liu, J. G., Sun, W. H., Chen, W., & Zhu, N. H. (2015). Multi-band local microwave signal generation based on an optical frequency comb generator. Optics Communications, 338, 90-94.
5. Carpintero, G., Guzmán, R. C., Gordón, C., Kervella, G., Chitoui, M., & Van Dijk, F. (2015). Photonic integrated circuits for radio-frequency signal generation. Journal of Lightwave Technology, 34(2), 508-515.
6. J. Ohtsubo, Semiconductor Lasers: Stability, Instability and Chaos (Springer, 2006).
7. A. Murakami, K. Kawashima and K. Atsuki, "Cavity resonance shift and bandwidth enhancement in semiconductor lasers with strong light injection," in IEEE Journal of Quantum Electronics, vol. 39, no. 10, pp. 1196-1204, Oct. 2003, DOI: 10.1109/JQE.2003.817583.
8. C. Sun *et al.*, "Modulation Characteristics Enhancement of Monolithically Integrated Laser Diodes Under Mutual Injection Locking," in *IEEE Journal of Selected Topics in Quantum Electronics*, vol. 21, no. 6, pp. 628-635, Nov.-Dec. 2015, Art no. 1802008, doi: 10.1109/JSTQE.2015.2478817.
9. Li, Y., Yang, N., Xie, Y., Chu, W., Zhang, W., Duan, S., & Wang, J. (2019). Basic phase-locking, noise, and modulation properties of optically mutual-injected terahertz quantum cascade lasers. Optics Express, 27(3), 3146-3160.

Smartphone Interrogation of Narrow Bragg Gratings in G.652 Fibre

Marcos Madeira[1], *Álvaro* Santos[1], *Valmir* Oliveira[2], and *Hypolito* Kalinowski[1*]

[1] Dept. Eng. Telecomunnications, Universidade Federal Fluminense 24210-240 Niterói, Brazil
[2] Multi-Foton Laboratory, Universidade Tecnológica Federal do Paraná 80230-901 Curitiba, Brazil

Abstract. Transmission and reflection spectra of FBG in the visible region are seen and measured with a smartphone coupled portable apparatus. Standard G.652 optical fibre is used, operating in a few-mode regime.

1 Introduction

Along recent years attention was given to the properties of Fiber Bragg Grating (FBG)[1] with spectrum in the visible region [2, 3]; it was also shown that the spectra from such gratings can be detected by a commercial hand-held spectrometer, even when illuminated from the LED flashlamp source of a smartphone [4, 5]. Such findings are promising for the interrogation of FBG sensors by a low-cost smartphone-based spectrometer, almost continuously connected to the global World Wide Web. The connectivity will permit data transfer from distributed measuring points to a central processing unit, and it will also allow the use of cloud computing in the analysis of obtained data.

FBG spectra using a smartphone have already been measured [6, 7]. The first paper describes the interrogation of a chirped-FBG (FWHM ~ 6 nm) inscribed in multi-mode G.651 fibres and their use as a strain sensor, whereas the second one describes the smartphone measurement of narrow FBG (FWHM ~ 0.2 nm) produced in standard G.652 fibre optic; the last authors reported the shift of the central wavelength with temperature, replicating the FBG use as a thermal sensor.

It is worth noticing that the authors of [6] state that FBG spectra in the visible should not be observed in single-mode fibres with the apparatus' LED source. However, previous work has already demonstrated such spectra both in special single-mode fibres (designed for the visible) and in standard G.652 fibres operating as few mode fibre in the visible wavelengths [2]. The last condition was also reported in [7], using the LED light source of the smartphone. The present work complements the measurement techniques of FBG in standard G.652 fibres with use of a smartphone either in transmission or reflection set-ups.

2 Materials and Methods

FBG with the spectra in the visible region were inscribed using the under the phase-mask method with light from an *Ar:F*–excimer laser, Coherent Xantos XS-500 at 193 nm, repetition rate 250 Hz, pulse energy 2.0 mJ. Conventional ITU G-652 fibre optics (NA ~ 0.14) were used, previously hydrogenated at 100 atm for one week. As previously pointed, the G.652 fibre operates as a few-mode fibre at visible wavelengths [2], a roughly estimate shows 7 modes at 670 nm. The obtained FBG have Bragg wavelengths at 593 nm and 671 nm and their line-widths are in the range ~ 0.2 – 0.3 nm, measured by an OSA (Yokogawa 6373) just after inscription

A hand-held spectrometer (Thorlabs, CCS100/M) is used to obtain reference spectra, using light either from a red LED (Thorlabs M660F1) or from the LED lamp in the smartphone. To connect the optical source to the grating and to observe the reflected signal, a fibre coupler (Thorlabs TW670R5F2) is used – this device is also used when transmission experiments are performed, using the straight channel of the coupler. When using the red LED, a buffer fibre serves to couple the SMA connector of the LED to the FC/PC of the fibre coupler input pigtail.

The smartphone spectrometer is based on a replica mould transmissive diffraction grating (1000 lines/mm, Rainbow Symphony) mounted in a designed attachment unit to the smart- phone. The unit is 3-D printed with ABS plastic, and the prototype is seen on Fig. 1. The cylindrical holes permit to fix FC/PC adapters to connect the optical fibre collecting light from the LED and to deliver light to the CCD camera; the small rectangular side-located slot is used to insert a mount with the diffraction grating. The angled cylinder is designed to allow the visible region of the spectrum to fall on the smartphone camera. The grating mount also holds an entrance slit, whose width was measured by optical interference to be 80 μm. The used smartphone is an ASUS 550KL model, whose camera has 4128 × 3096 pixels. The recorded photographs are obtained with an ISO setting of 50.

* Corresponding author: hjkalinowski@id.uff.br

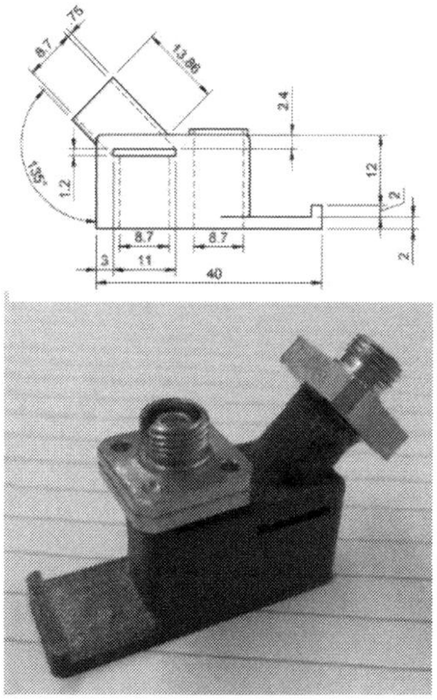

Fig. 1. Top: Schematic design of the smartphone clipped spectrometer. Bottom: 3-D printed smartphone coupling device incorporating a transmissive diffraction grating and fibre connectors.

The prototype device is calibrated in wavelength using two laser pointers (green and red) coupled to an input fibre. As for the spectral response of the CCD, it is replicated the procedure proposed by [8].

3 Results and Discussion

Figure 2–left shows a photograph taken by the used prototype of the transmission spectrum of one FBG when illuminated by a red-light LED. It is possible to discern the shadowed region associated to the forbidden band of the grating. That photo is clear evidence that, even using simple components in the instrument design, a non-apodized FBG in standard optical fibre can be measured by a smartphone. The shadow of the photonic band gap attributed to the FBG is stronger than that presented by [6] as the actual picture is taken with a intense light source. It is worth mentioning that it is possible to see the dark region inside the red light at naked eye, when looking the light coming out from the diffraction grating and projected on a blank screen (in a dark room). This can be of use in applications where a human certification of the sensing link integrity is needed but it is also a nice demonstration for a FBG showroom. The right side of Figure 2 shows the smartphone photograph of the reflection spectrum of a grating at 593 nm, where the faint spot associated to the reflected band is still discernible in the orange-red region (the photographs do not have the same scale), now using the LED flashlamp of the smartphone.

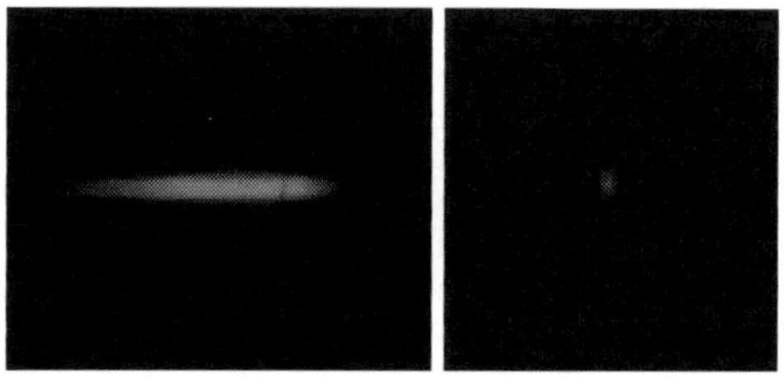

Fig. 2. Left: Photograph of the transmitted spectrum of a FBG ©Asian J. Phys. (with permission). Right: Image of the reflection spectra of another FBG.

The procedure to obtain the spectra corresponding to the smartphone photographs has been detailed in [7] (a usual optic labs urban legend states that *"If you can see it, you can measure it"*). The spectral measurements of FBG in transmission mode are presented on the graphs in Figure 3–left as compared to the measurement made with the commercial hand- held spectrometer. The vertical scales are arbitrary and not directly related, due to the different characteristics of the measuring devices.

The spectrum measured with the smartphone prototype displays clear broadening of the optical bands, caused by the lower resolution of the diffraction grating and by the difference in the entrance slit width. However, the obtained signal to noise ratio is good and permit the grating identification with ease. The transmission dip of the grating is about 20% of the nearby background baseline, which is clearly better than that obtained by [6] with the chirped FBG.

The reflection spectrum of another FBG is displayed on the inset graph of Figure 3, when the grating is illuminated by the smartphone LED source. For this experimental configuration the light intensity is lower and this is shown in the graph by the noisy background and a few spikes. Further measurements show better signal to noise by increasing the exposure time of the CCD.

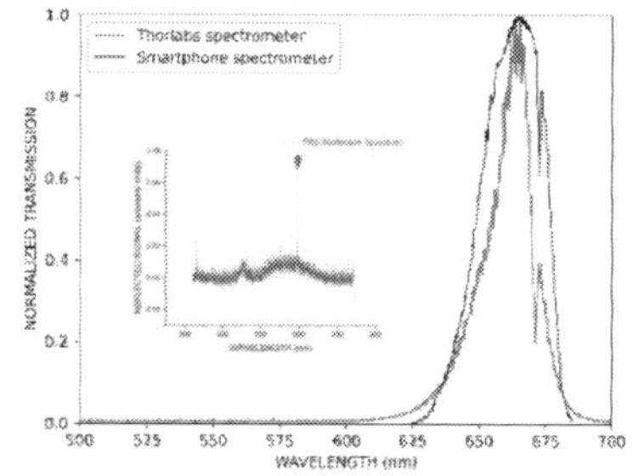

Fig. 3. Transmission spectrum of a 671 nm FBG measured with the current prototype as compared with a reference spectrum obtained by commercial spectrometer. The inset displays a raw reflection spectrum of a 593 nm FBG measured by the prototype.

It can be seen on the graph in Fig. 4 – top that the obtained signal to noise ratio with the smartphone can be comparable to that given by the commercial spectrometer measurement. In this condition the exposure time of the camera was set to a few hundred milliseconds, in order to replicate the integration time of the commercial spectrometer used for comparison (however, the vertical scales have no direct comparison).

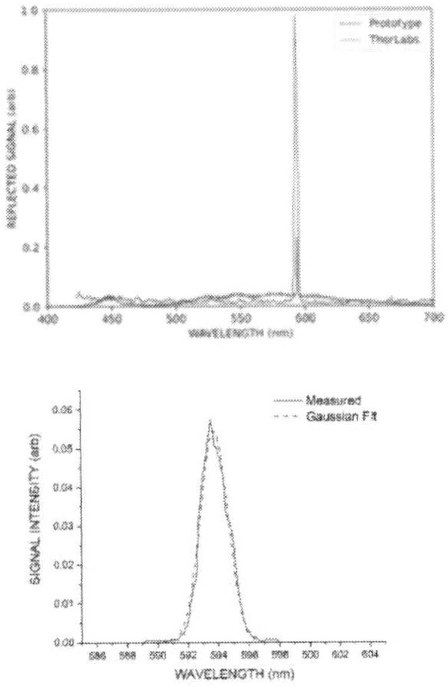

Fig. 4. Top: Reflection spectrum of a 593 nm FBG measured with the current prototype as compared with a reference spectrum obtained by commercial spectrometer. Bottom: The graph displays a detail of the central region as compared to the applied Gaussian fit.

The graph on the bottom of Fig. 4 shows a detailed view of the FBG reflection spectrum as compared to a Gaussian fit. The adjusted Gaussian width is 1.9 nm, close to the experimentally observed FWHM of 2.1 nm, which depicts the mentioned broadening caused by the diffraction grating and slits used in the apparatus. For comparison, the FWHM of the reference spectrum taken with the commercial spectrometer results as 0.84 nm. Although the current resolution can impose some constraints in the use of the prototype for a measuring application, it can be improved using a diffraction grating with higher line per mm count and a better designed entrance slit. Also, FBG measurements rely on peak detection techniques, which can be applied even in the presence of the broadening. It shall be noticed that the lower NA of the used fibre reduces the beam spread after the diffraction grating and helps to have a slightly better resolution than that of [6].

4 Conclusion

Spectra of standard, narrow, FBG are measured by using a diffractive attachment to a smartphone apparatus, both in transmission and reflection configurations. Albeit displaying a broadening due to the measuring conditions, it is possible to use the setup for sensing purposes.

Considering data presented in [9], only $\sim 3-6\%$ of incoherent light source should be transmitted along a few ($\sim 6-8$) mode fibre, which could be used to support the statement of [6] that the LED flashlamp on smartphones should not permit the measurement of FBG spectra in single-mode fibres. However, the actual measurements, as well those of [2, 7] show clearly that the mentioned limitation can be relaxed when the standard ITU G.652 fibre (single-mode at $\lambda \gtrsim 1260$ nm) operates at visible wavelengths, thus allowing the use of smartphones as interrogators for FBG devices with the LED flashlamp as an optical source. As compared to the modal equilibrium length, the short length fibre used in laboratory experiments can probably induce a higher non equilibrium power distribution, which allow propagation of an adequate power fraction in the lower order modes. This is supported by measurements of very narrow FBG in ITU G.651 graded index fibre [2], and consistent with single- or few-mode propagation.

The use of a smartphone-based interrogator for FBG sensors in the visible region will permit low-cost measurements with geo-localization and access to the communications network [8]. Recently the acquisition and signal processing procedures of the previous measurements were integrated in a single software application running in Android smartphones [10]. These developments can disseminate FBG measuring applications with reduced equipment & software costs.

Authors acknowledge scholarships and financial support from Brazilian agencies CNPq (404612/2023-7) and FAPERJ. The used gratings were inscribed at the Multi-FOTON facilities of UTFPR, Curitiba (Brazil).

References

1. K.O. Hill, Y. and Fujii, D.C. Johnson, B.S. Kawasaki, Photosensitivity in optical fiber waveguides: Application to reflection filter fabrication. Appl. Phys. Letters **32**, 647– 649 (1978)
2. P.L. Inácio, I.S.V. Suahele, I. Chiamenti, V. Oliveira, H.J. Kalinowski, Characterization of femtosecond-inscribed fiber Bragg gratings in the visible spectral range. Appl. Opt. **56(3)**, 510–514 (2017).
3. P.L. Inácio, C.C. Moura, I. Chiamenti, V. Oliveira, H.J. Kalinowski, Thermal sensitivity increase of RFBG in the visible range, in *Proceedings of 3rd International Conference on Applications of Optics and Photonics*, M.F.M. Costa, ed. (Proc. SPIE 10453, 2017), pp. 1045325-1–4.
4. G.C. Mastrapa, G.C. Guimarães, P.L. Inácio, V. Oliveira, H.J. Kalinowski,"Fibre Bragg gratings in the visible: towards low-cost detection," in *Optical Sensing and Detection V*, F. Berghmans and A.G. Mignani, eds. (Proc. SPIE, 10680, 2018), pp. 106800N-1–4.
5. G.C. Mastrapa, G.G. Guimarães, H.J. Kalinowski, "Pushing Fiber Bragg Gratings into the Visible," in *PHOTONICS 2018: International Conference on*

Fiber Optics and Photonics, New Delhi (India), (Proc. PHOTONICS 2018, TA1-11, 2018).

6. A.A. Markvart, L.B. Liokumovich, I.O. Medvedev, N.A. Ushakov, Smartphone-based interrogation of a chirped FBG strain sensor inscribed in a multimode fiber. J. Lightwave Technol., **39(1)** 282–289 (2021)
7. G.C. Mastrapa, A.A.C.O. Santos, G.G. Guimarães, H.J. Kalinowski, Smartphone reading of standard Bragg gratings in single mode optical fiber. As. J. Phys., **30(5)** 809–813 (2021)
8. M.A. Hossain, J. Canning, K. Cook, A. Jammalipour, Portable smartphone optical fibre spectrometer, in *24th International Conference on Optical Fibre Sensors, Curitiba, Brazil.* H.J.Kalinowski, J.L. Fabris, W.J. Bock, eds. (Proc. SPIE, 9634, 2015) pp. 81—84. .
9. A.W. Snyder, C. Pask, Incoherent illumination of an optical fiber. J. Opt. Soc. Am.**63(7)** 806–812 (1973)
10. M.V.C. Madeira and H.J. Kalinowski, Smatphone spectrometer for FBG interrogation, in *Latin American Workshop on Optical Fiber Sensors,Campinas, Brazil.*

POF Sensor for Dynamic Physiological Evaluation in a Dog-Assisted ASD Therapy: A Case Study

María Gaitán-Padilla[1,*], *Daniela* Farfal[2], *Patricio* Barria[2], *Marcelo Eduardo* Vieira Segatto[1], *Carlos A.* Cifuentes[3], *Maria Jose* Pontes[1], and *Camilo A. R.* Diaz[1]

[1]Telecommunications Laboratory (LabTel), Federal University of Espírito Santo, Vitória, Brazil
[2]Rehabilitation Center Club Leones Cruz del Sur, Punta Arenas, Chile
[3]Bristol Robotics Laboratory, University of the West of England, Bristol, UK

Abstract. Optical fiber sensors have been validated for physiological measurement as a potential device for smart textiles and wearable sensors. Also, developing sensors with Polymeric Optical Fiber (POF) shows promising applications in dynamic and real scenarios, as low-cost alternatives. Furthermore, the functioning of POF sensors can be used in no-skin contact applications, such as in highly skin-sensitive children with Autism Spectrum Disorder (ASD) during therapies. On the other hand, Dog-Assisted Therapy (DAT) is a stress-reducing method and has been studied to identify its influence on improving therapies and stress-related events. This work presents a case study in three weekly DAT sessions focused on speech therapy in a subject with ASD. A validated low-cost POF sensor was used in this first dynamic physiological measurement scenario. The analysis of five Heart Rate Variability (HRV) variables was made to compare the participant's physiological status before and during the DAT. The obtained pulse signal showed the dynamic movements affecting the signal but being able to obtain the pulse peaks. The physiological comparison showed a difference in the mean peak-to-peak time interval when the child was interacting with the dog than previous to the interaction in Session 1 and Session 2 (p-value<0.05). Also, an improvement in phonological consciousness was obtained, increasing by eight points after the three weeks of therapy. This case study showed the potential use of POF sensors in dynamic environments, and for wearable or smart textiles for children with ASD.

1 Introduction

Autism Spectrum Disorder (ASD) is a complex neurodevelopmental condition characterized by persistent challenges in social interaction, communication, and repetitive behaviors [1]. The prevalence of ASD has been increasing globally, making it a significant public health concern [2]. Individuals with ASD often experience a range of comorbidities, including anxiety, sensory sensitivities, and difficulties with motor coordination, which can complicate their daily functioning and quality of life [3, 4].

Conventional therapies for ASD typically involve a combination of behavioral interventions, educational strategies, and sometimes pharmacological treatments [5, 6]. Applied Behavior Analysis (ABA) is one of the most widely used approaches, focusing on reinforcing positive behaviors and reducing harmful or undesired ones [7]. Despite the effectiveness of these therapies in many cases, they often require significant time, resources, and specialized training, which can be barriers to access for some families [8].

Speech therapy is a critical component of the therapeutic regimen for many individuals with ASD [5]. This intervention aims to improve verbal and non-verbal communication skills, which are often significantly impacted in this population. Speech therapists work on language development, articulation, social communication skills, and alternative communication methods when necessary [9]. Enhancing communication abilities can greatly improve the overall quality of life and social integration for individuals with ASD.

Dog-Assisted Therapies (DAT) have emerged as a complementary approach to traditional ASD interventions [10]. These therapies leverage the effects of the interactions between humans and dogs to promote physical, emotional, and social well-being [11]. Interaction with therapy dogs has been shown to reduce anxiety, improve social interactions, and provide sensory comfort to individuals with ASD [12]. The non-judgmental nature and unconditional affection of therapy dogs can create a safe and motivating environment for therapeutic activities.

In this case, therapy dogs can be useful in therapies of stress management strategies, and the impact of therapy dogs can be quantified through the analysis of the Autonomic Nervous System (ANS). The ANS is crucial in regulating physiological responses to stress, when dysregulation of the ANS, often observed in individuals with ASD, can manifest as heightened stress responses, impacting overall health and well-being [10]. Understanding and monitoring ANS activity can provide valuable insights into the stress levels and physiological state of individuals

*e-mail: maria.padilla@edu.ufes.br

with ASD, which is essential for tailoring effective therapeutic interventions [13]. However, in Chile, there are few canine assistance centers, and most studies of this type of therapy do not perform monitoring during the session due to the movement and activity during the therapies. Performing stress measurements at specific points of the session limits the confidence of the monitoring information because the ANS is highly regulated and changes in time.

Heart Rate Variability (HRV) is a key indicator of ANS function and overall cardiovascular health. HRV measures the variation in time intervals between heartbeats and is influenced by the balance between the sympathetic and parasympathetic systems of the ANS [14]. Higher HRV is generally associated with better adaptability and resilience to stress, while lower HRV indicates increased stress and potential health risks [15]. Monitoring HRV can offer a non-invasive and objective measure of stress and autonomic regulation in individuals with ASD [16].

Advancements in sensor technology have revolutionized the ability to monitor physiological parameters in real-time [17]. However, while accurate, there are uncomfortable sensors for children with ASD, who tend to avoid unfamiliar items and devices that come in direct contact with their skin [18, 19]. Optical fiber sensors, in particular, Polymeric Optical Fiber (POF) sensors, offer several advantages, including low cost, high sensitivity, biocompatibility, and the ability to function in dynamic environments [20]. These sensors can be integrated into wearable devices, providing continuous and non-intrusive monitoring of physiological signals such as heart rate and HRV. Their application in therapeutic settings holds great potential for enhancing the understanding and management of ASD during DATs. In addition, the indirect measurement of respiration and pulse due to chest movement and vibration is an advantage toward commercial sensors in the case of children with ASD, who can reject unknown devices due to high sensitivity in the skin.

This work aims to evaluate the effect of DATs on HRV in a case study of a child with ASD, utilizing a proposed wearable POF sensor in a dynamic environment. By integrating advanced sensor technology with therapeutic interventions, this study aims to provide new insights into the physiological impacts of DAT and contribute to the development of more effective, personalized therapeutic strategies for individuals with ASD.

2 Methodology

The methodology described in this study was approved by the ethics committee of the Special School Club Leones Cruz del Sur and is based on analyzing six segments and five HRVs as the physiological monitoring of a participant with ASD, in rest and during a DAT, with a POF-based sensor. This section describes the sensors' operation principle, the case study, the DAT protocol, the experimental setup, and the signals and features for the analysis.

2.1 POF-based sensor

The proposed POF sensor [21], is made of SH4001 Simplex Optical Fiber Cable with Polyethylene Jacket, with a 1.0 mm Core Optical Fiber diameter, and 2.2 mm Polyethylene Jacket diameter (Mitsubishi Chemical Co., Japan). The sensor is based on the optical intensity variation concept, where a constant wavelength and intensity Light-Emitting Diode (LED) is connected through a POF to a photodetector configured with a trimmer to read as best as possible the light intensity variation [20]. The sensor has a light-sensitive area to generate a variation in the light intensity that can be detected with a photodetector [22]. A 3D-printed piece was designed to maintain the coupling between two ends of POF, having one end fixed to the 3D-printed piece connected to the LED and another free-moving end connected to the photodetector; shown in Figure 2.

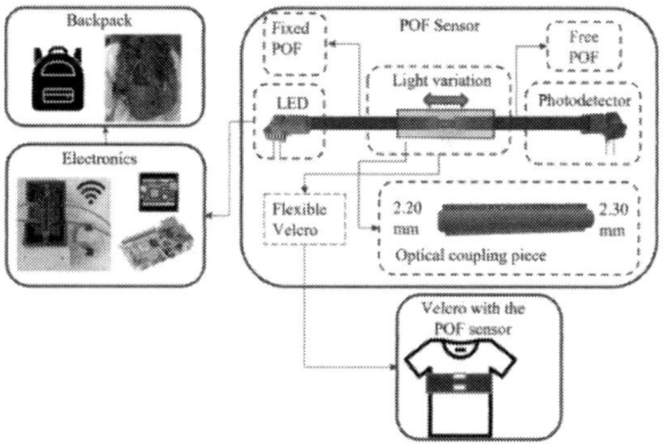

Figure 1. Diagram of the POF sensor functioning and the interrogation portable system.

2.2 Participant description

The participant is a 10-year-old boy with a diagnosis of Autism Spectrum Disorder (ASD), a user of the Rehabilitation Center Club de Leones Cruz del Sur, Punta Arenas, Chile. The participant has no respiratory, cardiac, or cardiovascular disease that could generate a high-risk situation during the session. He has no notable fear of animals, has no allergy to dogs, does not report having had any violent episodes towards an animal, and performs therapies with the dog periodically. The therapy dog has three years of experience assisting speech therapists in the institution, has all the requirements, and is a co-therapist at the rehabilitation center.

2.3 Experimental setup

The sessions were conducted in a therapy room adapted and pre-organized for the sessions, the relaxing activity material was located on a therapy table to continue with the routine of the conventional sessions. The sensor was adapted to the participant's preferences (his therapy t-shirt and backpack) and located in the room, as shown in Figure 2. The therapist ensured the participant's comfort to continue with the session. Two cameras were distributed

Table 1. Sessions flux for the procedure conducted during the case study, showing the goal of the session, and the control and DAT activities.

Session	Goal	Relax activity	DAT activity
1	To achieve identification of the initial phoneme	Cognitive control game: Set in a table with phonemic activities, with interactive cards, and boards	"Search the ball" game: Pictograms with phonetical indications glued in tenis balls asking to search the ball to the dog and asking to identify the initial phoneme to the participant
2	To achieve identification of the final phoneme		"Pressing buttons" game: Buttons indicated to press to the dog with audios with words to identify final phonemes and classifying them by the participant
3	To achieve phonemic synthesis		"Crazy cubes" game: A tower of cubes with letters are set in the middle of the room and it is indicated to the dog to take down the tower and it is asked to the participant to spell the expected word and identify the complete phonemic challenge

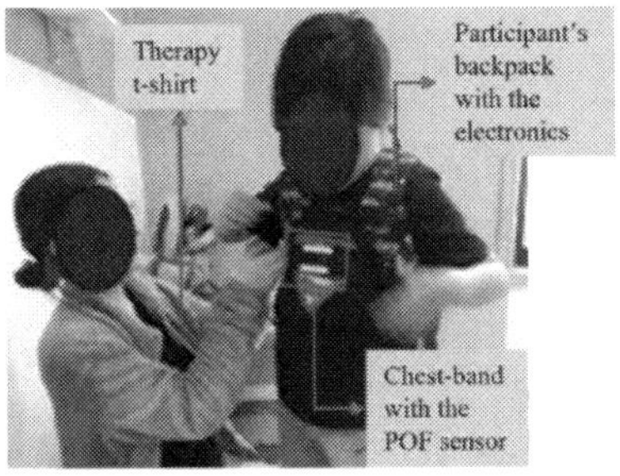

Figure 2. Location of the sensor in the participant at the beginning of the session, adapting the original sensor to the participant's preferences.

in the room to record the whole session and to monitor the sensor's functioning and control the times during the therapies, this helped to identify disconnections of the sensor by the engineer located outside the room. There was a computer at the main table to make it easier for the therapist to conduct the activities correctly and see the time of the session. The initial part of the session was conducted without the dog as the control time of therapy, and the second part was done with the dog assisting with the personalized phonetical activities for each session.

2.4 DAT procedure

This study was conducted as a case study with a pre and post-test evaluation of the standardized Phonological Awareness Screening Test (PECFO). This test was conducted also to develop a personalized therapy plan for the three sessions based on the main speech difficulties and limitations of the participant. Three DAT sessions were done, one session per week, and after this time, the PECFO test was applied again. The flow of the sessions is shown in Table 1. Before starting the study, the protocol was explained to the parents and the informed consent for participation was signed. In addition, a sensor familiarization was performed in a previous therapy to introduce the device to the participant. During the session, a resting phase

Table 2. Variables extracted from the pulse signal of the sensor with the description and processing.

Var	Description	Processing
HR	Heart rate	Counting the peaks of each 30 s window after the pulse peak detection multiplied by two
mIBI	Mean of the interbeat intervals	Mean of the detected peaks intervals of every height 30 s window
stdIBI	Standard deviation of the interbeat intervals	Standard deviation of the detected peaks intervals of every 30 s window
I2B	Mean of every two consecutive peaks	Mean difference of every two detected peaks intervals for each 30 s window
I10B	Mean of every ten consecutive peaks	Mean difference of every ten detected peaks intervals for each 30 s window

was performed to ensure that the participant was not fatigued, agitated, or stressed. The POF sensor was instrumented on the participant, and the sensor was checked for proper functioning with an interval of one minute. A baseline resting phase without the dog in the room was conducted with the control activity, and then, the therapy activity was performed with the dog. Afterward, the sensor was removed and the sessions were approximately 36 minutes, Figure 3 shows an example of the therapy room during the DAT activity with the sensor.

2.5 Processing for analysis

The results obtained from the sensors during the control and DAT activities were processed, graphically, and statistically compared, and analyzed with the HRV variables concerning the baseline to evaluate the influence of

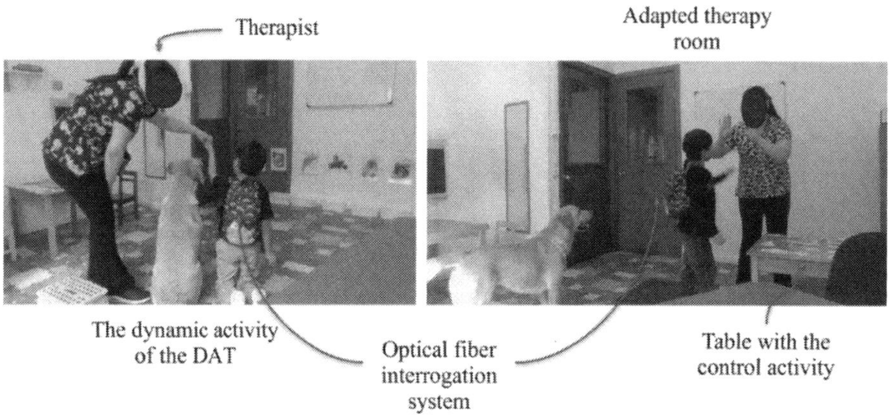

Figure 3. The therapy room as an example of the setup showing an activity during Session 2, the dog, the participant, the therapist, and the interrogation system of the sensor as a portable device.

the DAT program on the participant with ASD. Also, the PECFO test was compared before and after the therapy program.

For the POF data processing, the subject's first and last 30-second segments were omitted, and the control and DAT times were divided according to the video annotations. To confirm the frequency bands of the obtained signal with the POF, the relaxation segment was analyzed, the raw data was plotted, and the fast Fourier transformation (FFT) was performed to see the frequency bands and confirm the filtering process. The frequency bands for the biosignals correspond to the normal frequencies of heart and respiratory rates, as have been defined previously [23], from 0.1 to 0.8 Hz for the respiration and from 1 to 3 Hz for the pulse. According to the FFT with and the bands found in the literature [24, 25], a Butterworth filter was implemented for both signals. After the filtering process, the signals were verified by plotting them, as shown in Figure 4, and the Findpeaks function of Matlab was used to collect the peaks in both cases and extract the HRV variables described in Table 2.

3 Results and discussion

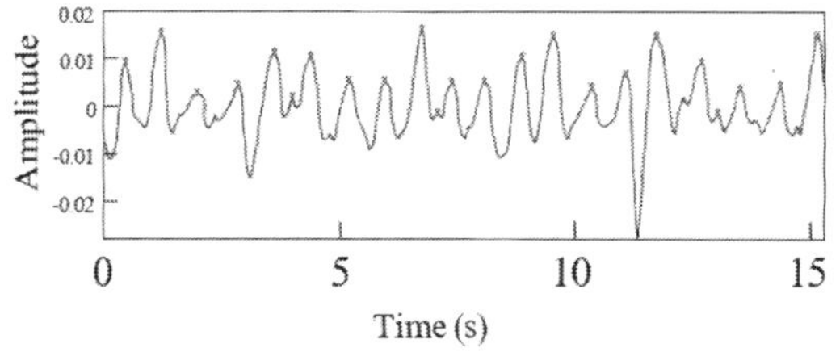

Figure 4. Pulse signal of a 15 s segment during a DAT activity of the participant.

Figure 4 illustrates the pulse signal obtained during the therapy sessions, highlighting the dynamic nature of the signal due to movements. Despite these fluctuations, the pulse peaks were successfully identified, allowing for analysis.

The physiological data, focusing on HRV, revealed a significant difference in the mean peak-to-peak time interval between sessions when the child interacted with the dog compared to before the interaction. Specifically, there was a notable change when comparing the mean peak-to-peak interval with the Wilcoxon statistical test (p-value < 0.05) during Session 1 and Session 2, as shown in Table 3. This suggests that the interaction with the dog had a measurable impact on the child's physiological state, potentially indicating reduced stress or increased relaxation during the therapy sessions.

The significant difference in the mean peak-to-peak time interval during dog interaction, as evidenced by the p-value < 0.05 in the Wilcoxon statistical test, supports that DATs can affect physiological responses. The reduction in the peak-to-peak interval indicates a potential calming effect or decreased stress levels, which aligns with existing literature on the therapeutic benefits of animal-assisted interventions [10]. This finding is particular given the dynamic nature of the pulse signal, which underscores the robustness of the physiological measurements despite potential sources of variability.

Also, with the DAT and the relation with the physiological analysis, an improvement in the PECFO assessment in the phonological consciousness was obtained. The results demonstrated an improvement in the PECFO score, with an increase of eight points following the three weeks of therapy, as shown in Figure 5. This improvement suggests that the DAT may have positively influenced the child's phonological awareness, aligning with the physiological findings.

The observed increase in PECFO scores by eight points post-therapy further corroborates the physiological findings. Improved phonological consciousness, as measured by PECFO, suggests that the therapy not only had an immediate physiological impact but also contributed to cognitive and developmental progress. This enhancement in phonological awareness may be attributed to the therapeutic environment facilitated by

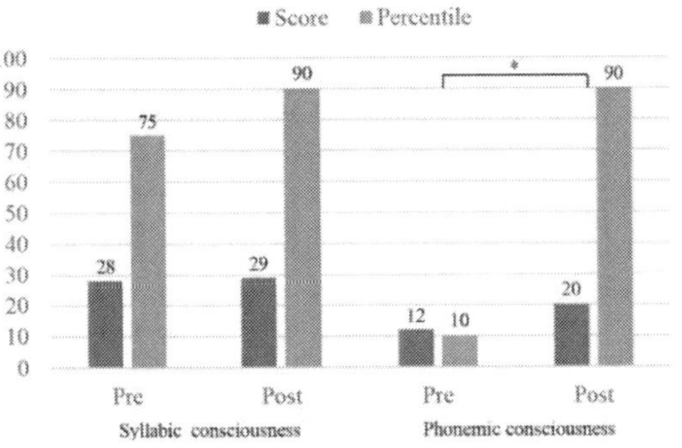

Figure 5. Pre and post-test results of the PECFO test of the participant, showing the statistical difference (*) obtained in the phonemic consciousness metric.

the presence of the dog, which could have created a more engaging and supportive atmosphere for the child.

The combined evidence from both physiological and cognitive assessments highlights the multifaceted benefits of dog-assisted therapy. The physiological data suggests a reduction in stress or anxiety, while the PECFO improvement indicates potential cognitive and educational benefits. These results align with prior studies that have reported positive outcomes from animal-assisted interventions, reinforcing the potential of such therapies in supporting both the emotional and developmental aspects of children.

Table 3. Results of the Wilcoxon statistical comparison between the control activity and the DAT in the three sessions, indicating p-values < 0.05 with *.

Var	Session 1	Session 2	Session 3
HR	0.91	0.54	0.90
mIBI	0.01 *	0.01 *	0.16
stIBI	0.02 *	0.12	0.42
I2B	0.05 *	0.61	0.22
I10B	0.39	0.18	0.70

4 Conclusion

This case study showed the potential use of optical fiber sensors in dynamic environments such as DATs as a no-skin contact device, and for wearable or smart textiles, mainly in children with ASD. In addition, it is a potential tool to identify the real impact and benefits of dog-child interactions in therapies. Future work will improve the POF-sensor's response in terms of sensitivity and performance in dynamic settings and it will be applied in other ASD therapies, to promote stress awareness and control strategies by analyzing HRV variables. While the results are promising, further research with larger sample sizes and diverse populations is needed to validate these findings. Future studies could explore the long-term effects of dog-assisted therapy on both physiological responses and cognitive development. Additionally, incorporating the evaluation of respiratory signals alongside pulse measurements could provide a more comprehensive understanding of the physiological impact of the therapy, potentially revealing how different aspects of physiological arousal are influenced by animal-assisted interventions.

5 Acknowledgement

This work is supported by FAPES (EXTRATO ATA DA 6ª REUNIÃO CCAF/2022 – "CPID 2.2"). Camilo A. R. Diaz acknowledges the financial support of FAPES (459/2021, 2022-97F12), CNPq (310668/2021-2, 404111/2023-8), and IEEE RAS SPARX and María Gaitán-Padilla acknowledges the financial support of FAPES (490/2024).

References

[1] L. Wing, The definition and prevalence of autism: A review, European child & adolescent psychiatry **2**, 61 (1993).

[2] J. Baio, Prevalence of autism spectrum disorder among children aged 8 years—autism and developmental disabilities monitoring network, 11 sites, united states, 2014, MMWR. Surveillance Summaries **67** (2018).

[3] B. Zablotsky, M.D. Bramlett, S.J. Blumberg, The co-occurrence of autism spectrum disorder in children with adhd, Journal of Attention Disorders **24**, 94 (2020).

[4] M. Micai, L.M. Fatta, L. Gila, A. Caruso, T. Salvitti, F. Fulceri, A. Ciaramella, R. D'Amico, C. Del Giovane, M. Bertelli et al., Prevalence of co-occurring conditions in children and adults with autism spectrum disorder: A systematic review and metaanalysis, Neuroscience & Biobehavioral Reviews **155**, 105436 (2023).

[5] E. Hollander, E. Anagnostou, Clinical manual for the treatment of autism (American Psychiatric Pub, 2007)

[6] R. Hurwitz, R. Blackmore, P. Hazell, K. Williams, S. Woolfenden, Tricyclic antidepressants for autism spectrum disorders (asd) in children and adolescents, Cochrane Database of Systematic Reviews (2012).

[7] H.S. Roane, W.W. Fisher, J.E. Carr, Applied behavior analysis as treatment for autism spectrum disorder, The Journal of pediatrics **175**, 27 (2016).

[8] A. Genovese, M.G. Butler, Clinical assessment, genetics, and treatment approaches in autism spectrum disorder (asd), International journal of molecular sciences **21**, 4726 (2020).

[9] M.H. Charlop-Christy, M. Carpenter, L. Le, L.A. LeBlanc, K. Kellet, Using the picture exchange communication system (pecs) with children with autism: Assessment of pecs acquisition, speech, social-communicative behavior, and problem behavior, Journal of applied behavior analysis 35, 213 (2002).

[10] R.E. Griffioen, G.J. van Boxtel, T. Verheggen, M.J. Enders-Slegers, S. Van Der Steen, Group changes in cortisol and heart rate variability of children with down syndrome and children with autism spectrum disorder during dog-assisted therapy, Children 10, 1200 (2023).

[11]F. Martin, J. Farnum, Animal-assisted therapy for children with pervasive developmental disorders, Western journal of nursing research **24**, 657 (2002).

[12]L.A. Redefer, J.F. Goodman, Brief report: Petfacilitated therapy with autistic children, Journal of autism and developmental disorders 19, 461 (1989).

[13]M. Espinosa, L. Escobedo, SATORI: a support tool for young adults with autism to self-manage stress and anxiety, in X Latin American Conference on HCI (2021)

[14]S.K. Schäfer, F.R. Ihmig, K.A. Lara H, F. Neurohr, S. Kiefer, M. Staginnus, J. Lass-Hennemann,

[15]T. Michael, Effects of heart rate variability biofeedback during exposure to fear-provoking stimuli within spider-fearful individuals: Study protocol for a randomized controlled trial, Trials 19 (2018).

[16]F. Shaffer, J.P. Ginsberg, An overview of heart rate variability metrics and norms, Frontiers in public health (2017).

[17]R. Castaldo, L. Montesinos, P. Melillo, C. James,

[18]L. Pecchia, Ultra-short term HRV features as surrogates of short term HRV: A case study on mental stress detection in real life, BMC medical informatics and decision making 19 (2019).

[19]M. Gjoreski, M. Luštrek, M. Gams, H. Gjoreski, Monitoring stress with a wrist device using context, Journal of biomedical informatics 73 (2017).

[20]N.E.J. Asha, R. Khan et al., Low-cost heart rate sensor and mental stress detection using machine learning, in 5th International Conference on Trends in Electronics and Informatics (IEEE, 2021).

[21]C. Kyriacou, R. Forrester-Jones, P. Triantafyllopoulou, Clothes, sensory experiences and autism: is wearing the right fabric important?, Journal of Autism and Developmental Disorders 53, 1495 (2023).

[22]A.G. Leal-Junior, C.A. Diaz, L.M. Avellar, M.J. Pontes, C. Marques, A. Frizera, Polymer optical fiber sensors in healthcare applications: A comprehensive review, Sensors 19 (2019).

[23]M. Gaitán-Padilla, M. Múnera, M. Pontes, M. Segatto, C. Cifuentes, C. Diaz, Development of a polymeric optical fiber sensor for stress estimation: A comparative analysis between physiological sensors, IEEE Sensors Journal (2024).

[24]J. Casas, A. Leal-Junior, C.R. Díaz, A. Frizera, M. Múnera, C.A. Cifuentes, Large-range polymer optical-fiber strain-gauge sensor for elastic tendons in wearable assistive robots, Materials 12 (2019).

[25]D.L. Presti, D. Bianchi, C. Massaroni, C. Coricciati, A. Rainer, S. Silvestri, A. Gizzi, E. Schena, Optimization and characterization of a 3d-printed wearable strain sensor for respiration and heartbeat measurements, Measurement p. 114377 (2024).

[26]L. Zhichao, Z. Xi, S. Taoping, M. Jiahe, Heartbeat and respiration monitoring based on fbg sensor network, Optical Fiber Technology 81, 103561 (2023).

[27]A.G. Leal-Junior, C.R. Díaz, C. Leitão, M.J. Pontes, C. Marques, A. Frizera, Polymer optical fiber-based sensor for simultaneous measurement of breath and heart rate under dynamic movements, Optics & Laser Technology 109 (2019).

A Comparative Study of Surface Plasmon and Tamm Plasmon Polaritons for Hydrogen Sensing

Miguel A. S. Almeida[1,2*], *João* P. M. Carvalho[1], *Isabel* Pastoriza-Santos[3], *José* M. M. M. Almeida[1,4], *Luís* C. C. Coelho[1]

[1] INESC TEC – Institute of Systems and Computer Engineering, Technology and Science, and Department of Physics, Faculty of Sciences, University of Porto, 4169-007 Porto, Portugal
[2] Department of Engineering Physics, Faculty of Engineering, University of Porto, 4200-465 Porto, Portugal
[3] CINBIO – Center of Research in Nanomaterials and Biomedicine, Campus Universitario Lagoas, Universidad de Vigo, 36310, Vigo, Pontevedra, Espanha, and SERGAS-UVIGO – Galicia Health Service, 36312 Vigo, Pontevedra, Espanha
[4] Department of Physics, School of Sciences and Technology, University of Trás-os-Montes e Alto Douro, 5001-801 Vila Real, Portugal
*Corresponding author: miguel.a.almeida@inesctec.pt

Abstract. Due to the exponential increase in energy consumption and CO_2 emissions, new sustainable energy sources have emerged, and hydrogen (H_2) is one of them. Despite all the advantages, H_2 has high flammability, so constant monitoring is essential. Two optical techniques were numerically studied and compared with the goal of H_2 sensing: surface plasmon polaritons (SPP) and Tamm plasmon polaritons (TPP). The H_2-sensitive material used was palladium (Pd) in both techniques. The SPP structure was found to have more sensitivity to H_2 than TPP, 23 and 5nm/4vol% H_2, respectively. However, the latter has lower FWHM, with the minimum of the band showing reflectivity near 0%. In addition, TPP also uses more cost-effective materials and can be interrogated at normal incidence with depolarized light. The potential of using each of these optical techniques for H_2 sensing was demonstrated.

1 Introduction

In the last few years, it has been observed an exponential increase in energy consumption and carbon dioxide (CO_2) emissions due to using fossil fuels, leading to significant ecological issues. In this way, alternative and sustainable energy sources have emerged, where green hydrogen (H_2) is one of the most promising fuels. This energy source has several advantages, such as double the mass-energy density of traditional fuels and the fact that it can be produced, storable, transportable, and utilized without CO_2 emissions. However, H_2 has some disadvantages, like its flammability and explosivity with low ignition energy at concentrations between 4 and 75 vol%. In this way, the correct monitoring of possible leaks is essential for the safer use of this fuel [1].

Several types of sensors have been developed to address these challenges. Optical sensors, in particular, offer advantages over other types of sensors. They can operate in harsh environments with immunity to electromagnetic interference, enable remote measurement, support multiplexing, and allow for miniaturization. Consequently, new optical principles should be explored to obtain high-sensitivity and reliable optical sensors [2].

Among several techniques, Surface Plasmon Polaritons (SPP) and Tamm Plasmon Polaritons (TPP) are two techniques that can be used for H_2 detection. The first, currently used in different applications, is well-known and is based on the excitation of collective electron oscillations (plasmons) at the interface between a metal and dielectric material due to the interaction with an evanescent wave. This excitation occurs at specific resonant conditions that depend on the light' wavelength, incident angle, and dielectric constants of the metal and dielectric material, as well as the external medium [3,4]. On the other hand, TPP is not yet commonly used for sensing. This plasmonic technique is based on the electromagnetic resonance between a distributed Bragg reflector (DBR) and a metallic film, excited with normal incidence. In this case, the resonant conditions depend on the light's wavelength, each layer's thickness, and the dielectric constants of the materials and external medium [5,6]. A layer composed of palladium (Pd) is used as sensing material to H_2 in both techniques. This element has a high ability to absorb H_2, forming palladium hydride (PdH_x), and during the transition from Pd to PdHx, the optical properties of the material change, changing the resonant conditions in both techniques, allowing H_2 detection [7].

An SPP structure composed of Au, TiO_2, and Pd was numerically characterized and optimized and then compared with a TPP structure composed of a DBR with TiO_2/SiO_2 and Pd layer that was also optimized. The advantages and disadvantages of each optical principle will be further discussed.

2 Numerical results

The numerical results presented were obtained using a Python routine based on Transfer Matrix Method (TMM). The optical properties of Au, SiO_2, and TiO_2 were extracted from the database *refractiveindex.info* [8], and the properties of Pd and PdH_x from the article of Subramanian *et al.* [9]. For each technique, the thickness of each layer was optimized by analyzing the band's spectral position, sensitivity (S), full-width-at-half-maximum (FWHM), and figure-of-merit (FOM) of each structure. This last parameter is obtained by S divided by FWHM, while S is the wavelength shift between the plasmonic band with Pd and PdH_x completely saturated with H_2.

2.1 Surface plasmon polaritons

SPP waves are obtained between a metallic and dielectric layer. In this case, the plasmonic material is Au, while the dielectric layer is TiO_2. Above these two layers, a thin layer of Pd is used to achieve H_2 sensitivity. The first step of this study is the influence of the Pd layer on the SPP band. For this, an optimized structure of 50nm of Au and 50nm of TiO_2 was used. The SPP band for this structure is present around 725nm, as observed in Fig. 1-a). Maintaining the metallic layer at 50nm and increasing the thickness of TiO_2 to 60nm, the SPP band is red-shifted to around 795nm (dashed yellow curve in Fig. 1-a)). This behavior is related to the high refractive index (RI) of TiO_2, which increases the real component of the plasmon propagation constant. On the other hand, in addition to the wavelength shift, higher dielectric thickness also contributes to higher FWHM.

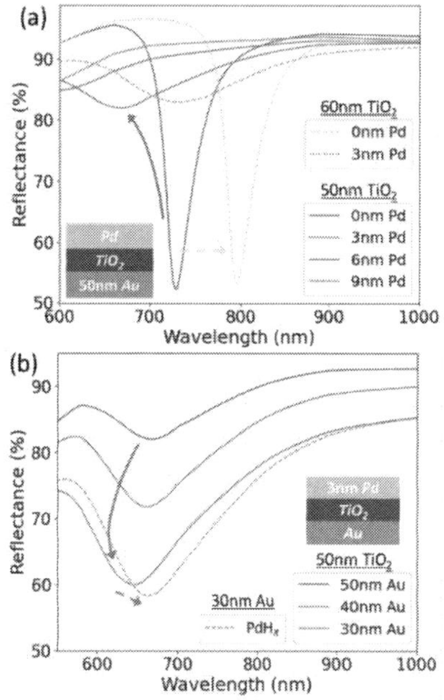

Fig. 1. Reflection spectra with SPP bands for different combinations of thicknesses on the SPP structure ($Au/TiO_2/Pd$). (a) Influence of Pd thickness and TiO_2, and (b) influence of Au thickness and PdH_x.

Fig. 1-a) also shows the SPP bands with different Pd thicknesses. The inclusion of this layer and the increase in the thickness contribute to an increase in the minimum of the band and, consequently, an increase in the FWHM. For thicknesses higher than 3nm, the SPP disappears due to the strong absorption of Pd. Considering this observation and the fact that 3nm of Pd is currently defined as the minimum that can be used to obtain layer uniformity and H_2 sensitivity, this thickness will be used in the subsequent analysis. With 3nm of Pd over the SPP structure, it's also possible to shift the SPP band to higher wavelengths using thicker layers of TiO_2 (see dashed blue curve in Fig. 1-a)).

Despite the possibility of obtaining an SPP band with an Au layer of 50nm, the obtained band is too shallow due to the high absorption of the structure. Considering the thin thickness of the Pd, reducing the Au thickness decreases the structure absorption. Fig. *1*-b) shows the results of that reduction from 50nm to 30nm, where a better definition is obtained with a slight blue-shift of the band. Again, the band's position can be rectified with the dielectric thickness. In Fig. *1*-b), the influence is also possible to observe the influence of H_2 on the resonance band. Comparing the orange curve with the dashed line, where the difference is only the transformation of Pd into PdH_x, it is possible to observe a red-shift and a decrease in the minimum value of the band. This behavior is related to the passage from metallic Pd to dielectric and, consequently, increases in RI.

From these results, the quantities S, FWHM, and FOM were calculated for different thicknesses of Au and TiO_2, while maintaining the Pd layer with 3nm. The obtained results are represented in Fig. 2. Sensitivity, FWHM, and Figure. Regarding the sensitivity, it is possible to observe an increase in higher thicknesses of TiO_2. Considering the influence of TiO_2, this result indicates that sensitivity is higher in higher wavelengths. This can be related to the variation of permittivity values of Pd, which is higher at higher wavelengths. Nevertheless, based on the FWHM graph, higher TiO_2 also contributes to higher FWHM and, consequently, a lower band definition. Regarding the Au thickness, in the study interval, it has a lower influence on performance compared with TiO_2. Finally, the FOM reflects the behaviors previously observed, where the FWHM has a strong impact. In this way, the optimized configurations have lower thicknesses of Au and TiO_2. In particular, the configuration with higher FOM is the one with 25nm/60nm/3nm of $Au/TiO_2/Pd$, respectively, achieving a sensitivity of 23nm/4vol% H_2 and an FOM of $0.20(4vol\%\ H_2)^{-1}$.

2.2 Tamm plasmon polaritons

In the TPP, electromagnetic resonance occurs between a DBR and a metallic film. The DBR is a periodic structure composed of alternating materials with different RI that reflect specific wavelengths due to constructive interference. The central reflected wavelength, λ_0, defines the thickness of each layer, d, given by $\lambda_0/4RI$ [5]. As an example, in Fig. 3 is represented a DBR centered at 1200nm and composed of 3 periods of TiO_2 and SiO_2 with 130.4 and 205.5nm ($n_{TiO_2}=2.3$, and $n_{SiO_2}=1.46$ at 1200nm), respectively. To excite the Tamm modes, a spacer layer should be used as a spacer above the DBR followed by the metallic layer. In Fig. 3, it is possible to observe three different Tamm bands that correspond to the use of 150, 200, and 250nm of SiO_2 as a spacer layer and 50nm of Pd as the

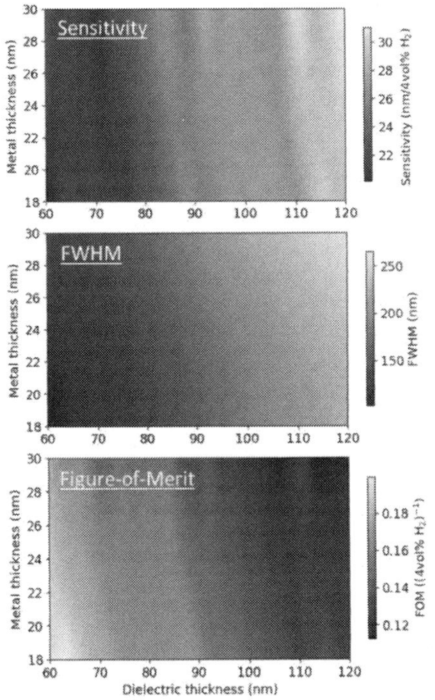

Fig. 2. Sensitivity, FWHM, and Figure-of-Merit of SPP structures with different thicknesses of Au/TiO_2 and 3nm of Pd.

metallic layer. Comparing these three bands, it's easy to find a relation with the dielectric layer in the SPP structure. As previously, higher thicknesses on the last layer in this structure induce higher red-shifts inside the DBR stopband. To obtain shifts beyond the DBR stopband, the DBR layers' thickness should be adjusted. Considering these observations, the influence on reflection spectra of the last layer and Pd thickness and the number of periods were analyzed. The DBR was considered fixed around 1200nm and the thickness of the DBR layers was not changed during the studies.

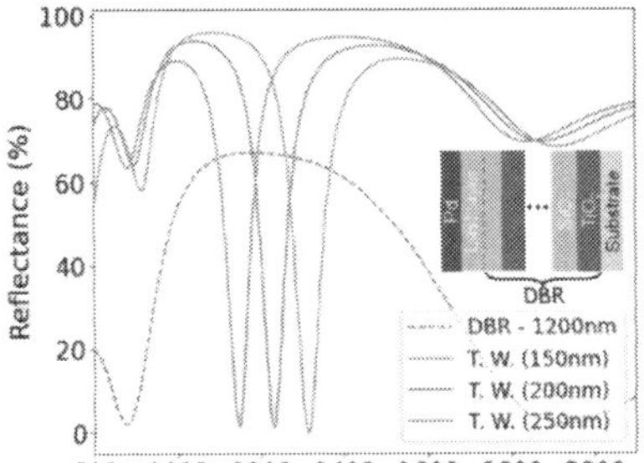

Fig. 3. Reflection spectra with the characteristic stopband of a simple DBR around 1200nm and three Tamm bands excited inside the stopband using a layer of SiO_2 and 50nm of Pd.

The first analysis is represented in Fig. 4 and corresponds to the influence of the number of periods and last layer thickness considering the Pd layer with 50nm. In these graphs, the highly reflective zone at yellow color, between 1000 and 1600nm, corresponds to the DBR stopband, while the blue shade corresponds to the Tamm wave. As previously mentioned, for all the number of periods, the increase of the thickness of the last layer induces a red-shift of the band with a quasi-linear relation. Regarding the number of periods, this parameter strongly influences the DBR stopband, where a higher number of periods allows a better definition of the stopband, with higher reflectivity and well-defined borders. Regarding the Tamm band, the number of periods can influence the FWHM and minimum value of the band. Concerning the first characteristic, comparing the three graphs, it is possible to constate that a greater number of periods contribute to a lower FWHM and better definition of the Tamm band. However, when this number is too high, the DBR reflectivity becomes appreciable, and the minimum value obtained on the Tamm wave increases. This can be seen in the graph obtained for four periods, where the minimum value is between 20 and 40%, while the minimum value for three periods is lower than 20%. Considering the FWHM and minimum value, the optimized number of periods is three.

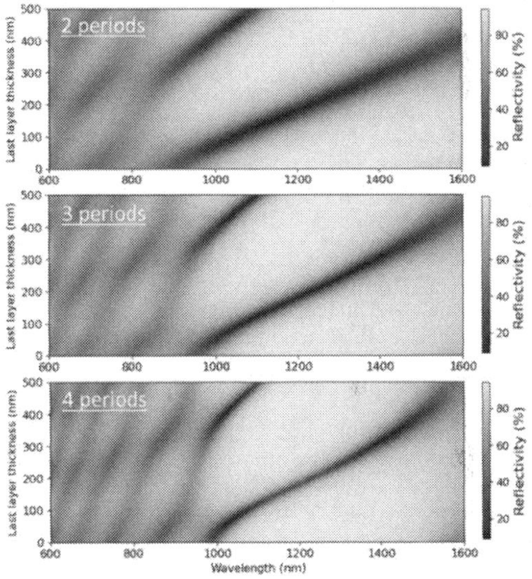

Fig. 4. Study about the influence of the last layer thickness and number of periods on the final spectra of a Tamm structure using 50nm of Pd.

In addition to the last layer thickness, the influence of the metal thickness on the Tamm resonance was also analyzed, and the results are represented in Fig. 5. The conclusions related to the influence of periods are the same as previously. With a higher number of periods, it is possible to obtain a lower FWHM. However, for structures with more than 3 periods, the minimum of the band increases, and the Tamm band starts to lose its definition. In this way, the optimized number of periods for this type of structure and final application is again three. Regarding the metal layer thickness, this layer is essential in Tamm structures for the confinement of light due to its negative dielectric constant [6]. In this way, when the metallic film is too thin, the effective dielectric permittivity of the metal/air system will not be negative enough, so higher thicknesses are needed. For example, in Fig. 5, for the structure with three periods and thicknesses below 20nm, it is not possible to excite

Tamm modes. Above that thickness, the Tamm mode can be excited, where higher thicknesses correspond to higher light confinement. In terms of the spectral position of the Tamm band, this parameter does not have any influence, being this characteristic only defined by the last layer thickness.

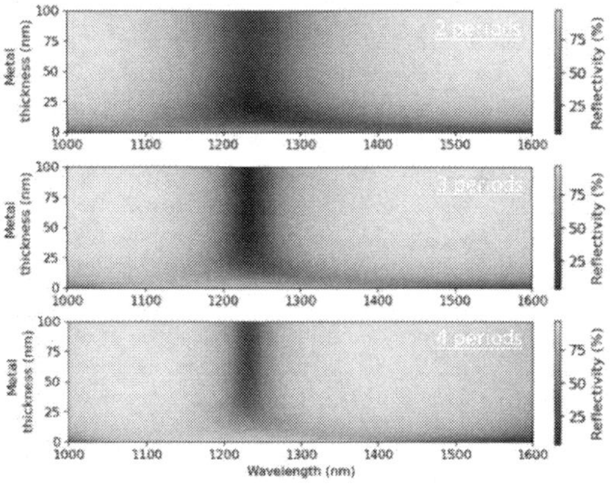

Fig. 5. Study about the influence of the metal layer thickness and number of periods on the final spectra of a Tamm structure using 200nm of SiO_2 as a spacer layer.

Considering these observations, a similar study was conducted on the SPP structure in terms of the quantities S, FWHM, and FOM on the Tamm structure, and the results are represented in Fig. 6. Sensitivity, FWHM, and Figure. Regarding the sensitivity, it is possible to observe higher sensitivity when lower thicknesses of Pd and higher thicknesses of SiO_2 are used. This occurs due to the higher penetration of the field into the metallic layer. When higher Pd thickness is used, higher is the electric field confinement on the last dielectric layer and lower the penetration on the Pd, reducing the sensitivity. However, as previously observed and as seen in Fig. 6. Sensitivity, FWHM, and Figure, the lower thickness of Pd results in higher FWHM, so thicker layers of Pd are needed to obtain a well-defined Tamm band. The graph of the FOM resumes the results of the other two graphs. This graph shows us that the optimized configurations, with higher FOM, need higher thicknesses of Pd and intermediate thicknesses of SiO_2. This intermediate thickness corresponds to the Tamm band centered on the DBR stopband. To find the optimized configurations, this work's authors consider the Pd film's expected time response, where thicker layers correspond to higher response time. Considering that FOM is very stable for thicknesses higher than 60nm of Pd with 200nm of SiO_2, and considering the time response, the optimized configuration found was with these same values, that is, with 60nm and 200nm of Pd and SiO_2 as the last layer. With this configuration, it is possible to obtain a sensitivity of 5nm/4vol% H_2 and a FOM equal to $0.05(4vol\% H_2)^{-1}$.

2.3 Direct comparison between techniques

To directly compare both techniques, the optimized bands of each without and with H_2 were plotted with the same scale in the x- and y-axis, and the result is represented in Fig. 7. Analyzing the shape of the bands, it's clear that the TPP band has a better definition and

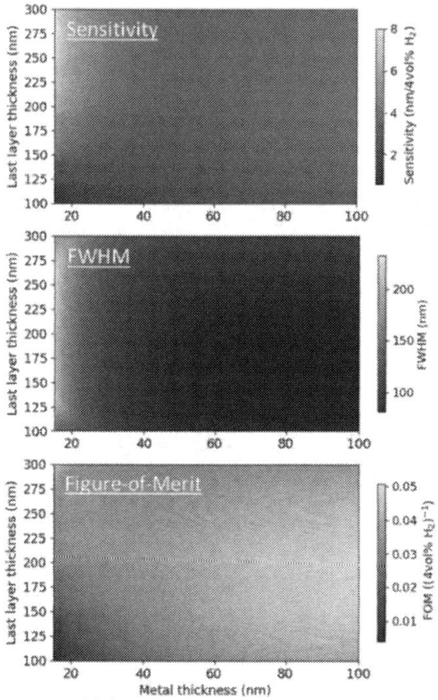

Fig. 6. Sensitivity, FWHM, and Figure-of-Merit of TPP structures with different thicknesses on the last layer (SiO_2) and metallic layer (Pd) with DBR central wavelength considered constant at 1200nm.

lower FWHM than the SPP band. While the latter has a minimum of around 52% of reflectivity and a maximum of around 80%, the first has a minimum of almost 0% and a maximum of 95%. Thus, in terms of interrogation, it will be easier to implement the TPP structure. However, comparing the wavelength shift with the presence of H_2, the SPP band has more than 4 times the shift presented on the TPP band, which is an important advantage for the SPP.

In practical terms, namely in terms of production, the SPP uses expensive plasmonic materials, like Au or Ag, while the TPP uses cost-effective materials. In addition, the SPP structure needs very thin layers of Pd (3nm). This can be an advantage, allowing faster response times; however, in terms of production, it can require expensive and accurate deposition techniques and can decrease the reproducibility of the structures. On the other hand, the thicker layers needed for the TPP structures increase the production time and the use of material. Regarding interrogation, the TPP can also have some advantages over the SPP. While the latter is excited with p-polarized light and a sensitive incidence angle, the TPP can be excited with depolarized light and normal incidence. In addition, the spectral position of the SPP band is in the visible range and far away from the preferable range of optical communications (around 1500nm). Adjusting the DBR thicknesses makes it possible to easily enable the TPP band to reach that range without losing performance, as occurs in the SPP. Considering all these characteristics, it is possible to confirm that SPP has more sensitivity, but TPP also has a lot of advantages over this traditional method that can be important in real applications.

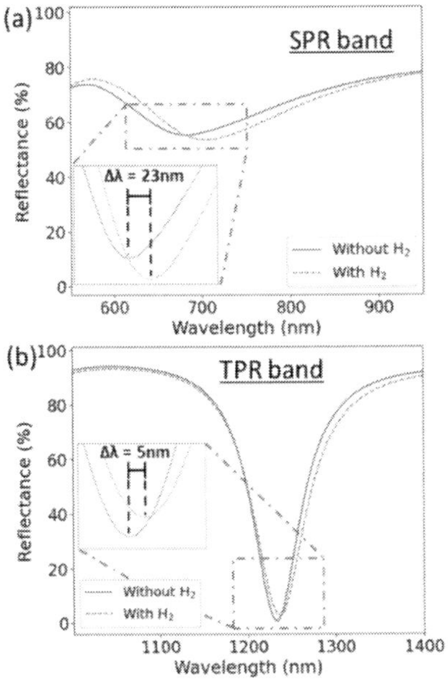

Fig. 7. Direct comparison between (a) SPP and (b) TPP optimized bands, with the x- and y-axis at the same scale.

3 Conclusion

Hydrogen flammability requires constant monitoring, where optical sensors can make a strong contribution. In this way, this work presents a comprehensive and systematic numerical study to compare the traditional SPP technique with the TPP, a less-known sensing technique. Pd was the material used in this work to obtain H_2 sensitivity. In the SPP structure, it has been found that the plasmonic layer's thickness strongly influences the FWHM, while the dielectric layer enables the spectral tunability of the band. In this case, very thin layers of Pd (3nm) must be implemented, which can allow fast response times. On the other hand, TPP can also be tunable using the thickness of the dielectric layers. However, the Pd layer needs to be thicker to allow the confinement of the electromagnetic field. Comparing the optimized bands of both techniques, the SPP has a higher sensitivity, but TPP has a lower FWHM. However, in real applications, TPP can have several advantages, such as the possibility of obtaining TPP bands in the infrared spectrum and using normal and depolarized light, which can facilitate the interrogation of the structures. In conclusion, this work demonstrates how these optical methods can be used for H_2 sensing and its advantages.

Acknowledgements: This work is financed by Component 5 - Capitalization and Business Innovation, integrated in the Resilience Dimension of the Recovery and Resilience Plan within the scope of the Recovery and Resilience Mechanism (MRR) of the European Union (EU), framed in the Next Generation EU, for the period 2021 - 2026, within project ATE, with reference 56. Miguel Almeida acknowledges the support from FCT research grant 2023.00890.BD. Luís Coelho acknowledges the support from FCT research contract grant CEECIND/00471/2017.

References

1. M. H. McCay, S. Shafiee, Hydrogen: An energy carrier. Future Energy: Improved, Sustainable and Clean Options for Our Planet, Elsevier, (2020). https://doi.org/10.1016/B978-0-08-102886-5.00022-0
2. Y. nan Zhang, H. Peng, X. Qian, Y. Zhang, G. An, Y. Zhao, Recent advancements in optical fiber hydrogen sensors. Sens Actuators B Chem 244 393–416 (2017). https://doi.org/10.1016/j.snb.2017.01.004
3. B. Ai, Y. Sun, Y. Zhao, Plasmonic Hydrogen Sensors. Small 18 (2022). https://doi.org/10.1002/smll.202107882
4. V. Yesudasu, H.S. Pradhan, R.J. Pandya, Recent progress in surface plasmon resonance based sensors: A comprehensive review. Heliyon 7 (2021). https://doi.org/10.1016/j.heliyon.2021.e06321
5. Y. Chen, Z. Yang, M. Ye, W. Wu, L. Chen, H. Shen, S. Ishii, T. Nagao, K. Chen, Tamm Plasmon Polaritons Hydrogen Sensors. Advanced Physics Research 2 (2023). https://doi.org/10.1002/apxr.202200094
6. C. Kar, S. Jena, D. V. Udupa, K.D. Rao, Tamm plasmon polariton in planar structures: A brief overview and applications. Opt Laser Technol 159 (2023). https://doi.org/10.1016/j.optlastec.2022.108928
7. A.K. Pathak, S. Verma, N. Sakda, C. Viphavakit, R. Chitaree, B.M.A. Rahman, Recent Advances in Optical Hydrogen Sensor including Use of Metal and Metal Alloys: A Review. Photonics 10, 122 (2023). https://doi.org/10.3390/photonics10020122
8. RefractiveIndex.Info, Refractive index database. https://refractiveindex.info/ (accessed March 1, 2023).
9. S. Subramanian, K. Kumar, A. Dhawan, Palladium-coated narrow groove plasmonic nanogratings for highly sensitive hydrogen sensing. RSC Adv 10, 4137–4147 (2020). https://doi.org/10.1039/C9RA08101A

Numerical evaluation of strain transfer model for steel-reinforced optical fiber cable embedded in a cylindrical concrete beam with two void inclusions

Mira Kabbara[1,2*], *Xavier* Chapeleau[1,2], *Qinghua* Zhang[1,2], and *Frédéric* Bourquin[3]

[1]COSYS-SII, Université Gustave Eiffel, 44340 Bouguenais, France
[2]I4S, Inria, 35042 Rennes, France
[3] Université Gustave Eiffel, 77420 Champs-sur-Marne, France

Abstract. In distributed optical fiber sensors, strain transfer is usually described through a simplified one-dimensional equation, derived from continuum mechanics. This equation, in which a parameter known as the strain-lag parameter takes into account the cable's geometric and mechanical characteristics, establishes a relationship between the measured longitudinal strain profile within the optical fiber and the real strain profile occurring in the host material. In the case of steel-reinforced optical fiber cables, appreciated for their resistance to breakage during on-site instrumentation, a notable discrepancy between the measured and actual strain profiles is revealed especially in the presence of a strain gradient, indicating that the ability to transfer strain from the host material to the optical fiber is restrained using this type of cable. This paper assesses numerically the strain transfer model for steel-reinforced optical fiber sensors in the presence of a strain gradient generated by two void inclusions in a concrete beam. The good accuracy of the strain transfer model is observed by the comparison with a 3D finite element simulation. However, the result points out the critical necessity of precisely determining the strain-lag parameter.

1 Introduction

In recent years, distributed optical fiber sensors, particularly those whose technology is based on Rayleigh backscattering, have become increasingly popular for Structural Health Monitoring due to their ability to provide real-time, high-resolution measurements along the entire fiber length [1-3]. These sensors are especially effective for detecting unlocalized and invisible cracks. Optical fibers, typically made of several protective coatings to prevent breakage, provide a blurred measurement of the real strain profile in the monitored structure. This phenomenon, referred to as strain transfer, can lead to misinterpretation of strain profiles, influenced by the type of optical fiber cable used. Efforts have been made to model this phenomenon, usually with a one-dimensional second-order differential equation that incorporates the cable's properties through a parameter known as the strain-lag parameter

* Corresponding author: mira.kabbara@univ-eiffel.fr

[4-6]. Indeed, this parameter depend on the mechanical and geometric properties of the cable, such as the radius, Young's modulus and Poisson's ratio of each layer and stiffness coefficient between consecutive layers. Achieving full strain transfer is particularly challenging for optical fiber sensors with multiple coating layers and steel reinforcement, especially in the presence of strain gradients. Despite this, steel-reinforced cables are beneficial for harsh environments and complex installations [7]. Most existing research studies simplify strain profiles as linear or bilinear functions as in [8]. This paper, however, addresses a more complex scenario with an arbitrary strain gradient, applying the strain transfer model proposed in [9] to steel-reinforced optical fiber sensors. First, a 3D finite element simulation on a cylindrical concrete beam with two void inclusions is conducted, and the resulting strain profiles are used to validate the model. Then, a study was done on the impact of erroneous beta values on the model's validity, as this parameter is influenced by the cable's properties, which are often unknown and can lead to potential inaccuracies.

2 Strain transfer model applied to a steel-reinforced optical fiber sensor

For this study, a representative model of a common steel-reinforced cable on the market is chosen and it is supposed to embedded into a concrete structure of length L. The steel-reinforced cable is composed of a glass fiber with a diameter of approximately 10 µm, surrounded by an optical cladding. This optical cladding typically has a diameter of125 µm. To increase the cable's durability, three protective layers are added: a steel coating that is further enclosed by two polymer layers. To simplify, the core and cladding are considered as a single layer, referred to as the core. About that, the system in question consists of a four-layered model as the one described in [9] and for which a formula is given for the calculation a strain lag parameter, $\boldsymbol{\beta}$. To remind, the strain profiles, $\boldsymbol{\varepsilon_f}$ and $\boldsymbol{\varepsilon_m}$,in the optical fiber and in the host material respectively, are linked by:.

$$\begin{cases} \varepsilon''_f(z) - \beta^2 \varepsilon_f(z) = -\beta^2 \varepsilon_m(z) \\ \varepsilon_f(0) = \varepsilon_0;\ \varepsilon_f(L) = \varepsilon_L \end{cases} \quad (1)$$

Where z represents the coordinate along the length of the cable.

3 Validation approach for the proposed model: computing the solution of the strain transfer model using Finite difference method

The strain-lag parameter, determined through the expression provided in [9], along with the longitudinal strain outputs obtained at discrete points from a 3D Finite Element simulation on the host material level (used as input of equation (1)), are employed to solve the second-order differential equation in (1) using the Finite Difference method, as elaborated in the subsequent discussion.

The space discretization into n+1 points, labelled as z_i, is outlined as follows:

$$z_i = ih \ where \ h = \frac{L}{n} \ and \ 0 \le i \le n \quad (2)$$

The Finite Difference method is employed to approximate the second derivative of ε_f in equation (1):

$$\epsilon''_f(z_i) = \frac{\varepsilon_f(z_{i+1}) - 2\varepsilon_f(z_i) + \varepsilon_f(z_{i-1})}{h^2} \ with \ 0 \le i \le n-1 \quad (3)$$

Substituting this approximation into (1) yields:

$$a\ \varepsilon_f(z_{i+1}) + b\ \varepsilon_f(z_i) + a\varepsilon_f(z_{i-1}) = -\beta^2 \varepsilon_m(z_i) \quad (4)$$

$$where \ a = \frac{1}{h^2} \ and \ b = -\frac{2}{h^2} - \beta^2$$

The equation (4) can be expressed as a linear system represented as follows:

$$AX = b \quad (5)$$

where

$$A = \begin{pmatrix} 1 & 0 & 0 & 0 & 0 & \dots & \dots & 0 \\ a & b & a & 0 & 0 & \dots & \dots & 0 \\ 0 & a & b & a & 0 & \dots & \dots & 0 \\ 0 & 0 & a & b & a & 0 & \dots & 0 \\ \vdots & \vdots & \ddots & \ddots & \ddots & \ddots & \ddots & \vdots \\ 0 & 0 & \dots & 0 & a & b & a & 0 \\ 0 & 0 & \dots & \dots & 0 & a & b & a \\ 0 & 0 & \dots & \dots & 0 & 0 & 0 & 1 \end{pmatrix},$$

and

$$X = \begin{pmatrix} \varepsilon_f(z_0) \\ \varepsilon_f(z_1) \\ \varepsilon_f(z_2) \\ \vdots \\ \varepsilon_f(z_{n-1}) \\ \varepsilon_f(z_n) \end{pmatrix} \text{ and } b = \begin{pmatrix} \varepsilon_0 \\ -\beta^2 \varepsilon_m(z_1) \\ -\beta^2 \varepsilon_m(z_2) \\ \vdots \\ -\beta^2 \varepsilon_m(z_{n-1}) \\ \varepsilon_L \end{pmatrix}$$

Solving the linear system in equation (5) yields ε_f at discrete points, representing the calculated strain profile in the optical fiber. A comparison is then made between the results of the 1D model and 3D simulation outcomes at the optical fiber level. Subsequently, a criterion for error is defined to assess the 1D model's accuracy. The described validation approach is detailed in Fig.1.

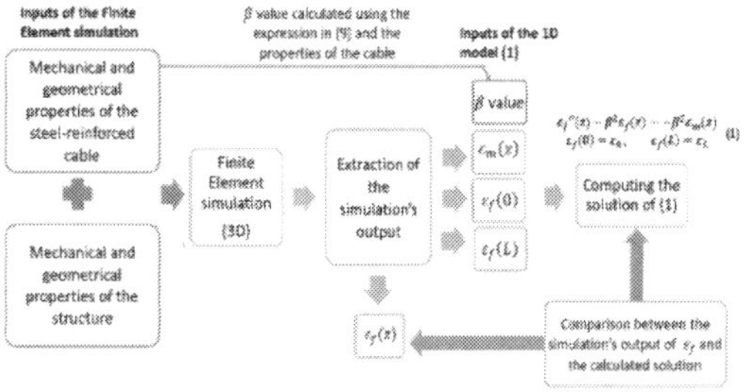

Fig. 1. Diagram representing the inputs and outputs of the 1D model.

4 Description of the three-dimensional simulation: a cylindrical beam including two inclusions with a steel-reinforced cable embedded at the center

A simulation is performed on a cylindrical concrete beam measuring 600 mm in length and 100 mm in radius, incorporating two void inclusions, as illustrated in Fig. 2. These void inclusions are characterized as annular regions resembling rings, defined by their geometric parameters. Each inclusion features an outer radius of 7 cm and an inner radius of 5 cm. The presence of these void inclusions can significantly affect the mechanical behavior of the concrete beam, influencing stress distribution within the structure and effectively creating a strain gradient. The beam is subjected to a tensile force of 125 kN. A 1-meter steel-reinforced optical fiber cable, with a radius of 1 mm, comprising an optical fiber and two polymer coatings encasing a steel coating, is embedded at the center of the structure. Based on the cable properties provided in Table 1 and the strain lag parameter expression given in [9], the strain lag parameter is found to be $\beta = 8.5\ m^{-1}$.

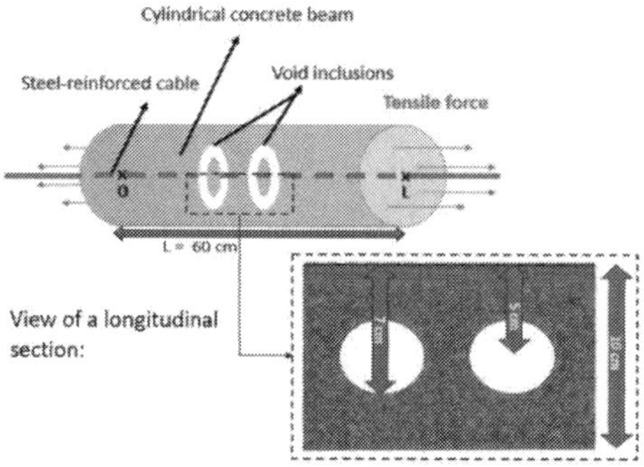

Fig.2. Cylindrical concrete beam with two inclusions and steel-reinforced cable in the center.

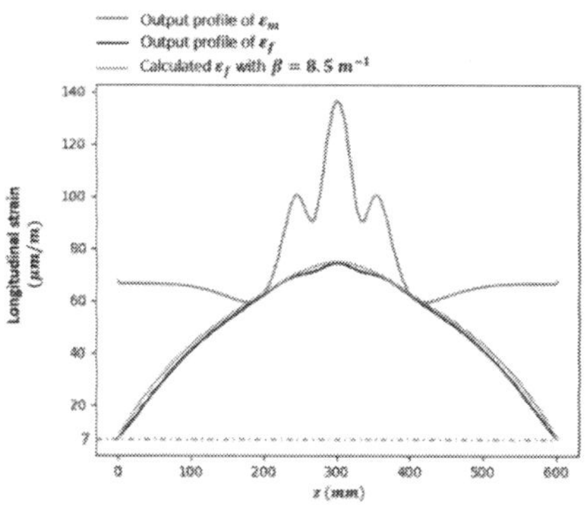

Fig.3. Simulation's outputs of the strain vs the 1D model's solution using appropriate boundary conditions (ε_0=ε_L=7 μm/m) and β = 8.5 m^{-1}

Table 1. Geometric and mechanical properties of the optical fiber cable's components.

Layer	Radius (mm)	Young Modulus (GPa)	Poisson 's ratio	Interfacial stiffness coefficient (Mpa/mm)
Optical fiber	0.06	70	0.3	10
Coating 1	0.12	1	0.4	10
Coating 2	0.77	200	0.3	10
Coating 3	1	4	0.4	10

5 Strain transfer model validity with appropriate boundary conditions and β value

The results obtained from the earlier discussed numerical simulation serve to validate the model proposed in (1). Within Fig.3, the longitudinal strain profiles of both the optical fiber ε_f and the host material ε_m of the 3D finite element simulation are represented by the black and blue curves, respectively. It is crucial to emphasize the smoothing effect of the optical cable on deformation measurements, as highlighted by the significant difference between the two extracted profiles. Employing the ε_m profile at discrete points alongside the previously determined beta value $\beta = 8.5$ m^{-1} as inputs, the differential equation (1) is solved using the Finite Difference method, as previously discussed. Boundary conditions are derived from the simulation's output of longitudinal strain at the optical fiber level, specifically denoted as ε_0=ε_L=7 µm/m. The resulting solution, which portrays the computed longitudinal strain in the optical fiber, is illustrated by the orange curve in Fig.3. A comparison is then made between this profile and the longitudinal strain output of the 3D model in the optical fiber. The error between the simulation's output and the calculated strain profile in the core remains below 3 µm/m. This strong agreement between the outputs of the 3D and 1D models at the core level points out the model's accuracy for the case studied.

6 Assessment of the strain transfer model validity under varied strain-lag parameter's values

While the outcomes appear suitable, it's crucial to examine how the value of the beta parameter might affect the reliability of the model since this parameter is linked to the mechanical and geometric characteristics of the sensor, which are often unknown consequently the beta value. Therefore, it becomes necessary to investigate the influence of using an erroneous beta value in the mechanical transfer model and to determine the acceptable margin of error required to achieve accurate results using the 1D model. In this context, equation (1) is solved for various beta values, utilizing the strain profile at the concrete beam level obtained from the numerical simulation and ensuring appropriate boundary conditions are applied (ε_0=ε_L=7 µm/m). Subsequently, for each value of beta, the error between the calculated profile and the simulation's strain profile in the optical fiber is assessed at each point. As evaluation criteria, we consider that an error of less than 4 µm/m at a point implies that the model yields acceptable results. The selected beta value is deemed suitable also if the extent of invalid areas does not surpass 100 mm from either end and provides valid results within the *z*-interval [100,500]. Fig.4 where the model is considered valid, i.e., according the chosen criteria, for a range of β value. It is demonstrated that the values of β between 7.4

and 9 m^{-1}, delimited as the green zone in Fig.4, provide validity of the model throughout at least two-thirds of the entire cable's length. However, values outside this interval yield inaccurate results, particularly in the center of the cable and at both ends.

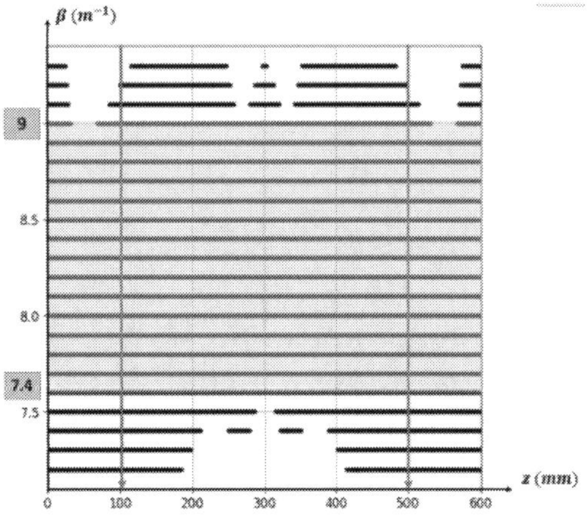

Fig.4. Zones of validity of the 1D model with several values of β

7 Conclusion

In conclusion, this paper evaluated through a numerical simulation the mechanical strain transfer model for steel-reinforced cable in the presence of a strain gradient generated by two inclusions in a cylindrical concrete beam. The results demonstrated the cable's smoothing effect on strain measurements. Additionally, the one-dimensional strain transfer model is validated, with errors remaining within acceptable limits, provided the strain-lag parameter was accurately estimated. However, accurately characterizing the cable remains a significant challenge due to insufficient information from manufacturers. Even minor errors in the strain-lag parameter, β, led to erroneous results.

Furthermore, a gap in the literature persists regarding the inverse problem of the strain transfer model, which focuses primarily on the direct problem, assuming known strain profiles within the host material. However, in on-site applications, this profile is often unavailable, requiring more in-depth studies.

References

1. Pierre Ferdinand, "The Evolution of Optical Fiber Sensors Technologies During the 35 Last Years and Their Applications in Structure Health Monitoring", 7th European Workshop on Structural Health Monitoring July 8-11, 2014. La Cité, Nantes, France.

2. Barrias A, Casas JR, Villalba S. A Review of Distributed Optical Fiber Sensors for Civil Engineering Applications. Sensors (Basel), May 2016

3. Brian J. Soller, Dawn K. Gifford, Matthew S. Wolfe, and Mark E. Froggatt," High-resolution optical frequency domain reflectometry for characterization of components and assemblies," Opt. Express 13, 666-674 (2005)

4. Farhad Ansari and Yuan Libo, "Mechanics of Bond and Interface Shear Transfer in Optical Fiber Sensors," Journal of Engineering Mechanics, vol. 124, no. 4, pp. 385-394, 1998.

5. Huaping Wang and Zhi Zhou, "Advances of strain transfer analysis of optical fiber sensors," Pacific Science Review, vol. 7677, pp.8–18, Sep. 2014.

6. Antoine Bassil, Distributed Fiber Optics Sensing for Crack Monitoring of Concrete Structures, PhD thesis, Université de Nantes, Nov. 2019.

7. Arianna Piccolo et al., "Mechanical Properties of Optical Fiber Strain Sensing Cables under γ-Ray Irradiation and Large Strain Influence," Sensors (Basel, Switzerland), vol.20, no. 3, p. 696, Jan. 2020

8. Xing Zheng, Bin Shi, Cheng-Cheng Zhang, Yijie Sun, Lei Zhang, Heming Han, "Strain transfer mechanism in surface-bonded distributed fiber-optics sensors subjected to linear strain gradients: Theoretical modeling and experimental validation", Measurement, Volume 179, 2021

9. Xavier Chapeleau and Antoine Bassil, "A General Solution to Determine Strain Profile in the Core of Distributed Fiber Optic Sensors under Any Arbitrary Strain Fields," Sensors, vol. 21, no. 16, pp. 1–33, Jan. 2021.

Development of a High-Precision, Portable and Automated Mobile Laser Scanner for the Recording and Digitation of Texture and Micro-marks in Archaeological and Heritage Stone

Pablo Pardiñas[1,2], *David* S. Canzobre[1,2], *Javier* Lamas[1,2], *Micaela* Gómez-Coronel[1], *Alberto* Ramil[1], *David M.* Freire Lista[3] and *Ana J.* López [1*]

[1]Laboratorio de Aplicacións Industriais do Láser, Campus Industrial de Ferrol, Universidade da Coruña,15471 Ferrol, Spain
[2]Antanorte S.L. Avenida Mestre Manuel Gomez Lorenzo, 30, Vedra, 15885, A Coruña, Spain
[3]Departament of Geology, UTAD Universidade de Trás-Os-Montes E Alto Douro. Quinta de Prados, 5001-801, Vila Real, Portugal

Abstract. The use of laser tools to digitise cultural heritage items is becoming more widespread but is generally limited to manual systems that rely on the user's movement or expensive automated equipment that is designed to scan large items but is not always able to accurately measure or reproduce specific details of small size. In this paper we present the novel high-precision system PALLAS, which allows the automatic digitisation of three-dimensional detailed surfaces. The equipment has been built using a set of motors and linear axes, control electronics and a line laser scanner that can be moved along the XY working plane. It is battery-powered for ease of use in the field and uses a standard laptop to communicate via Wi-Fi network. Custom software has been developed to automatically calculate paths, control the movement and store the profiles generated by the scanner. The software can also be used to reconstruct the surface from the set of scanner profiles, allowing the XYZ coordinates of any point on the surface to be measured and to obtain areal roughness parameters to characterise the finish and texture of the stone surface. Preliminary results obtained from masons' marks, rock engravings and petroglyphs are presented.

1 Introduction

Currently, multiple techniques exist for measuring or reproducing elements of cultural heritage. Among these, laser technology stands out as a particularly innovative and rapidly expanding approach. Like other optical techniques, such as photogrammetry [1], laser scanning offers the significant advantage of being contact-free. Its interaction with the sample relies on the reflection or incidence of light, ensuring no alteration to the artifact and allowing for measurements at a distance, even through transparent materials, within the laser's wavelength range. Another key benefit of laser scanning is its exceptional precision compared to conventional methods. Affordable equipment is available that can achieve measurements with accuracy down to hundredths of a millimetre. However, despite these advantages, most automated tools currently available for heritage conservation, particularly those suitable for field use, primarily focus on scanning large structures like facades or entire buildings [2, 3]. Smaller artifacts and intricate details, especially those with complex geometries, are often relegated to laboratory settings for high-precision digitation [4]. This involves either transporting the artifacts, which may be impractical or damaging, or manually scanning them in situ using handheld laser scanners or other techniques. Both approaches suffer from reduced precision and require a considerable time investment from conservation personnel. This issue becomes particularly critical when digitising heritage elements like petroglyphs. These archaeological markings, often found on exposed stone surfaces susceptible to erosion and located in difficult access areas, should ideally remain *in situ* [5]. Furthermore, their delicate nature and potential for faint markings require high-precision scanning to ensure accurate documentation. In order to address this challenge, this study presents a novel, low-cost, portable system designed and built by the authors. This system enables in-situ digitation of cultural heritage elements with laboratory-equivalent precision, operating autonomously to minimise user effort and maximise efficiency.

2 Methodology

In the design and manufacturing process of the equipment, the main criteria related to its use have been taken into account, specifically: light weight for easy transport, size suitable for its use in confined spaces, capable of measuring at any angle, precision equivalent to that achieved in a laboratory setting, automatic functioning during data collection in both smooth

* Corresponding author: ana.xesus.lopez@udc.es

surfaces and surfaces with protrusions or indentations, and low-cost or easily accessible components for assembly and operation, whenever possible.

2.1 Hardware

The hardware components that comprise the system are listed below, along with their respective functions. Upon completion of the prototype, a system with a mass of 10.4 kg and an effective scanning area of 500 mm × 160 mm was achieved. This system can be oriented in any direction and achieve a working speed of 5 mm/s while displacing a mass of up to 1 kg, with an XY plane displacement resolution of 0.05 mm, a repeatability of 0.1 mm and a surface measurement precision of up to 0.005 mm. Furthermore, the equipment is versatile and can operate when connected to any power supply system with an output of 12-24 VDC or 220 VAC.

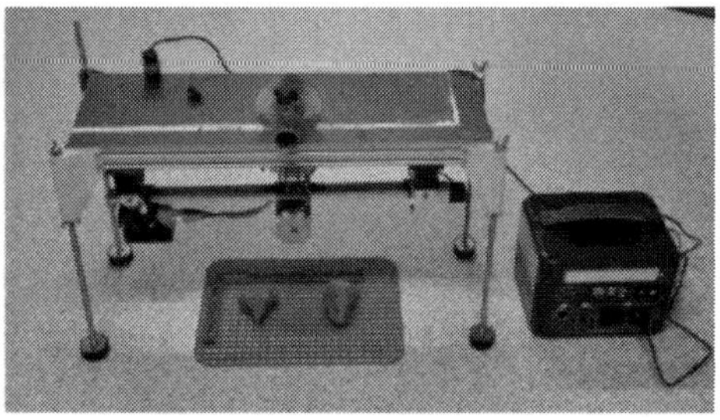

Fig. 1. Prototype of the PALLAS system

2.1.1 Structural frame

A rectangular structural frame was fabricated using custom-cut T-Slot aluminium profiles, which enabled a straightforward assembly process with a cost-effective and widely available material. Threaded holes were machined into each of the four corners, allowing for the optional installation of legs equipped with embedded interior nuts. These legs were designed and 3D-printed in PLA for integration into the prototype. The aforementioned nuts enable the anchoring or movement of a threaded rod through the leg, which features a rubber anti-vibration stop installed at the end oriented towards the sensor's measurement direction. These threaded rods, when installed, serve as a spacer and reference system, ensuring a consistent and known distance is always maintained between the equipment and the measurement surface. This configuration enables precise control over the measurement setup, allowing for accurate and reliable data collection. A mounting plate was secured to the rear face of the profiles, providing support for components associated with the control system. Additionally, an attachment element was mounted on the exterior of this plate, enabling the equipment to be secured to a tripod for measurements at varying heights and angles.

2.1.2 Motion system

The system consists of two motorised linear guides equipped with limit switches to determine the working area. Specifically, four magnetic reed-switch position sensors, model SMC D-A96, have been installed. For the X-axis movement, the low-cost, ultralight Fuyumotion FSK30J linear guide model, with a 500 mm stroke and an integrated low-power Nema11 stepper motor of 0.06 Nm, has been selected. A universal support, designed and 3D-printed in PLA with integrated magnets, has been installed on the carriage of this guide to enable the attachment of the scanning equipment and the activation of the position sensors related to the X-axis limits. The base of this guide has been coupled to both linear axes bolted to the infrastructure, which add rigidity and prevent potential bending of the linear guide, and the second linear guide, which has been bolted to the infrastructure to provide additional rigidity to the entire assembly and enable the movement of the first guide along the Yaxis. For this second guide, the low-cost Fuyumotion FSK40 model with a 160 mm stroke and a Nema23 stepper motor of 1.2 Nm has been chosen. Between the two linear guides, an anchoring spacer, designed and 3D-printed in PLA with integrated magnets, has been installed to enable the activation of the position sensors related to the Y-axis limits.

2.1.3 Control system

To carry out the control tasks, low-cost electronics that can be programmed with open-source tools have been selected. Specifically, an ESP32 microcontroller is used to send the signals that enable the control of the stepper motors, determine if the working area limits have been reached and send a signal that acts as an external trigger to control the timing of surface profile acquisition. In order to convert the signals into stepper motor movements, two drivers have been selected and installed: the Stepperonline DM320T, which controls the first linear guide, and the Stepperonline DM542T, which controls the second linear guide. Additionally, all the necessary electronics for signal conditioning and power supply have been incorporated.

2.1.4 Laser scanner

Auxiliary structures for holding various models of laser scanners on the previously mentioned universal support have been designed and 3D-printed using PLA. This level of adaptability allows for the selection of the most appropriate scanner to be used for the digitation process based on the required precision for the measurement of the surface. The functionality of the equipment has been tested with three commercial industrial-grade laser scanners, which operate by projecting a laser line onto the surface to be measured and then detecting the reflected light using a camera, to calculate the coordinates of the points forming the laser line in the scanner reference system based on the angle of refraction. Specifically, the models P+F VLE350-F280-B12-1100 (X:40-160 mm Z:60-350 mm, 960 points),

MicroEpsilon scanCONTROL 2500-25 (X:23.4-29.1 mm Z:53.5-78.5 mm, 640 points) and MEL M2DW-75/30 (X:30-40 mm Z:90-165 mm, 256 points) have been used. All these scanners have the capability to use an external trigger, which causes the scanner to obtain a profile when an activation pulse is sent, allowing for the programming of a sequence of measurements at a series of known coordinates. A critical angle of approximately 30 degrees has been identified, beyond which shadowing effects can occur, resulting in point loss, for holes, protrusions or indentations on the surface.

2.1.5 Auxiliary systems

A low-cost Wi-Fi router has been incorporated to enable communication between the control system, the laser scanner and a conventional laptop, allowing for the sending of control command sequences and the collection of surface data. Specifically, the N300 Mini Wireless Router model has been installed. A battery-based power system has also been included to allow for the use of the equipment outdoors in uninhabited areas. The VTOMAN Jump 600 has been selected for testing, offering a battery life of over 4 hours with intensive use. Finally, a tripod has been added to fix the equipment in the required position during the measurement process, allowing for the scanning of marks or elements on vertical surfaces and ceilings or at awkward angles. Specifically, the Nedo Heavy-Duty Wooden Tripod 200513 model has been chosen.

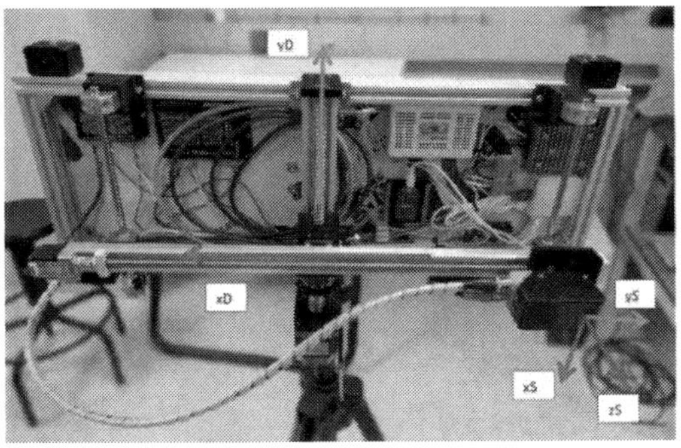

Fig. 2. Detail of the setup with linear guides and laser scanner reference axes overlaid

2.2 Firmware and software

The system's operation relies on open-source firmware installed on the control electronics and custom-developed software. This software suite enables path planning, command transmission to the laser scanner, reading, visualization, and storage of the digitized surface data. Furthermore, the software converts the acquired data into an open 3D representation format, promoting accessibility and interoperability

2.2.1 Control firmware

FluidNC, an open-source firmware, was selected for its ability to use G-code commands to control the linear guides' movements and trigger the laser scanner to capture profiles. This firmware, designed for the ESP32 microcontroller, features a graphical user interface for system management and status visualisation. It also enables remote command reception via a wireless network connection. Parameters defining the resolution, acceleration and maximum speed of each linear guide, as well as the digital inputs and outputs connected to magnetic sensors and the laser scanner, have been configured.

2.2.2 Custom software

A Python-based software has been designed and implemented to control the equipment via a wireless network connection, utilising a conventional low-end laptop. The software employs a communication socket to monitor the system status and transmit G-code commands, receiving responses accordingly. Upon establishing the communication channel and specifying the area to be measured, the application calculates the coordinates of each necessary path to scan the selected surface and generates the required command sequence. If the surface to be scanned exceeds the laser line width of the tool, multiple sweeps are generated to measure it completely, with overlapping regions according to a specified percentage. This overlap percentage is a programmable parameter whose optimisation can contribute to eliminating noisy points and shadow zones caused by the scanner's orientation relative to the piece, facilitating the merging of the different profiles obtained. The software calculates the coordinates and time intervals in which a uniform speed is maintained, based on the scanner's displacement speed and maximum acceleration configured in the firmware. During this interval, the tool's activation command is sent, enabling the transmission of a trigger sequence at a previously established frequency. Although the scanner's displacement speed can be selected based on the installed sensor model and its maximum working frequency, the software proposes a speed value that maintains a homogeneous distance between the obtained points that define the surface in the XY plane, facilitating subsequent data processing. During this interval, the application establishes a connection to receive profiles from the scanner and stores them in HDF5 format along with the linear guides' coordinates where the aforementioned profiles have been obtained, enabling grouping and conversion to the equipment's reference system coordinates.

After executing all planned paths, the software processes the stored points using various algorithms, eliminating outliers, merging sweeps and calculating the best approximation of height value for points in overlapping zones. The processed points are used to construct a triangle mesh, which is stored in PLY format. This open 3D format has been selected because it can be processed and visualised using multiple free or proprietary software tools in a straightforward manner. Additionally, it has the advantage of allowing the storage of colour information that can come from other sources. Furthermore, to facilitate the sampling task, the application allows for the visualisation of profiles as they are obtained and of the entire set using the Plotly library.

It is important to note that, in accordance with the current equipment features, when storing multiple scans, it is recommended to include auxiliary reference elements on the surface to optimise their merging, as they facilitate determining the relative position between them and allow for automatic processing through software. Furthermore, since the scanner always remains oriented in the same direction relative to the surface, if there are shadow or hole areas caused by surface protrusions or indentations that the overlap cannot resolve, it will be necessary to manually displace and rotate the equipment, perform a new sequence of sweeps that can add additional points and incorporate the previously mentioned ones. Using automated software in the conducted experiments, we achieved a percentage of valid points exceeding 90% (with an approximate average of 95%), which is greater than the required amount for a proper characterisation of the surfaces. In terms of time savings compared to manual scanning, a time savings of approximately 83% was calculated. It is important to note that the yield calculations were performed in comparison to tools previously developed to test the equipment, as no commercial software exists to perform the described set of tasks in a fully integrated manner.

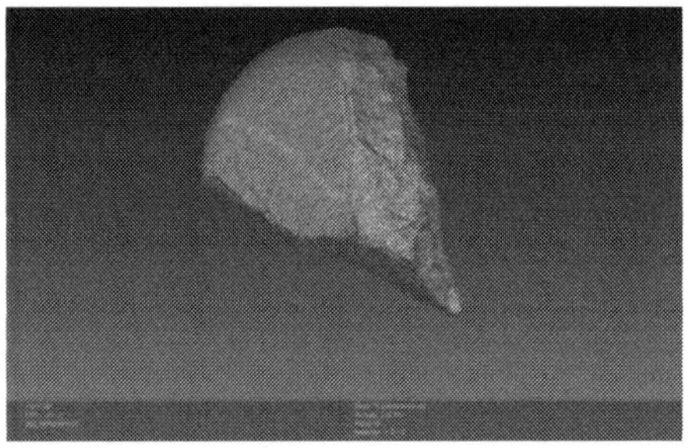

Fig. 3. Scanned hand axe saved in PLY format

3 Results

A series of laboratory tests have been conducted on the current prototype, with the dual purpose of calibration and verification of compliance with the predetermined design criteria. Following the successful outcomes of these tests, the equipment is presently undergoing field validation on heritage pieces located outdoors. It is noteworthy that preliminary results from ongoing field tests suggest that the automatic system yields measurements with enhanced precision, speed and accuracy compared to manual scanning and post-processing. While the approaches described in the referenced articles are adequate for obtaining statistical data on general roughness and determining the limits of medium-sized marks, the equipment presented in this work is necessary for a rapid and efficient processing of large areas with sub-millimetric precision. Further validation of the equipment and analysis of the results are necessary to complete the comparison.

The following text presents a series of results obtained during the tests already performed.

3.1 Calibration Grating

The test was conducted on a grid with known dimensions in order to determine the equipment's precision and perform a partial calibration of the equipment using different scanner models. The complete element has a size of 300 mm × 400 mm and a thickness of 9.5 mm, with gaps of 15.0 mm × 15.0 mm, the width of the bars being 2.5 mm.

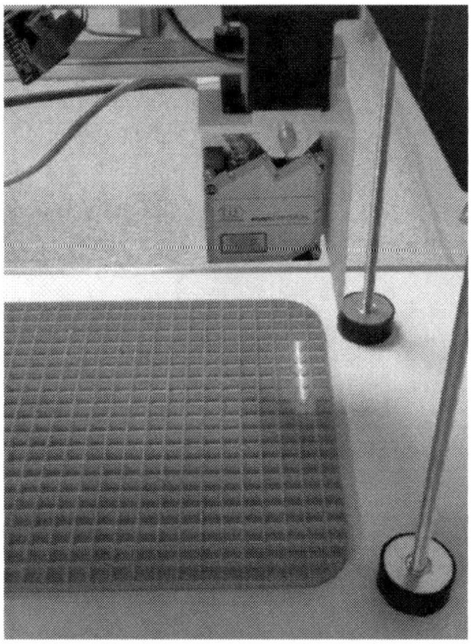

Fig. 4. Calibration grating scanning process

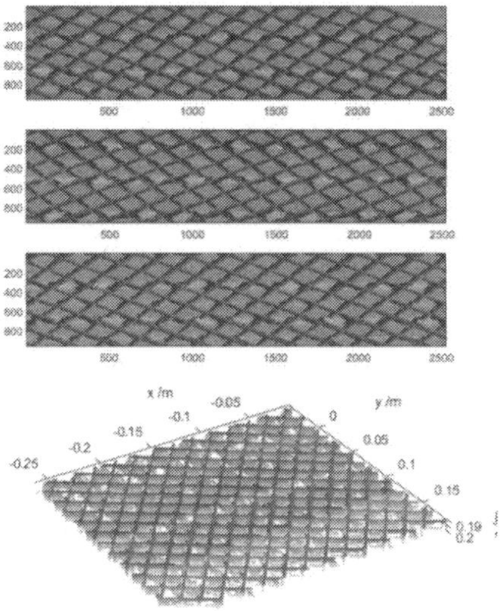

Fig. 5. Calibration grating measurements

3.2 Theatre Neutral Mask

The test was conducted on a mask similar to those used in classical theatre. It was selected due to its size, large enough to require multiple sweeps, its height and the level changes in the surface that allow for the completion of the equipment's calibration with different scanner models across the entire working range along the Z-axis. Furthermore, this example is considered representative of the types of surfaces commonly found in cultural heritage made of stone, due to its round shapes, the representation of a human face and the presence of cavities that create shadows. Additionally, it is considered an interesting example because it enables the validation of the system's capability to digitise individual pieces from other areas of cultural heritage with high precision.

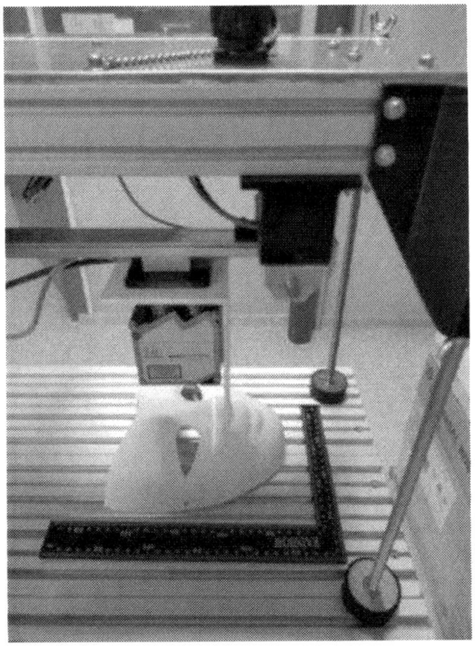

Fig. 6. Theatre neutral mask scanning process

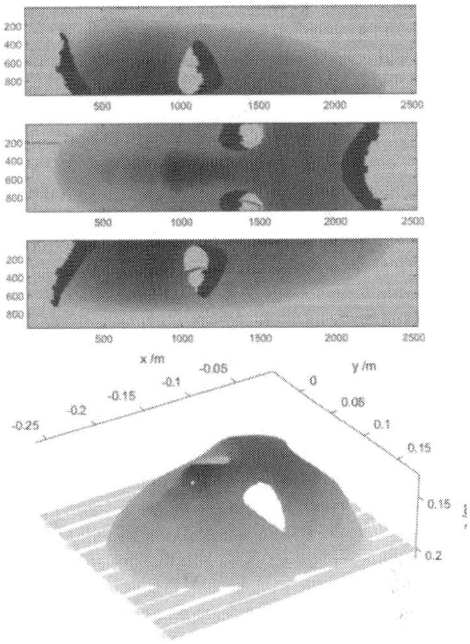

Fig. 7. Theatre neutral mask measurements

3.3 Reproduction of Granitic Stone Frieze

The test was conducted on a representative element of the stone heritage in the geographical area where the authors are carrying out most of the field tests. In this case, a granite piece was selected, as it is a very abundant material in the northwest region of the Iberian Peninsula, widely used in both archeologic stones with petroglyphs and cultural heritage buildings. Furthermore, it is a material that the authors have extensively studied in previous works and, due to its durability and heterogeneity, it presents a wide variety of wear marks, defects or interesting surface changes that should be recorded for conservation and treatment purposes. The example used is a reproduction of a simple frieze handcrafted by a stonemason, with a relief and depth equivalent to those commonly found on the façades of churches and manor houses. The surface was painted with blue graffiti paint and partially cleaned with a laser, emulating the restoration of the piece after being subjected to vandalism, to determine the level of detail that the equipment records before and after the treatment.

Fig. 8. Granitic stone frieze scanning process

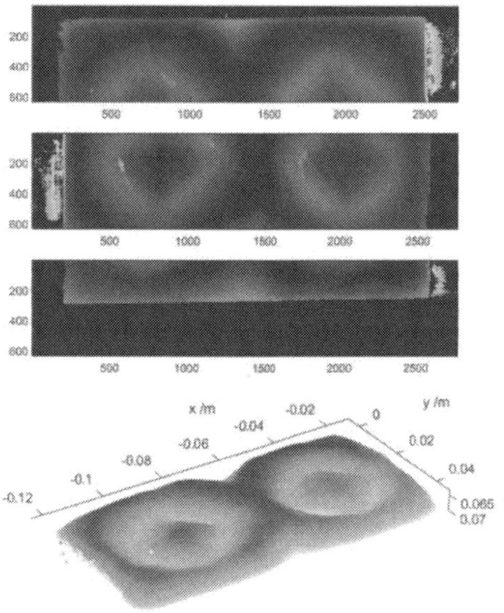

Fig. 9. Granitic stone frieze measurements

3.4 Rock with Engravings

A test was conducted on a fragment of a rock with engravings from the Valcamonica archaeological site due to its challenging nature for precise digitation *in situ* using conventional methods. The sample presents shallow marks and signs of wear caused by prolonged exposure to the environment. The marks are difficult to discern with the naked eye, not only due to erosion but also due to the material's discoloration. The use of the prototype enables the visualisation of all existing marks, which appear to represent a set of lines and an arrow, as well as measuring their depth and length. Additionally, the surface data obtained from the scanning process allows for the partial determination of the texture differences between the marked areas and the base material, providing supplementary information that can be useful for archaeologists when analysing the type of tool used for engraving or searching for similarities between these marks. Ultimately, this sample has validated the equipment's level of precision allows for satisfactory results in the use cases for which it was designed.

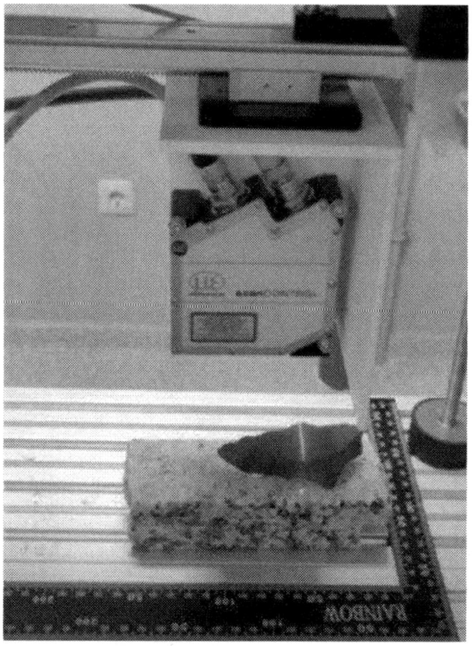

Fig. 10. Rock with engravings scanning process

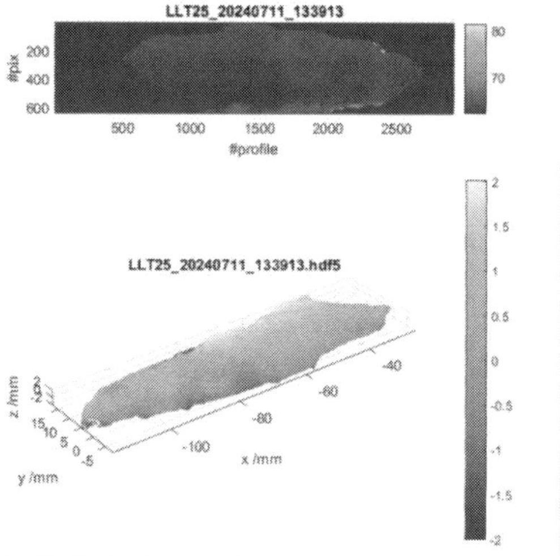

Fig. 11. Rock with engravings measurements

4 Conclusions

A novel method is presented that enables the acquisition of high-resolution data from archaeological stone *in situ* with laboratory-grade precision, using a device and software developed by the authors. It is based on a lightweight positioning system with two motorised linear axes XY that enable the scanning of the surface with different laser scanner models. A 3D model of the entire surface within the working area without holes can be generated by merging different scan sequences, but it requires auxiliary reference markers and manually rotating the device.

Future work will focus on enhancing the prototype's performance. While the current prototype functions correctly and executes its intended processes effectively, potential improvements include the incorporation of a rotating scanner, the addition of encoders to the motors and the extension of the software. These modifications are expected to improve usability and reduce the time required for surface scanning.

Acknowledgements: Work carried out under the Memorandum of Understanding between the Universidade da Coruña and the Centro Camuno di Studi Preistorici (Brescia, Italy). Work partially financed by PID2021-123948OB-100, funded by MCIN/AEI/ 10.13039/501100011033 and by the ERDF "A way of making Europe" of the European Union.

References

1. Höll, T. Holler, G. Pinz, A., A novel high accuracy 3D scanning device for rock-art sites. ISPRS - Int. Arch. Phot., Rem. Sen. Spa. Inf. Sci. **XL-5**. 285-291 (2014) https://doi.org/10.5194/isprsarchives-XL-5-285-2014
2. Vacca, G., Deidda, M., Dessi, A., Marras, M. . Laser Scanner Survey To Cultural Heritage Conservation And Restoration. ISPRS - Int. Arch. Phot., Rem. Sen. Spa. Inf. Sci. **XXXIX-B5** 589-594 (2012) https://doi.org/10.5194/isprsarchives-XXXIX-B5-589-2012
3. Masciotta, M. G., Sanchez-Aparicio, L. J., Oliveira, D. V., Gonzalez-Aguilera, D., Integration of Laser Scanning Technologies and 360o Photography for the Digital Documentation and Management of Cultural Heritage Buildings. Int. J. Arch. Her. **17(1)**, 56–75 (2023) https://doi.org/10.1080/15583058.2022.2069062
4. Fontana, R., Gambino, M. C., Greco, M., Pampaloni, E., Pezzati, L., Scopigno, R., High-resolution 3D digital models of artworks. Proc. SPIE. **5146**, 34-43 (2003) https://doi.org/10.1117/12.501248
5. Freire-Lista, D.M., Campos, B.B., Moreira, P., Ramil Rego, A. y Lopez Díaz, A.J., Building Granite Characterisation, Construction Phases, Mason's Marks and Glyptography of Nossa Senhora de Guadalupe Church, Mouçós e Lamares, Galicia-North Portugal Euroregion. Geoher. **15,** 24 (2023) https://doi.org/10.1007/s12371-023-00790-4

Refractive index and thickness analysis of planar interfaces by prism coupling technique

Paulo Lourenço [1,2*], *Yuri* Vygranenko[2], *Miguel* Fernandes[1,2], *Alessandro* Fantoni[1,2], and *Manuela* Vieira[1,2,3]

[1]IPL/ISEL/DEETC, Instituto Politécnico de Lisboa, Instituto Superior de Engenharia de Lisboa, Departamento de Engenharia Eletrónica e Telecomunicações e de Computadores, R. Conselheiro Emídio Navarro, 1949-014 Lisboa, Portugal
[2]CTS-UNINOVA and LASI, Caparica, Portugal
[3]NOVA School of Science and Technology, Caparica, Portugal

Abstract. Since 2022, various foundries are offering the manufacture of integrated photonic structures for the visible spectrum. As this technology continues to enter the market, there will be an increasing demand for accurate optical and dimensional characterization of these structures. To meet this demand, we have developed a highly precise optical characterization system based on the prism coupling technique, also known as m-lines spectroscopy, to investigate the optical properties of hydrogenated amorphous silicon nitride planar waveguides deposited by plasma-enhanced chemical vapor deposition. In this work, by combining visible spectroscopy with the prism coupling technique to excite modes that propagate resonantly in the waveguide via frustrated total internal reflection, using either parallel or perpendicularly polarized light beams, we were able to analyze the waveguide properties of silicon nitride thin films with an interfacial oxide layer. Furthermore, through numerical simulation of the bilayer structure, we calculate the waveguide's refractive index and thickness, and determine the characteristics of the interfaces in terms of refractive index and thickness.

1 Introduction

Approximately two decades after the midpoint of the last century, a reliable and highly precise technique for measuring the refractive index and thickness of transparent thin films [1]–[3] was developed. This technique commonly utilizes an optical setup known as the prism waveguide coupler. The basic components of this setup include a prism that is pressed against the thin film under analysis, which is typically deposited on an optical glass substrate. By controlling the pressure applied to the prism/film system, the air gap between the prism and the film can be adjusted, thereby influencing the coupling efficiency. The theoretical foundation of this method is based on the prism-waveguide coupler mode equations, which allow for the accurate determination of the refractive index and thickness of the transparent thin film.

Regarding the air gap dimension, the theoretical approach may be classified and treated as in the following waveguide systems:

- Slab waveguide mode equations: These consider an infinite air gap between the prism and the thin film [4], [5];
- Leaky waveguide mode equations: These assume the air gap is zero [6];
- Prism waveguide coupler mode equations: These allow for a variable air gap thickness, ranging from zero to infinity,

where the slab and leaky waveguide approaches represent specific cases of the more general prism waveguide mode equations.

To determine the refractive index and thickness of a transparent thin film, the synchronous directions corresponding to specific incidence angles at the prism's edge, which excite a propagating mode in the thin film via evanescent coupling, are experimentally obtained using a prism waveguide coupler setup. The thickness and refractive index of the transparent thin film can then be calculated by numerically solving the mode equations. In this work, we have combined the prism coupling technique with visible spectroscopy to determine the optical constants and thickness of thin films and developed an approach which enables the identification of the presence and thickness of an interfacial layer.

The Manifacier/Swanepoel method used in this study was initially developed by Manifacier et al. [7] and later refined by Swanepoel [8] for determining the optical constants and thickness of hydrogenated amorphous silicon thin films. This method, also known as the T-spectra method, is based on the transmission spectra of a transparent thin film under normal incidence. The complex refractive index and thickness of the sample are determined by analysing the interference fringes observed in the transmission spectra.

* Corresponding author: plourenco@deetc.isel.ipl.pt

Moreover, the T-spectra method provides insights into the optical uniformity of the transparent thin film. If the film is uniform, the interference effect produces a series of fringes in the transmission spectra, which can be used to determine the film's complex optical constants and thickness. However, the presence of an interfacial layer, or a graded refractive index within the film bulk, reduces the amplitude of these interference fringes and alters their peak values.

The refractive index and thickness results obtained in this work for the transparent thin film were verified using our software platform, Synopsys RSoft [9], through simulations of the structure discussed earlier and later in this document. This verification process also included determining the interfacial layer thickness by simulating the bilayer structure, and identifying the propagating modes for TE, and TM polarizations.

2 Experimental

The hydrogenated amorphous silicon nitride (a-SiN_x:H) thin films for this study were deposited on alkali-free glass substrates using a capacitively-coupled parallel-plate 13.56 MHz PECVD system. The gases used were SiH_4, NH_3 and H_2 at an SiH_4/H_2 dilution ratio of 1:9. The details on the deposition process have been reported elsewhere [10]. The optical transmission spectra were measured by a Shimadzu 3100 UV-Vis-NIR spectrophotometer.

Our optical setup for prism coupling measurements [11] comprises an angular actuator/motor, the Thorlabs PRM1-Z8, which is controlled by a Thorlabs K-Cube KDC101 controller connected to a PC. The photodetector placed adjacent to the prism's edge where the reflected beam is refracted. It is connected to a Rigol DM-3058 multimeter for signal measurements. By incorporating a polarizer, the Thorlabs LPVIS100, between the laser diode and the prism, we can monitor parallel and perpendicular polarized incoming beams, enabling the selective excitation of transverse magnetic (TM) or transverse electric (TE) modes.

The entire system is controlled via USB connection to the PC, which oversees both the angular displacement of the prism and data acquisition from the photodetector through the multimeter. A developed C# application on the PC manages various functions of the angular actuator/motor, and it also handles the display and storage of the photodetector's signal.

3 Results and Discussion

Fig. 1 shows the transmission spectra of the a-SiN_x thin film used in our study. The spectra were measured on the day of sample preparation and three years later when the sample was used for prism coupling experiments. Both spectra show similar interference fringes; however, the peaks of the later curve are above the transmittance of the glass substrates. This is indicative of films with a graded refractive index. We assume that the SiO_x layer grows permanently over time due to unavoidable nitride surface oxidation in air [12].

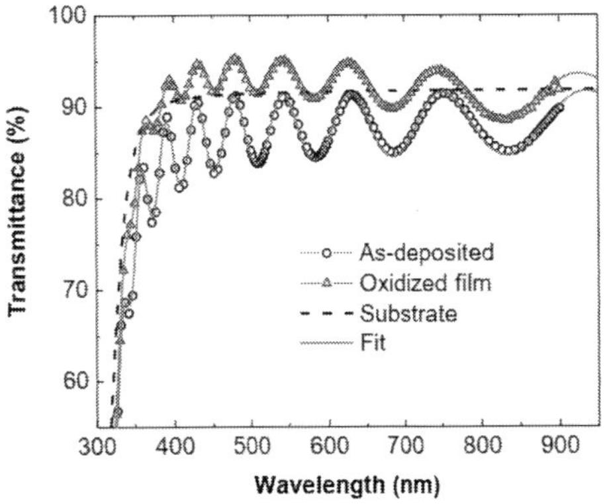

Fig. 1. Transmission spectra of the a-SiNx:H thin film measured immediately after deposition and 3 years later.

The film thickness of 1068 nm and the n&k dispersion curves shown in Fig. 2 were obtained from the spectrum of the as-deposited film using the Swanepoel method [8]. Similar optical constants for a-SiN_x:H have been reported elsewhere [13]. These data were utilized to simulate the fitting curves presented in Fig. 1. The Transfer Matrix Method (TMM) was employed to simulate the transmission spectrum of the glass/SiN_x/SiO_2 bilayer at n_{SiO2} = 1.5 [14,15]. The best fit was achieved with SiN_x and SiO_2 layer thicknesses of 1012 nm and 72 nm, respectively.

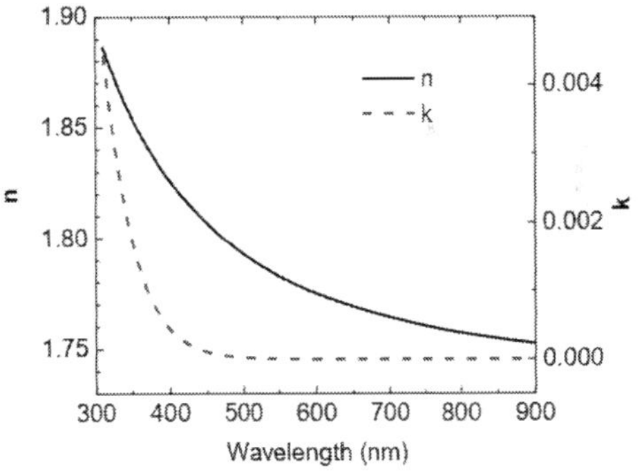

Fig. 2. Refractive index and extinction coefficient curves for an a-SiN_x:H thin film.

Fig. 3 displays the processed signals measured under TE and TM light polarizations at a wavelength of 656 nm, revealing three resonance peaks at both polarizations corresponding to mode orders 0, 1, and 2. The obtained modal indices, alongside the calculated refractive index (n_f) and film thickness (d) (combinations mode 0 & mode 1 and mode 0 & mode 2, have been considered), are listed in Table 1. The n_f values, ranging from 1.766 to 1.770, are consistent with the refractive index n(656nm) = 1.768 obtained by spectroscopy. The average thickness values are 1023 nm and 1012 nm for TM and TE modes, respectively, with the latter matching the thickness of the SiN_x layer calculated from the observed resonances.

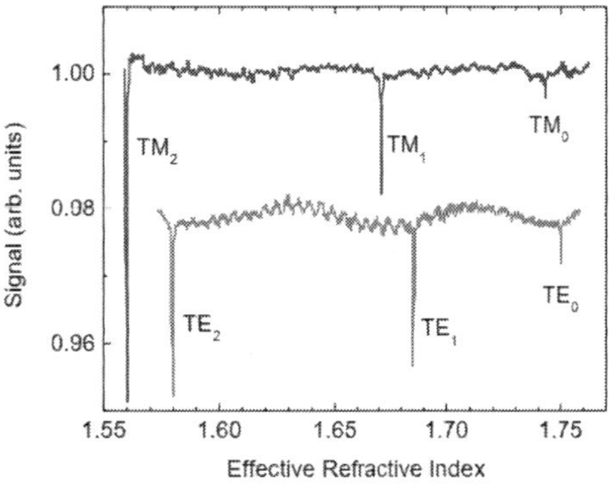

Fig. 3. Experimental dependences for excitation of TM and TE modes in the a-SiN_x:H waveguide.

Table 1. Measured modal refractive indices and calculated refractive index and film thickness.

Modes	0	1	2	n_f	d [nm]
TM	1.7425	1.671	-	1.767	1019
TM	1.7425	-	1.559	1.766	1027
TE	1.7496	1.685	-	1.770	1009
TE	1.7496	-	1.579	1.770	1014

The following equations were used to calculate n_f and d for TE and TM modes [1]:

$$\sqrt{\frac{n_f^2-N_i^2}{n_f^2-N_j^2}} = \frac{\arctan\sqrt{\frac{N_i^2-n_a^2}{n_f^2-N_i^2}}+\arctan\sqrt{\frac{N_i^2-n_s^2}{n_f^2-N_i^2}}+m_i\pi}{\arctan\sqrt{\frac{N_j^2-n_a^2}{n_f^2-N_j^2}}+\arctan\sqrt{\frac{N_j^2-n_s^2}{n_f^2-N_j^2}}+m_j\pi}, \quad (1)$$

$$d = \frac{\lambda}{2\pi\sqrt{n_f^2-N_m^2}}\left[\arctan\sqrt{\frac{N_m^2-n_a^2}{n_f^2-N_m^2}} + \arctan\sqrt{\frac{N_m^2-n_s^2}{n_f^2-N_m^2}} + m\pi\right] \quad (2)$$

$$\sqrt{\frac{n_f^2-N_i^2}{n_f^2-N_j^2}} = \frac{\arctan\frac{n_f^2}{n_a^2}\sqrt{\frac{N_i^2-n_a^2}{n_f^2-N_i^2}}+\arctan\frac{n_f^2}{n_s^2}\sqrt{\frac{N_i^2-n_s^2}{n_f^2-N_i^2}}+m_i\pi}{\arctan\frac{n_f^2}{n_a^2}\sqrt{\frac{N_j^2-n_a^2}{n_f^2-N_j^2}}+\arctan\frac{n_f^2}{n_s^2}\sqrt{\frac{N_j^2-n_s^2}{n_f^2-N_j^2}}+m_j\pi}, \quad (3)$$

$$d = \frac{\lambda}{2\pi\sqrt{n_f^2-N_m^2}}\left[\begin{array}{c}\arctan\frac{n_f^2}{n_a^2}\sqrt{\frac{N_m^2-n_a^2}{n_f^2-N_m^2}} + \\ \arctan\frac{n_f^2}{n_s^2}\sqrt{\frac{N_m^2-n_s^2}{n_f^2-N_m^2}} + m\pi\end{array}\right], \quad (4)$$

where $k_0 = 2\pi/\lambda$, λ is the radiation wavelength, n_a the refractive index of air, n_s the refractive index of substrate, N_m the modal index, and m is the mode order.

Equations (1) to (4) are valid for the optically uniform film, and their use for a film with a graded index interface may result in significant errors. To address this, we simulated the glass/SiN_x/SiO_2 structure using RSoft's mode solver. Fig. 4 displays the calculated modal refractive indices for TE and TM modes as a function of the SiO_2 layer thickness. The modelling was conducted while keeping the combined thickness of the SiN_x/SiO_2 bilayer constant at 1000 nm. We then processed these data using equations (1 – 4) to determine the values of n_f and d. The obtained refractive index values matched exactly with the refractive index of the SiN_x layer used in the mode solver, while the thickness varied depending on the SiO_2 layer thickness. Figure 5 shows the relationship between the SiN_x core layer thickness and the SiO_2 layer thickness. In this figure, the reference line (dash-dot line) indicates the model parameter for the SiN_x layer thickness. This suggests that the presence of an interface layer can lead to an overestimation of the core layer thickness. This procedure predicts an error of approximately 23 nm when the interface layer is 72 nm thick. Additionally, the difference between the TM and TE thicknesses increases linearly at a rate of 0.13, predicting a thickness difference of 9.4 nm at a SiO_2 thickness of 72 nm, which aligns well with experimental results. The SiO_2 layer thickness can also be estimated in reverse.

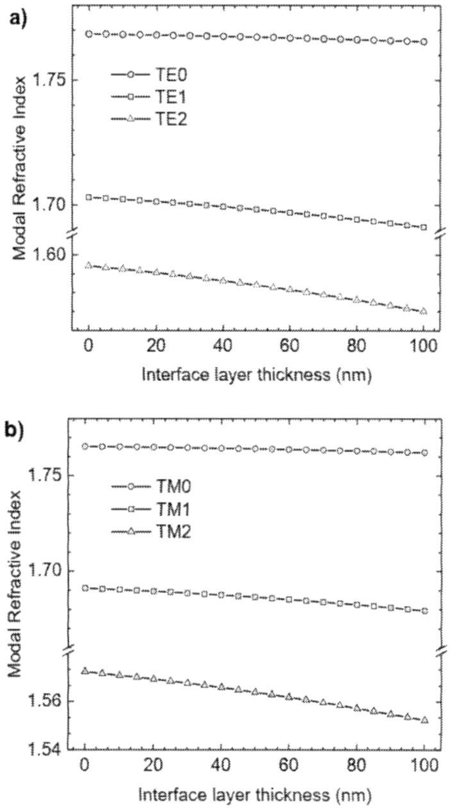

Fig. 4. Simulated modal refractive indices for (a) TE and (b) TM polarizations versus the SiO_2 layer thickness.

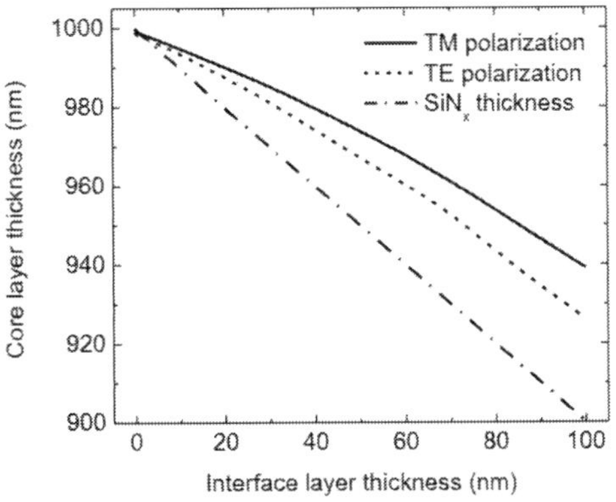

Fig. 5. Calculated SiN_x core layer thickness versus the SiO_2 layer thickness. The reference line (dash-dot line) indicates the SiN_x layer thickness in the numerical model.

4 Conclusions

We have applied the prism coupling technique to thin films with a capping interface layer. Numerical simulations demonstrate that the refractive index of the a-SiN_x:H film with an interfacial oxide layer can be accurately determined using the propagation equations, which are obtained for a single-core waveguide with abrupt interfaces. However, the thickness of the film may be overestimated depending on the thickness of the oxide layer. This thickness overestimation can be corrected by numerically modelling the bilayer structure and analysing the experimental data.

Acknowledgments

This research was supported by Portuguese national funds provided by the Portuguese FCT program, Centre of Technology and Systems (CTS) UIDB/00066/2020/ UIDP/00066/2020, by FCT project ASER-META 2022.07694.PTDC and by IPL project IPL/IDI&CA2023/ DETPREC_ISEL. A special word of appreciation to BSc. Jorge Fidalgo for his contributions in this work, namely concerning the design and implementation of the mechanical fixtures in the optical setup.

References

[1] P. K. Tien, R. Ulrich, and R. J. Martin, "Modes of propagating light waves in thin deposited semiconductor films," *Appl. Phys. Lett.*, vol. 14, no. 9, pp. 291–294, 1969, doi: 10.1063/1.1652820.

[2] P. K. Tien and R. Ulrich, "Theory of Prism–Film Coupler and Thin-Film Light Guides," *J. Opt. Soc. Am.*, vol. 60, no. 10, pp. 1325–1337, 1970, doi: 10.1364/josa.60.001325.

[3] R. Ulrich, "Theory of the Prism–Film Coupler by Plane-Wave Analysis," *J. Opt. Soc. Am.*, vol. 60, no. 10, p. 1337, Oct. 1970, doi: 10.1364/JOSA.60.001337.

[4] R. Ulrich and R. Torge, "Measurement of Thin Film Parameters with a Prism Coupler," *Appl. Opt.*, vol. 12, no. 12, p. 2901, 1973, doi: 10.1364/ao.12.002901.

[5] D. Marcuse, *Theory of Dielectric Optical Waveguides*. Elsevier, 1991.

[6] R. T. Kersten, "A new method for measuring refractive index and thickness of liquid and deposited solid thin films," *Opt. Commun.*, vol. 13, no. 3, pp. 327–329, 1975, doi: 10.1016/0030-4018(75)90112-1.

[7] J. C. Manifacier, J. Gasiot, and J. P. Fillard, "A simple method for the determination of the optical constants n, k and the thickness of a weakly absorbing thin film," *J. Phys. E.*, vol. 9, no. 11, pp. 1002–1004, Nov. 1976, doi: 10.1088/0022-3735/9/11/032.

[8] R. Swanepoel, "Determination of the thickness and optical constants of amorphous silicon," *J. Phys. E.*, vol. 16, no. 12, pp. 1214–1222, Dec. 1983, doi: 10.1088/0022-3735/16/12/023.

[9] RSoft, "Synopsys RSoft Solutions." https://www.synopsys.com/optical-solutions/rsoft.html (accessed Jul. 30, 2024).

[10] G. Lavareda *et al.*, "Dependence of optical properties on composition of silicon carbonitride thin films deposited at low temperature by PECVD," *J. Non. Cryst. Solids*, vol. 551, no. July 2020, p. 120434, Jan. 2021, doi: 10.1016/j.jnoncrysol.2020.120434.

[11] P. Lourenço *et al.*, "Determining Thin Film Characteristics by Prism Coupling Technique," In Doctoral Conference on Computing, Electrical and Industrial Systems, pp. 337–347. Cham: Springer Nature Switzerland, 2024.

[12] S. I. Raider, "Surface Oxidation of Silicon Nitride Films," Journal of The Electrochemical Society, vol. 123, no.4, 560, 1976, doi:10.1149/1.2132877.

[13] G. Lavareda, Y. Vygranenko, A. Amaral , P. Brogueira, "Effect of dehydrogenation on optical constants of silicon nitride thin films", Optical Materials, vol. 145, no. 8, 114480, Nov. 2023, doi: 10.1016/j.optmat.2023. 114480.

[14] C.C. Katsidis, D.I. Siapkas, General transfer-matrix method for optical multilayer systems with coherent, partially coherent, and incoherent interference, Appl. Opt., vol. 41, pp. 3978–3987, 2002, doi: 10.1364/AO.41.003978.

[15] Y. Vygranenko, G. Lavareda, M. Fernandes, M. Vieira, "Characterization and modeling of a-Si:H-based n-i-p photodiodes with protocrystalline silicon p-layers," Thin Solid Films, vol. 784, no. 14, Nov. 2023, doi: 10.1016/j.tsf.2023.140072.

Monitoring Reinforced Concrete Structures Using Iron Thin Film Coated Optical Fibre Sensors

Pedro M. da Silva[1,2*], *João P. M. Carvalho*[1], *João P. Mendes*[1], *José M. M. M. de Almeida*[1,3], *and Luís C. C. Coelho*[1]

[1] INESC TEC – Institute for Systems and Computer Engineering, Technology and Sciences, and Faculty of Sciences, University of Porto, Rua do Campo Alegre, 4169-007 Porto, Portugal
[2]Faculty of Engineering, University of Porto, Rua Dr. Roberto Frias, 4200-465 Porto, Portugal
[3]Department of Physics, School of Sciences and Technology, University of Trás-os-Montes e Alto Douro, 5001-801, Vila Real, Portugal

Abstract. Structural health monitoring (SHM) of reinforced concrete structures (RCS) is crucial for mitigating the consequences of their deterioration. By identifying and addressing the issues early, SHM helps reduce environmental impact, safeguard lives, and enhance economic resilience. Rebar corrosion is a leading cause of early RCS decay and optical fibre sensors (OFS) have been employed for its monitoring. Reflection optrodes using optical fibres where the tip is coated with iron (Fe) thin films offer a robust, long-lasting and straightforward solution. This study investigates the tracking of spectral changes during the Fe thin film corrosion, which has been neglected in the literature, in favour of tracking reflection changes from thin film spalling. A multimode fibre tip, coated with a thin Fe layer embedded in concrete, allows spectral changes to be observed during corrosion. A 100 nm thick Fe film was deposited using radio frequency magnetron sputtering on polished fibre tips. Corrosion was induced by applying salted water drops and allowing the fibre tip to dry. Corrosion monitoring was successful for both air-exposed and cement-embedded tips, with results compared to reflection simulations of Fe, Fe_2O_3, and Fe_3O_4 thin films. This study supports monitoring at different wavelengths, enhancing robustness, cost-effectiveness and earlier detection.

1 Introduction

Reinforced concrete structures (RCS) are ubiquitous in our society; they measure human development and are fundamental to connecting, housing, and protecting humanity. Its compressive and tensile strengths, availability, longevity (50-100 years service life), and embodied energy and carbon make it stand out amongst other building materials [1,2]. Still, its usage incurs an environmental cost due to the cement and steel (rebars) industries, which are significant sources of emissions and have a substantial carbon footprint [3]. Considering the proliferation of RCS infrastructure over the past century, the coming decades will be marked by the need to repair, improve and build new infrastructure. A necessity that will challenge efforts to curb current emissions [4]. Monitoring infrastructures will be vital in finding the causes of RCS decay – such as cement paste carbonation, alkali-silica reaction, chloride ion ingress, and rebar corrosion – and mitigating its social, economic and environmental impact. Among these, rebar corrosion stands as a principal cause of RCS decay. Corrosion diminishes rebars tensile strength and, consequently, the overall tensile performance of the RCS. As the rebars corrode, the larger volume of the corrosion products causes cracks in the RCS, reducing its compressive strength and increasing its porosity thereby facilitating further decay [5].

Optical fibre sensors (OFS) have been employed to monitor rebar corrosion in RCS. Gratings-based OFS, specifically Fibre Bragg Gratings (FBG), were positioned near the rebars [6–8]. As the rebars expand due to the formation of Fe oxides, the period of the grating changes, influencing the optical response of the OFS. The use of FBG for the physical monitoring of RCS has been extensively studied in the literature and applied in the field [9,10]. These commercially viable sensors have a long service life (a requirement for monitoring infrastructure that lasts decades) and can be multiplexed. However, these gratings are typically inscribed in single-mode fibres (SMF), operating in the infrared, usually in the C band, where optical absorption is at a minimum (0.2 dB/km) [11]. This allows the system to work with standardized components used in optical communications, though this benefit comes with an associated financial cost. Given the difficulties of coupling light into the SMF, detecting in the infrared and the small changes in the Bragg wavelength, expensive interrogation systems and optical components are required.

Alternatively, a reflection OFS relies on the change in reflection of a Fe (or Fe-C) thin film at the fibre tip caused by corrosion, which is then used to infer the

* Corresponding author: pedro.m.madeira@inesctec.pt

corrosion of surrounding rebars [12,13]. These sensors are predicated on the spalling of the thin film occurring has its corrosion progresses. As the thickness of the thin film decreases, the amount of light being reflected also decreases. This approach was successfully deployed in the field [13], but its implementation through SMF, albeit closer to the visible spectrum (850 nm), continues to require expensive interrogation systems. Moreover, current implementation of reflection OFS neglect some key aspects: mechanical changes to the thin film heavily influence reflection, but do not necessarily mean corrosion has occurred. The influence of Fe oxides in the reflection spectrum is, also, not considered. Fe oxidation is accompanied by spalling, but for an embedded fibre tip coated with Fe, the cement paste will hold much of these Fe oxides in place.

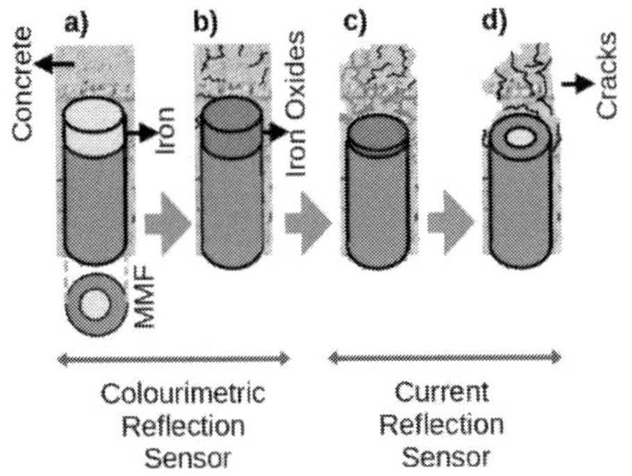

Fig. 1. Different stages of Fe thin film corrosion, detection scope of the current reflection OFS for RCS and the colorimetric reflection sensor proposed in this work. a) Fe thin film in healthy concrete. b) Increase in porosity and formation of cracks, with much of the corrosion products remaining near the fibre tip. c) Increased porosity and formation of cracks allow the spalling of the thin film, and d) complete escape of Fe and its oxides from the vicinity of the fibre tip.

Spalling of Fe and its oxides does occur, but in a later stage of RCS decay, Fig. 1, where increased porosity and formation of large cracks allow them to leave the vicinity of the fibre tip. Ignoring the absorption and reflection changes of Fe oxides along with the mechanical changes in the thin film, increases uncertainty and reduces the pre-emptive potential of these sensors in monitoring the onset of rebar corrosion, as seen in Fig. 1.

In the present work, the capabilities of a reflection OFS for monitoring rebar corrosion are expanded by exploiting colour changes that occur at the fibre tip due to the formation of the different Fe oxides. This approach increases the sensors resiliency to the heterogeneity of Fe corrosion. To do so, a multimode fibre (MMF) has a thin film of Fe formed at the fibre tip through radio-frequency (RF) magnetron sputtering. By employing an MMF, the price tag of the optical components required for these sensors is drastically reduced due to the facility of coupling light into MMF. Additionally, detectors and light sources operating in the visible range of the spectrum are more affordable and readily accessible. The thickness of Fe thin films was chosen through thin film reflection simulations, which are also employed for estimating changes in the reflected spectrum of light due to Fe_2O_3 and Fe_3O_4 formation near the fibre tip. These simulations are compared with real-world oxidation of Fe thin films, and the uncertainty of using reflection OFS without monitoring spectral changes is evaluated.

2 Methodology

2.1 Fe Thin Film Simulations for Oxidation

Thin film simulations assess changes in reflection between the fibre core (substrate) and the external medium, when a thin film is present between them. For a single thin film layer, equation 1 was used [14]:

$$E_0^- = \frac{r_1 + r_2 e^{-2i\delta_1}}{1 + r_1 r_2 e^{-2i\delta_1}} E_0^+ \qquad (1),$$

$$\text{with } r_1 = \frac{n_0 - n_1}{n_0 + n_1}, \qquad r_2 = \frac{n_1 - n_2}{n_1 + n_2},$$

$$\delta_1 = \frac{2\pi}{\lambda} n_1 \, d_1 \cos \phi_1,$$

where n_0, n_1 and n_2 is the refractive index of the fibre core, the thin film and air, respectively. λ is the wavelength of the light transversing the thin film, d_1 is the thickness of the thin film, and ϕ_1 the angle of incidence. E_0^- is the incident light intensity and E_0^+ is the reflected light intensity with reflectance (R) being deduced through them [14]. For several layers of thin films, equation 1 is iteratively solved from the outside in [14]. A Python script was used in this work to implement these equations to achieve simulations of several Fe oxides on the fibre tip

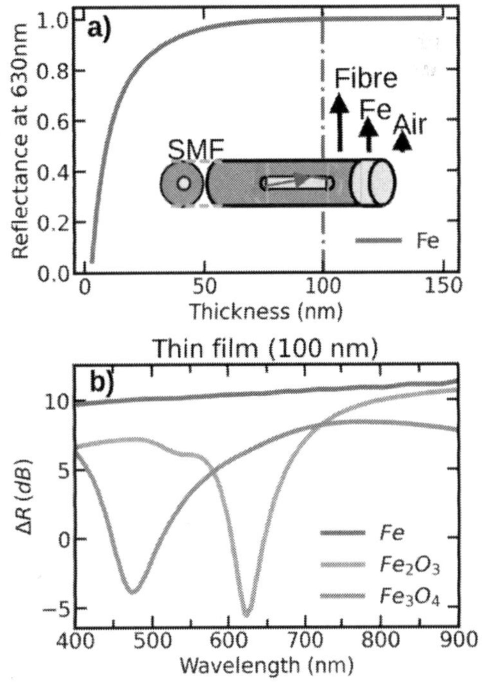

Fig. 2. a) Change in reflectance of light into the fibre core from different thicknesses of Fe, as represented in the inset. b) Reflectance change (ΔR) between fibre tips with Fe, Fe_2O_3 and Fe_3O_4 with bare fibre as a reference.

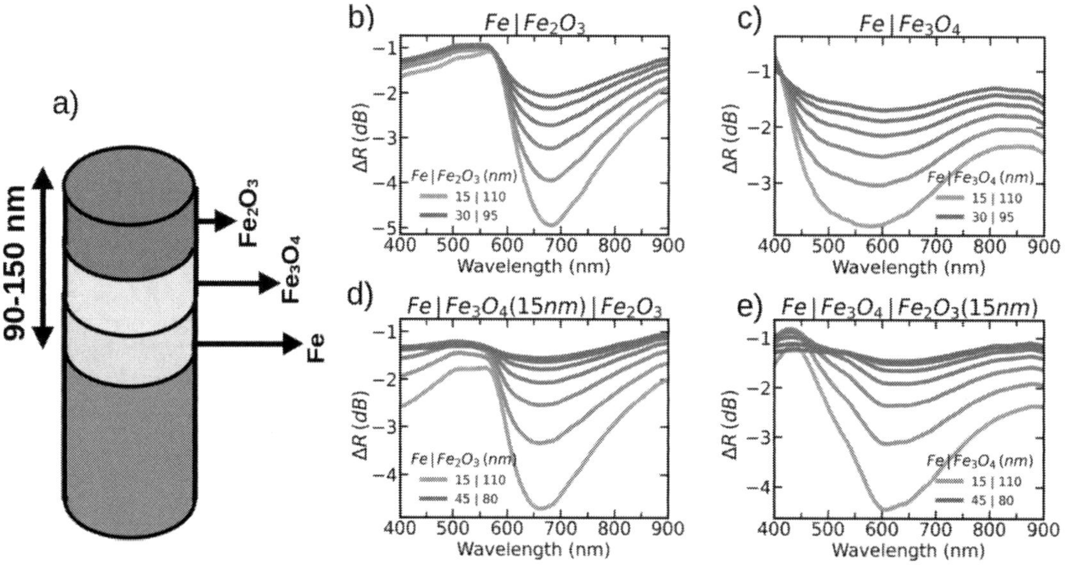

Fig. 3. a) Parameters of the simulated structure through thin film reflection. b) Simulation of two thin film structures of Fe being converted into Fe_2O_3 and c) Fe into Fe_3O_4. Simulation of three thin film structures of Fe being transformed into Fe_3O_4 and Fe_2O_3, d) with a fixed 15 nm thick layer of Fe_3O_4 and e) with a fixed 15 nm thick layer of Fe_2O_3.

Employing equation 1 for a simulation of a single-layer thin film is sufficient to estimate the reflectance of light from a Fe thin film, Fig. 2a. A 100 nm thick Fe thin film was chosen because there is barely any transmission beyond this point, which means no colorimetric changes can be observed beyond this thickness. Using less than 100 nm means fewer Fe oxides that could contribute to a colorimetric change. This colorimetric change is on display in Fig. 2b, where single-layers simulations are used to plot the change in reflectance ($\Delta R = 10\log_{10}R/R_0$) between Fe or Fe oxides coated fibre tips (R) and a bare fibre tip in air as reference (R_0).

At the onset of corrosion, Fe is slowly transformed into Fe_2O_3 and Fe_3O_4 [5]. Provided the cement matrix remains without significant cracks, the corrosion products of Fe will form in place, Fig. 1. For this reason, simulations are done considering multiple layers of thin films composed of Fe and its oxides. The formation of Fe oxides is highly heterogeneous, both spatially and temporally. Simulation of thin films are used to reduce the complexity of studying Fe oxide formation, yet they still reveal the significant heterogeneity present in the process. One should consider that there could be a volume buildup at the fibre tip with the formation of Fe oxides, but some of these products could escape and permeate through cement paste pores. Corrosion initiates from the external surface of Fe and progresses inward. Formation of Fe_3O_4 occurs in environments lacking oxygen [5]. Thus, the parameters for simulation are, Fig. 3a:

- Final structure of Fe, Fe_2O_3 and/or Fe_3O_4 is between 90-150 nm.
- Changes in reflectance (ΔR) are always compared to 100 nm of Fe in a fibre tip.
- When present, Fe is always next to the fibre core.
- Fe_2O_3 is the outermost layer, succeeding Fe_3O_4.

Figures 3b and 3c display two thin film systems, Fe/Fe_2O_3 and Fe/Fe_3O_4. They show an Fe thin film being corroded from 30 to 15 nm, with Fe_2O_3 and Fe_3O_4 increasing from 95 to 110 nm. Figure 3d shows the corrosion of Fe into Fe_2O_3 with a transition layer of Fe_3O_4 in between. Figure 3e portrays the formation of Fe_3O_4 from Fe, followed by a protective layer of Fe_2O_3.

2.2 Experimental Setup

The experimental setup can be seen in Fig. 4, wherein an MMF (FT600UMT, Thorlabs, Dortmund, Germany) with a core diameter of 600 μm, has a Fe thin film produced through RF Sputtering, over which water with NaCl with a concentration of 3.5% and 10% is used to induce corrosion.

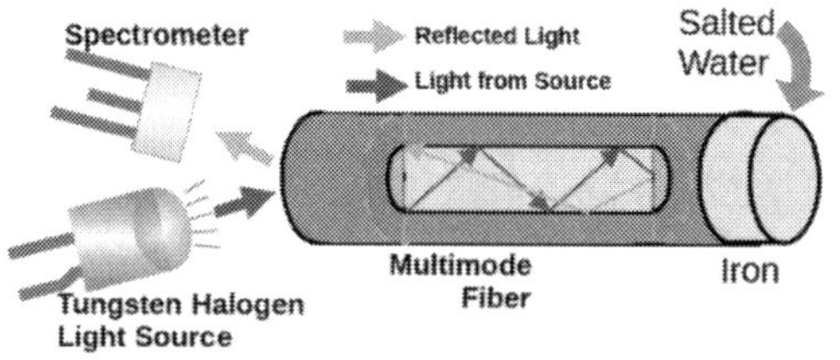

Fig. 4. Rebar corrosion probing tip - MMF coated with a thin film of iron. Interrogation occurs with a tungsten halogen light source and a spectrometer connected to the probing fibre with an MMF bundle. Corrosion is induced by water with NaCl.

A tungsten halogen light source (HL-2000, Ocean Optics, Dunedin, FL, USA) and a spectrometer (USB4000, Ocean Optics, Dunedin, FL, USA) are connected to the fibre through a fibre bundle composed of two 200 µm core diameter MMF (FT200UMT, Thorlabs, Dortmund, Germany). Using the procedure described in [15], fibre tips coated with Fe, are dipped in freshly mixed cement paste to cover and protect the thin film. The fibre tips, with and without cement paste, had a drop of salt water added and then left to dry. The process is repeated until the Fe in the fibre tip is eroded. Changes in the spectrum are ultimately compared with simulations of the structures of Fig 3a.

When working with the bare fibre tip coated with Fe, without the cement paste, better reproducibility is achieved by reducing the number of drops and increasing the salinity. This approach prevents the mechanical force of applying a drop of water from spalling Fe oxides from the tip.

3 Results & Discussion

3.1 Reflection changes from Fe-coated fibre with salted water.

Corrosion of Fe thin film coating the fibre tip without cement paste was produced in two ways: water with 3.5% and 10% NaCl. The lower concentration requires more drops to induce corrosion in the thin film of Fe, which leads to the spalling of the corrosion products, given that there is not a cement paste matrix to hold these in place.

An MMF coated with a thin film of Fe, Fig. 5a, has seven drops of water (3.5%) added over 115 hours, which reduced the amount of reflected light throughout the entire spectrum, Fig. 5c.

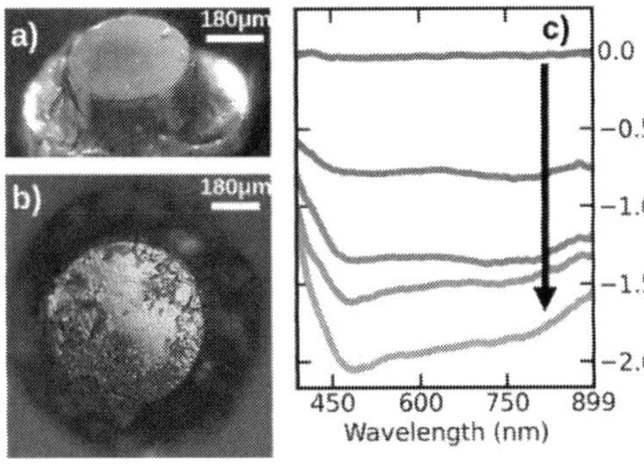

Fig. 5. a) Thin film of Fe sputtered into a polished MMF tip. b) Changes to the thin film from adding drops of water with a NaCl concentration of 3.5% and c) the accompanying ΔR of the Fe thin film at a multimode fibre tip.

Only a slight feature in the spectrum is observed between -1dB and -2dB at 450 nm, indicating little to no corrosion occurred. As much can be confirmed in the image of the MMF tip after the drops were added, Fig. 5b. Significant topological changes were made in the Fe thin film, but no significant corrosion occurred. Nevertheless, the signal change is as much as -2dB between the reflection for the starting thin film and the current one. This uncertainty between overall ΔR and corrosion requires more broad parameters of corrosion to be chosen when spectral changes are not being monitored, such as the ΔR of a bare fibre in air (-8dB) and in water (-20dB) [12].

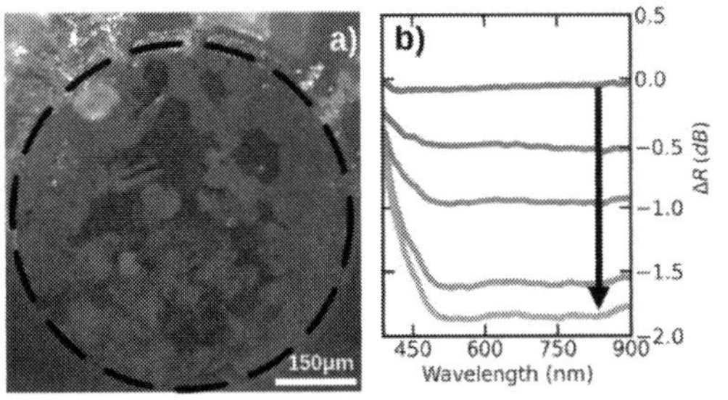

Fig. 6. a) Thin film of Fe sputtered at the polished tip of a multimode fibre after corrosion. Dashed line represents the limits of the fibre core. b) ΔR of the Fe thin film throughout its corrosion.

Fig. 6 displays the thin film after corrosion was induced with seven drops of salted water with a 3.5% concentration of NaCl over 113 hours and the ΔR that accompanied it. Much of the corrosion products have spalled away from the fibre, Fig. 6a, resulting in a weak colorimetric signal, Fig. 6b. It is essential to consider that the Fe-coated fibre tips of Fig. 5 and Fig. 6 underwent the same procedure of induced oxidation but are in different stages of corrosion while producing a similar change in reflection intensity (-2dB). This highlights the uncertainty associated with corrosion, with this monitoring, and the need for improvements.

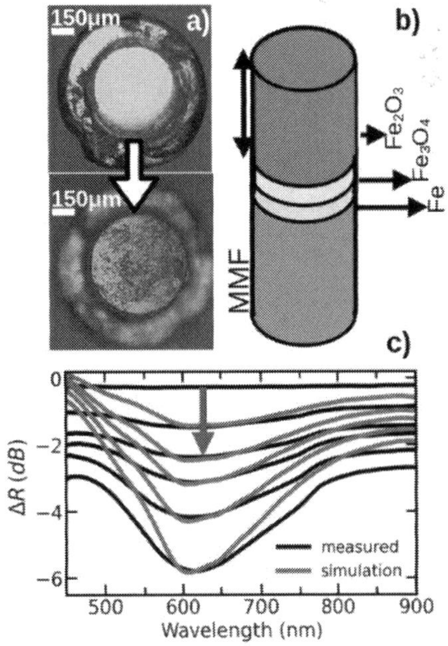

Fig. 7. a) Corrosion of a Fe thin film sputtered into a core of an MMF with a 10% NaCl solution. b) Structure used to simulate the ΔR of the thin film due to corrosion. c) Measured and simulated ΔR of the Fe thin film due to corrosion.

With a higher concentration of salt (10%), only two drops were used over 22 hours, allowing the corrosion to occur in place (Fig. 7a). By remaining in the vicinity of the fibre tip, a large colorimetric response is preserved, Fig. 7. The structure used to simulate the thin film corrosion is represented in Fig. 7b, and the resulting ΔR is juxtaposed to the measured one, Fig. 7c, with corrosion being ascertained with a ΔR, as low as, between -1dB and -2dB. The simulated structure measures 125 nm in thickness, with Fe being corroded from 24 nm into 10 nm, growing the Fe_2O_3 from 86 nm to 100 nm. A 15 nm transition layer of Fe_3O_4 is placed between Fe and Fe_2O_3. Corrosion is monitored with ΔR between -1dB and -6dB, with a pronounced decrease between 550 and 700 nm. This spectral response ensures the ΔR is accompanied by thin film corrosion, which is crucial when there is strong variability from mechanical changes and spalling, Fig. 5 and 6, respectively. Thus, an earlier warning can be attained for ΔR below -8dB, a point over which cracks and enlarged porosity allow the thin film to spall outwards, indicative of severe decay.

3.2 Reflection changes from Fe-coated fibre tip embedded in cement with salted water.

Six drops of salted water with 3.5% NaCl are added to the fibre tip coated with a thin film of Fe embedded in cement paste, Fig 8a, over 120 hours. By cracking the cement paste and looking at the corroded thin film, it is possible to see darker shades (dark orange) of the corrosion product that permeated the cement paste, Fig 8b.

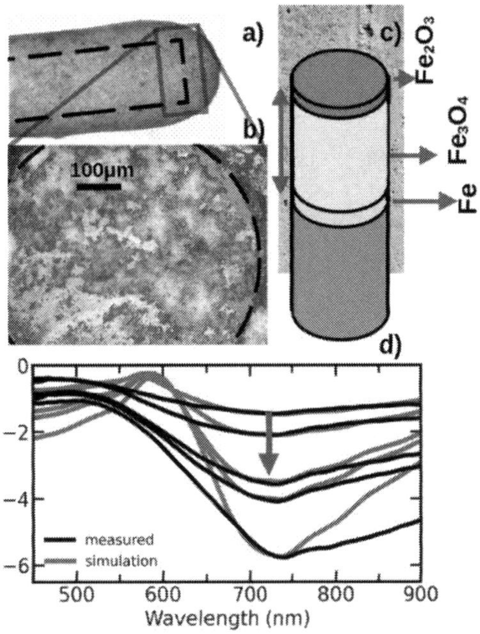

Fig. 8. a) MMF tip with Fe thin film embedded in cement paste. b) Corroded Fe thin film with 3.5% NaCl solution. c) Structure used to simulate the $\boldsymbol{\Delta R}$ of the thin film corrosion. d) Measured and simulated $\boldsymbol{\Delta R}$ of the Fe thin film due to corrosion.

As well as a bright blue corrosion product that formed close to the optical fibre. The simulated structure in Fig. 8c has a total thickness of 142 nm. It shows Fe corroding from 33 nm to 13 nm, with Fe_3O_4 growing from 95 nm to 115 nm. A 15 nm protective layer of Fe_2O_3 is placed over Fe_3O_4. The measured ΔR and the simulation are compared in Fig. 8d.

A strong colorimetric change is seen in Fig. 8, portraying that much of the corrosion products remained in place due to the cement paste, unlike Fig. 6, where a similar number of drops spalled the thin film away from the fibre tip. Considering a probing wavelength of 850 nm [12], the heterogeneity of Fe corrosion is another source of variability in an approach solely focused on monitoring the ΔR of the thin film in a single wavelength. Looking at the measured ΔR of Fig. 7 and 8, both reach ΔR close to -6 dB, but at 850 nm one is at -3dB and the other at -5dB. This showcases the need to probe several wavelengths to monitor the onset of corrosion.

4 Conclusion

Reflection optical fibre sensors based on thin film-coated fibre tips interrogating around a single wavelength have proven a simple and robust way of monitoring the corrosion conditions that might affect the rebars. However, through simulation and experiment, uncertainties from the heterogeneity of the corrosion process and mechanical changes that might afflict the thin film were evaluated. These delay the time frame of action of the sensor to when the section of cement paste being monitored by the fibre tip is starting to be exposed to the outside elements. A colorimetric approach employing a multimode optical fibre coated with a thin film of Fe, not only dispels much of the uncertainty, it provides a low-cost solution that can be used to monitor changes in reflectance as low as -1 dB, as opposed to the -8 dB or higher of a simple reflection method. Thin film reflection simulations were employed to better understand the corrosion process and its influence on reflectance and were successfully compared with real-world reflectance changes. Future work on the relationship between thin film corrosion, rebar corrosion and concrete deterioration is a crucial next step for real world application.

Acknowledgements

This work is co-financed by Component 5 - Capitalization and Business Innovation, integrated in the Resilience Dimension of the Recovery and Resilience Plan within the scope of the Recovery and Resilience Mechanism (MRR) of the European Union (EU), framed in the Next Generation EU, for the period 2021 - 2026, within project ATE, with reference 56 and by National Funds through the Portuguese funding agency, FCT - Fundação para a Ciência e a Tecnologia, within the PhD grant UI/BD/152695/2022 and the research contract CEECIND/00471/2017.

References

[1] G. P. Hammond and C. I. Jones, *Embodied Energy and Carbon in Construction Materials*, Proceedings of the Institution of Civil Engineers - Energy **161**, 87 (2008).

[2] F. Pacheco-Torgal, *1 - Introduction*, in *Eco-Efficient Repair and Rehabilitation of Concrete Infrastructures*, edited by F. Pacheco-Torgal, R. E. Melchers, X. Shi, N. D. Belie, K. V. Tittelboom, and A. Sáez (Woodhead Publishing, 2018), pp. 1–12.

[3] N. Darraj and G. Bender, Carbon Neutral Energy Intensive Industries, Brief, UNECE, 2022.

[4] P. M. Da Silva, J. P. Mendes, L. C. C. Coelho, and J. M. M. M. De Almeida, *Real-Time Monitoring of Cement Paste Carbonation with In Situ Optical Fiber Sensors*, Chemosensors **11**, 449 (2023).

[5] J. P. Broomfield, *Corrosion of Steel in Concrete: Understanding, Investigation and Repair*, Third edition (CRC Press, Boca Raton London New York, 2023).

[6] J. Mao, J. Chen, L. Cui, W. Jin, C. Xu, and Y. He, *Monitoring the Corrosion Process of Reinforced Concrete Using BOTDA and FBG Sensors*, Sensors **15**, 8866 (2015).

[7] H. Zhou, Y. Wang, Z. Zhou, and C. Ma, *Effects of Bonding on Fiber Bragg Grating (FBG) Sensor for Monitoring Steel Corrosion Condition in Concretes*, Measurement **222**, 113565 (2023).

[8] J. Gao, J. Wu, J. Li, and X. Zhao, *Monitoring of Corrosion in Reinforced Concrete Structure Using Bragg Grating Sensing*, NDT & E International **44**, 202 (2011).

[9] I. Robles Urquijo, A. Quintela Incera, S. Van Vaerenbergh, D. Inaud, and J. M. Lopez Higuera, *Risks and Opportunities of Using Fibre Optic Sensors for Long Term Infrastructure Health Monitoring Systems in an 18 Year Old Installation*, in *International Conference on Smart Infrastructure and Construction 2019 (ICSIC)* (ICE Publishing, Cambridge, UK, 2019), pp. 623–630.

[10] P. Moyo, J. M. W. M. W. Brownjohn, R. Suresh, and S. C. C. Tjin, *Development of Fiber Bragg Grating Sensors for Monitoring Civil Infrastructure*, Engineering Structures **27**, 1828 (2005).

[11] *Bend-Insensitive Single Mode Fiber (CCC1310-J9)*, (2024).

[12] C. Leung, K. Wan, and L. Chen, *A Novel Optical Fiber Sensor for Steel Corrosion in Concrete Structures*, Sensors **8**, 1960 (2008).

[13] K. T. Wan and C. K. Y. Leung, *Durability Tests of a Fiber Optic Corrosion Sensor*, Sensors **12**, 3656 (2012).

[14] O. S. Heavens, *Optical Properties of Thin Solid Films* (Dover Publ, New York, 1991).

[15] P. M. Da Silva, L. C. C. Coelho, and J. M. M. M. De Almeida, *Measuring Water Vapor Sorption Hysteresis of Cement Paste through an Optical Fiber Sensor*, Chemosensors **11**, 123 (2023).

An affordable optical detection scheme for LSPR sensors

Rima Mansour[1,2,3,4], *Catarina Serafinelli*[2,3,5], *Alessandro Fantoni*[2,3]*, *and Rui Jesus*[2,4]

[1]Department of the Computer Science, Nova School of Science and Technology, 2829-516 Caparica, Portugal
[2] Lisbon School of Engineering (ISEL)/IPL, Rua Conselheiro Emídio Navarro, nº1, 1959-007 Lisboa, Portugal
[3]CTS—Centre of Technology and Systems and Associated Lab of Intelligent Systems (LASI), 2829-516 Caparica, Portugal
[4]NOVA LINCS, Nova School of Science and Technology, 2829-516 Caparica, Portugal
[5]Department of Electrical and Computer Engineering, Nova School of Science and Technology, 2829-516 Caparica, Portugal

* Corresponding author: afantoni@deetc.isel.ipl.pt

Abstract. Biosensing technologies are essential for advancing healthcare by enabling rapid point-of-care (POC) testing and diagnosis, potentially saving lives. Biosensors, such as Localized Surface Plasmon Resonance (LSPR) sensors with gold nanoparticles (AuNPs), show promise in early disease diagnosis due to their simple structure and high sensitivity. However, their commercialization is limited by high production costs and the need for precise optoelectronic systems. This article proposes an affordable optical detection scheme for LSPR sensors named BioColor. BioColor uses a color CMOS camera to capture images of light passing through the sensor elements, which are plasmonic papers composed of AuNPs. Variations in the refractive index over the sensor's surface cause changes in the color of the transmitted light. This color change can be detected through image processing algorithms and the detection results are visualized on the BioColor mobile app, providing instant automated access to sensing outcomes.

1 Introduction

Regardless of the tremendous improvements in biosensor technologies, there remain challenges to overcome, as the progressive medical community necessitates the development of small-sized, portable, and cost-effective biosensors allowing point-of-care diagnostic [1]. The biosensors based on Local Surface Plasmon Resonance (LSPR) of gold nanoparticles (AuNPs) are promising for advancing next-generation point-of-care (POC) devices due to their simple structure and good sensitivity [2,3].

Localized Surface Plasmon Resonance (LSPR) sensors utilize the interaction between light and the sensor array elements (AuNPs) to detect changes in their surrounding environment. The underlying sensing process is that each array element maximizes its specific LSPR at its own wavelength and acts like a band-stop optical filter for the specific wavelength. When exposed to a broad light, each element produces a different response, resulting in a certain color of the transmitted light. Changes in the refractive index of the sensor's surrounding medium lead to a shift in the rejection band of the sensing element, hence changing the color of the transmitted light [4].

The novel approach of BioColor is analyzing the color of the transmitted light using image processing algorithms, providing an affordable solution for the sensor's interrogation scheme. This study investigates the sensor's sensitivity to different mediums using the BioColor approach and explores how using different types of transmitted light,such as Warm LED and Halogen Lamp, breakcan enhance color variations and thus the sensor's sensitivity.

2 Preparation of the AuNPs plasmonic paper

All the chemicals and reagents used in this study were analytical grade and used as received without further purification. Tetrachloroauric acid (HAuCl4.3H2O), ethanol and glycerol were purchased from Aldrich. Tea extract has been obtained from commercial black tea leaves. The Tea-AuNPs are adsorbed on laboratory filter paper (Whatman grade No. 1, 180 μm thick).

The tea solution was prepared by mixing a weighted amount of dried black tea leaves (2.29 g) with 20 mL of distilled water followed by vigorous stirring for 15 min at room temperature. After, the mixture was filtered and used to prepare 6 mL of solution 10% (v/v) in water. Successively, 0.1 ml of 0.1 M HAuCl4.3H2O solution was added to the tea solution and stirred for 20 min. In the end, the color of the solution turned violet indicating the presence of AuNPs.

One strip of filter paper was immersed in the solution of the pre-synthesized Tea-AuNPs and left overnight at 4 °C Singameni [5]. The paper strip was taken out, washed with MilliQ water and, as a final step, it was dried for about 45 minutes at 44 °C inside the oven.

3 BioColor system

The BioColor system consists of four main components: (1) experimental setup that incorporates LSPR samples made of AuNP-composed plasmonic papers positioned between a Halogen Lamp located at the top and a CMOS camera at the bottom of the setup; (2) hosting database that stores captured images in the Firebase cloud service; (3) mobile app to display the sample images and the obtained sensing results; (4) API server which runs the image processing algorithms. The bioColor system architecture is illustrated in Fig. 1.

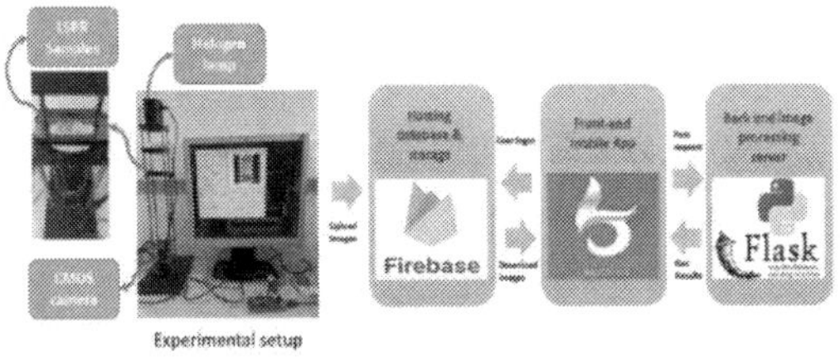

Fig. 1. BioColor system overview.

The user utilizes the CMOS camera interface to take pictures of the samples. The images are then uploaded to the Firebase database. In the mobile app, the authenticated user can view the uploaded images. Then, proceed with analyzing the selected images, which are sent to the API server for executing the image processing algorithm. The server, in turn, returns the obtained analysis results to the mobile app, providing the color difference of the samples.

3.1 Image processing

The image processing algorithms discussed in our previous paper [6] are executed by the BioColor API (Application Programming Interface) server. In our research, we presented a scientific approach for analyzing the color of the light passing through the LSPR sensor surface. First, a Region of Interest (ROI) detection algorithm is applied to remove unnecessary parts, followed by a ROI segmentation method. After that, to represent the sample in a single color, mean or dominant color methods are computed [7] in the LAB color space in the ROI.

The analysis process includes comparing the color of the sample under different mediums to validate the sensitivity of the LSPR sensor to refractive index variations. This comparison uses the CIEDE2000 equation 1, which defines the total color difference according to the International Commission on Illumination (CIE) [8].

$$\Delta E_{00} = \sqrt{\left(\frac{\Delta L^*}{K_L S_L}\right)^2 + \left(\frac{\Delta C^*}{K_C S_C}\right)^2 + \left(\frac{\Delta H^*}{K_H S_H}\right)^2 + R_T \left(\frac{\Delta C^*}{K_C S_C}\right)\left(\frac{\Delta H^*}{K_H S_H}\right)} \quad (1)$$

3.2 BioColor app

The BioColor mobile app features a user-friendly interface that simplifies image analysis and gives users information about color variations in the analyzed samples. The first pages of the app are related to the authentication process. After that, in the Home page users can view the captured images using the CMOS camera. By tapping in one image, the Sample page is presented with detailed information about the sample image, such as the capture date and the camera settings used during capture, see Fig. 2A. Then, the user can add the sample to the Analysis page where the user can select images for analysis and choose one of the two color comparison methods, "Mean" and "Dominant," see Fig. 2B. Finally, the Results page displays the obtained color difference results in a visualization chart, see Fig. 2C.

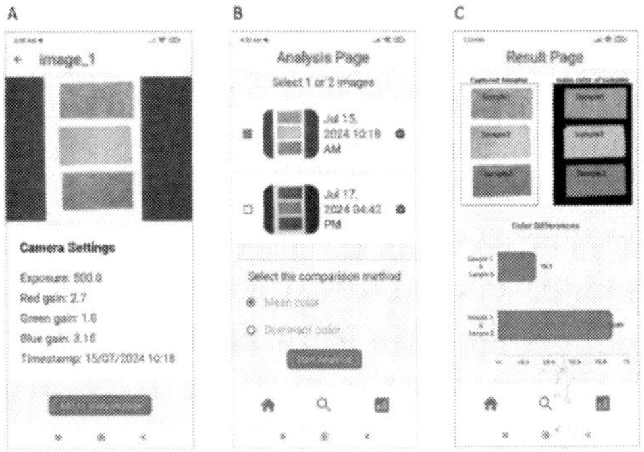

Fig. 2. BioColor app interface. (A) Sample page, (B) analysis page, and (C) results page.

4 Results & discussion

The study examines the response of plasmonic papers composed of AuNP to various solutions with increasing refractive index: (1) dry substrate, (2) substrate moistened with water, (3) substrate moistened with a mixture of glycerol and ethanol as sol.4, and (4) substrate moistened with glycerol. The substrates and the respective refractive indexes are detailed in Table 1.

Table 1. The refractive index of every medium.

Substrate	Refractive index
Dry plasmonic paper	1.00
Plasmonic paper in water	1.33
Plasmonic paper in sol. 4	1.44
Plasmonic paper in glycerol	1.47

The images are captured under two types of incident light: Halogen Lamp and Warm LED. Figures 3 and 4 show their respective spectra.

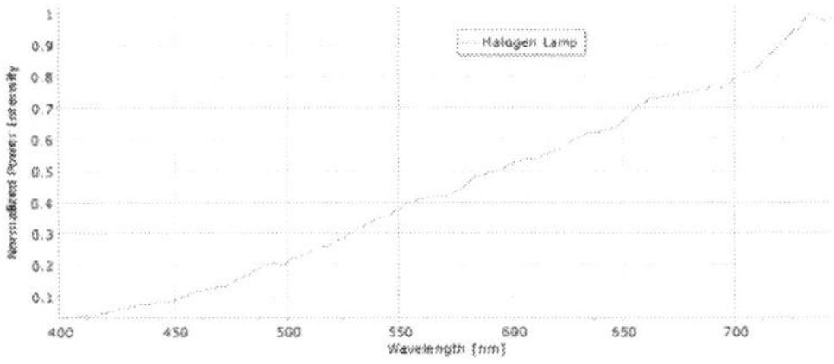

Fig.3. Spectra of the Halogen Lamp.

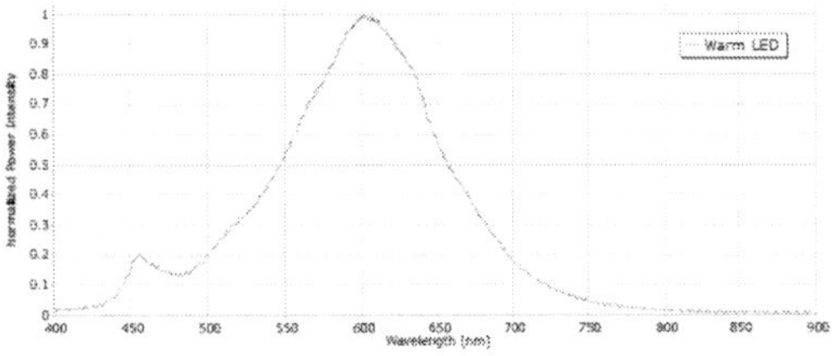

Fig.4. Spectra of the Warm LED.

The sample color is calculated using both the mean and dominant color methods. The mean color of a sample image is the sum of the pixel values for each channel L,A and B divided by the total number of pixels. The dominant color is the most frequent LAB color appearing in the image. The objective is to assess their effectiveness in detecting color variations and enhancing the sensor's sensitivity.

Fig. 5 shows the transmittance spectra of the dry plasmonic paper. The plasmonic resonance is located at about 540 nm, as could be expected from the LSPR theory [9].

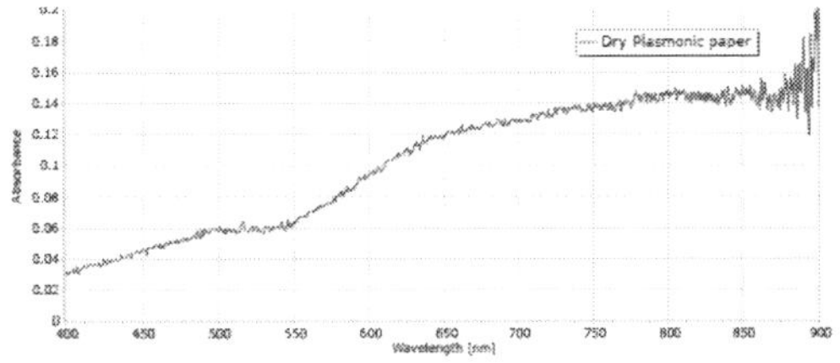

Fig.5. The absorbance spectra of the dry plasmonic paper.

To interpret the result of the color differences using CIEDE2000, a value close to '0' indicates similar colors, while higher values indicate increasing color differences. As shown in Table 2, the highest color difference occurred between the dry plasmonic paper substrate and the substrate immersed in glycerol (17.58). On the other hand, the smallest difference was between the substrate in solution 4 and the substrate in glycerol (0.73).

These results indicate that the color difference is larger when there is a greater variation in refractive index, and smaller when the variation is minimal. This supports the reliability and precision of the BioColor approach. This is valid specifically when employing the mean color method showing that the mean color is more sensitive to small color differences. The explanation is that mean color considers all the colors in the image, as small changes in pixel color will affect the mean color calculation. In contrast, the dominant color focuses only on the most prevalent color ignoring the presence of other colors in the image and the minor color differences.

In Table 3 the substrates were compared under the LED light. The comparison indicates that the color differences are larger compared to Table 2 where the Halogen Lamp was used. This means greater shifts within the visible spectrum of the LED light happened as it passed through the different substrates. Fig. 4 explains this by showing that the Warm LED emits a higher number of radiations, hence, more radiations are absorbed by the plasmonic paper leading to an increase in the color difference. Potentially enhancing the sensor's ability to detect changes in its surrounding medium.

Table 2. The Color difference results using the Halogen Lamp.

Substrate	Mean color	Dominant color
Dry vs. water	11.92	11.07
Dry vs. sol.4	17.44	20.81
Dry vs. glycerol	17.58	18.98
Water vs. sol.4	5.37	4.30
Water vs. glycerol	5.18	2.39
Sol.4 vs. glycerol	0.73	4.08

Table 3. The Color difference results using the Warm LED light.

Substrate	Mean color	Dominant color
Dry vs. water	20.41	21.64
Dry vs. sol.4	41.58	44.28
Dry vs. glycerol	42.60	45.10
Water vs. sol.4	19.62	21.54
Water vs. glycerol	20.91	22.55
Sol.4 vs. glycerol	1.90	1.86

5 Conclusion

This paper presented a new biosensor output reading technology based on image processing algorithms, which is less costly than traditional expensive spectrometers. The proposed BioColor approach analyzes the color of the light passing through LSPR sensor elements under different mediums with increasing refractive indices. The alterations of the refractive index in the surrounding medium correspond directly to a shift in the rejection band of the sensing element in this case the gold nanoparticle, leading to a change in the color of the transmitted light. Two color comparison methods, mean and dominant, were utilized to detect the color change. Based on the analysis, the mean color comparison method is recommended as it provides accurate detection of small color changes in the context of biomarkers, such as, pathogens interacting with the sensor's surface. It has been verified that the Warm LED light can enhance the sensor's sensitivity to different refractive index mediums compared to the Halogen Lamp. The outcomes of this work contribute significant value in understanding and optimizing the LSPR sensor performance for various applications, such as pathogen detection and medical diagnostics. Further research is being carried out to experiment with a broader range of samples, including preparing the sensor's surface with antibodies targeting actual pathogens.

Acknowledgments: This research was supported by Portuguese national funds provided by FCT – Fundação para a Ciência e Tecnologia, through the grants 2022.13579.BD and 2021.09347.BD, Center of Technology and Systems (CTS) UIDB/00066/2020 / UIDP/00066/2020, and by project IPL/IDI&CA2023/LUMINA ISEL.

References

1. S. K. Vashist, P. B. Luppa, L. Y. Yeo, A. Ozcan, & J. H. T. Luong. (2015). Emerging technologies for next-generation point-of-care testing. Trends in Biotechnology, 33(11), 692-705. https://doi.org/10.1016/j.tibtech.2015.09.001.
2. C. Serafinelli, A. Fantoni, E. Alegria, & M. Vieira. (2023). Recent progresses in plasmonic biosensors for point-of-care (POC) devices: A critical review. Chemosensors, 11(5), 303-316. https://doi.org/10.3390/chemosensors11050303.
3. B. Sepúlveda, P. C. Angelomé, L. M. Lechuga, & L. M. Liz-Marzán. (2009). LSPR-based nanobiosensors. Nano Today, 4(3), 244-251.https://doi.org/10.1016/j.nantod.2009.04.001.
4. Fantoni, A., Vygranenko, Y., Maçarico, A. F., Serafinelli, C., Fernandes, M., Mansour, R., Jesus, R., and Vieira, M. (2021). Arrayed graphene enhanced surface plasmon resonance for sensing applications. Optical Components and Materials XVIII. https://doi.org/10.1117/12.2583023.
5. Tadepalli, S., Kuang, Z., Jiang, Q., Liu, K. K., Fisher, M. A., Morrissey, J. J., Kharasch, E. D., Slocik, J. M., Naik, R. R., & Singamaneni, S. (2015). Peptide functionalized gold nanorods for the sensitive detection of a cardiac biomarker using plasmonic paper devices. Scientific Reports, 5, 16206. https://doi.org/10.1038/srep16206.
6. Mansour, R., Stojkovic, V., Vygranenko, Y., Lourenço, P., Jesus, R., and Fantoni, A. (2022). Colour and image processing for output extraction of a LSPR sensor. In Optics and Biophotonics in Low-Resource Settings VIII. https://doi.org/10.1117/12.2607463.
7. van der Walt, S., Schönberger, J. L., Nunez-Iglesias, J., Boulogne, F., Warner, J. D., Yager, N., Gouillart, E; Yu, T. (2014). Scikit-image: Image processing in python. PeerJ, 2. https://doi.org/10.7717/peerj.453.
8. Sharma, G., Wu, W., & Dalal, E. N. (2005). The CIEDE2000 color-difference formula: Implementation notes, supplementary test data, and mathematical observations. Color Research & Application, 30(1), 21-30. https://doi.org/10.1002/col.20070.
9. Fantoni, A., Fernandes, M., Vygranenko, Y., Louro, P., Vieira, M., Silva, R. P. O., & Alegria, E. C. B. A. (2018). Analysis of metallic nanoparticles embedded in thin film semiconductors for optoelectronic applications. Optical and Quantum Electronics, 50, 1-12. http://hdl.handle.net/10400.21/8577.

Lighting preferences of Iberian sausages for Spanish observers

Shino Okuda[1], *Masahito* Nagata[2], *Katsunori* Okajima[2], *Francisco José* Rodríguez-Pulido[3], *Francisco José* Heredia[3], *Rocío* Ortega-Velázquez[4], *María Dolores* Gaviño-Nadal[4], *Jesús* Casado-González[4], *Pedro Antonio* García[5], *María José* Rivas[6], *María* Tejada-Casado[6], *Ana* Yebra[6], and *Manuel* Melgosa[7]

[1]Doshisha Women's College of Liberal Arts, Kyoto, Japan
[2]Yokohama National University, Yokohama, Japan
[3]Food Color and Quality Laboratory, Faculty of Pharmacy, University of Seville, Seville, Spain
[4]Osborne Group. Sánchez Romero Carvajal, Jabugo S.A., Jabugo, Spain
[5]Department of Statistics and Operations Research, University of Granada, Granada, Spain
[6]Department of Optics, Faculty of Sciences, University of Granada, Granada, Spain
[7]Research Group FQM381 of the Andalusian Regional Government, University of Granada, Granada, Spain

Abstract. We report results of a visual experiment performed by 20 adult Spanish observers using calibrated images of 18 samples of sliced Iberian sausages under 15 illuminations, including 8 illuminants proposed by the International Commission on Illumination (CIE) and 7 light sources found in meat sections of supermarkets in Spain and Japan. Visual results from the pair comparison and magnitude estimation psychophysical methods used in our experiment showed a high correlation (Pearson's linear correlation coefficient r = 0.915). Illumination strongly influenced color preferences of sliced Iberian sausages. For example, using the magnitude estimation method, the average lighting preferences changed about 2.8/10 for the 15 illuminations tested. There were no statistically significant differences between lighting preferences for different kind of sausages. Exploring quality indices of white lights, we found that the correlated color temperature and gamut area index (Color Res. Appl. 33, 192-202, 2008) provided the highest linear correlations with visual preferences in our current experiment. Lights with correlated color temperatures below 3000 K or gamut area indices below 70 were not preferred. The CIE general color rendering and color fidelity indices provided poor predictions of visual preferences in our current experiment.

1 Introduction

In a first approach, color is an attribute of visual sensation generated by the interaction of light sources, objects and the human visual system [1]. Accordingly, a change in the spectral power distribution of the light source illuminating a given object may lead to a change in the color appearance of such object. Specifically, the use of different light sources may modify the visual appearance of foods [2], in such a way that proper food lighting can be useful to increase customer satisfaction and boost sales. In general, research about lighting preferences in meats is scarce, and it is unknown which are the main lighting parameters influencing people preferences [3-5]. In particular, as far as we know, researchers have not analyzed the influence of lighting on preferences of slices of Iberian cured meats, which constitutes the main goal of the current work.

From 2D spectroradiometric measurements of 18 commercial samples of sliced Iberian sausages from two manufacturers in Jabugo (Spain), we generated high-resolution color images of each sausage under 15 different illuminants or light sources. From these images, we studied lighting preferences for Iberian sausages in an experiment performed by 20 adult Spanish observers on a calibrated monitor using two different psychophysical methods. We also tested relationships between preferences in our visual experiment and some of the main color-quality indices currently used for white lights, like, for example, correlated color temperature (CCT) [6], Gamut Area index (GAI) [7], CIE general color rendering index (R_a) [8], and CIE general color fidelity index (R_f) [9].

2 Materials and methods

We selected 15 different illuminations, including some theoretical lights adopted by the International Commission on Illumination (CIE) [6] as prototypes, which are known as "illuminants", and a few light sources used in meat sections of some supermarkets in Spain or Japan. Specifically, the 15 illuminations tested are in the next 3 groups: I) The 3 CIE standard illuminants (A, D50 and D65). II) The 5 phosphor-converted blue LED illuminants proposed by the CIE and named LED-B1 to LED-B5 [6]. III) A set of 7 light sources found in meat sections of supermarkets in Spain or Japan, named S1 to S7, which spectral power distributions were measured by us using a portable Konica-Minolta CL-70F spectrometer. Table 1 lists the main characteristics of these 15 illuminations. Note that LED illuminants have CCTs in a relatively large range (2733 K – 6598 K) while the sources S1-S7 have CCTs in a considerably shorter range (1789 K – 3246 K), suggesting that warm lights have been usually preferred for meats in supermarkets. From Table I, the ranges for values of GAI [7] and R_a [8] (or R_f [9]) are 48-121 and 77-100, respectively. We note that R_a [8] and R_f [9] values are very similar for all LED illuminants, but not for sources S1-S7 (differences of up to 11 points).

Table 1. Main characteristics of 15 illuminants/sources used.

Illuminant/ Source	CCT (K)	GAI [7]	R_a [8]	R_f [9]
A	2856	53	100	100
D50	5001	88	100	100
D65	6503	98	100	100
LED-B1	2733	48	82	84
LED-B2	2998	55	83	84
LED-B3	4103	77	85	85
LED-B4	5109	83	77	77
LED-B5	6598	89	80	79
S1	1789	67	82	77
S2	2189	63	93	83
S3	2478	121	69	80
S4	2909	110	90	81
S5	3208	83	94	92
S6	3234	90	92	89
S7	3246	110	83	88

We selected 18 vacuum-packed envelopes of sliced pork Iberian sausages from two different manufacturers in Jabugo (Spain) named A and B for confidentiality (Table 2). These 18 samples included four different kind of Iberian pork sausages (ham, shoulder ham, loin, and *caña de presa*) with cuts made by hand (H) or machine (M). In our poster presentation we will show quality pictures of these 18 samples under the main CIE standard illuminant (D65), and one specific sausage sample under the 15 mentioned illuminations. We measured the spectral reflectance factors for very small regions in each one of the sausage samples using a 2D Topcon SR-5000 HWS spectroradiometer. The sausage samples were placed on a tilted grey surface and illuminated by two symmetrical NanGuang CN-600CSA LED sources, providing a nearly uniform illuminance of 4200 lx at the plane of the samples. The optical axis of the spectroradiometer was perpendicular to the sample and the distance from the head of the spectroradiometer to the sample was about 1 meter. From spectroradiometric measurements, bearing in mind the calibration of the EIZO ColorEdge CG2700X monitor used in our visual experiment, we generated realistic high-resolution images for each of the 18 sausage samples under the 15 illuminations indicated in Table 1. All these images had the same average equivalent illuminance (500 lx) to avoid the influence of illuminance level on visual preferences. We can assume that all the generated images were identical to the real samples under the selected illuminations.

Table 2. The 18 selected samples of pork Iberian sausages. Letters H and M mean cut by hand or machine, respectively.

Iberian Sausage	Number of samples	Manufacturer
Ham-H	3	A
Ham-M	3	A
Shoulder Ham-M	2	A
Loin-M	2	A
Caña de presa-M	2	A
Ham-H	2	B
Shoulder Ham-H	2	B
Loin-H	2	B

A panel of 20 observers born in Spain (50% men and 50% women), all of them with normal color vision from Ishihara test, non-vegetarian, and more than 35 years old (49 years old on average) participated in the visual experiment. Each observer performed 3 visual sessions in different days (40 minutes/session). The observer was in a dark laboratory, at a distance of about 60 cm from the monitor screen, and in each visual session used two psychophysical techniques, known as 'pair comparison' and 'magnitude estimation' methods, to perform visual preference tasks. In the pair comparison method, the observer must choose between two images appearing simultaneously in the monitor screen, corresponding to the same sausage sample under 2 different illuminations. In the magnitude estimation method, the observer sees only one image in the screen and must provide a score between 0 and 10, which mean 'I do not like at all this sausage sample, I'd never buy it' and 'I like a lot of this sausage sample, for sure I'd buy it', respectively. In both methods, to recover adaptation the observer sees a black screen between two successive visual assessments. The pair comparison method implies 105 assessments for each sausage sample (combinations of 15 illuminations by pairs), in such a way that this method is easy to perform but slow. Because of this reason, only 6 sausage samples were assessed using the pair comparison method. Regarding the magnitude estimation method, it is enough subjective but very fast, because only 15 assessments are made for each sausage sample. Because of this reason, observers assessed the 18 sausage samples using the magnitude estimation method. In each of the 3 visual sessions the observer assessed 2 samples using the pair comparison method and 6 samples using the magnitude estimation method. In the pair comparison method the final scores are the number of times each illumination is preferred to the 14 remaining ones, multiplied by a factor (10/14) in order to normalize the result to 10 for easier comparison with results from the magnitude estimation method. In overall, our current experiment implies 18000 visual assessments of lighting preferences, each observer performing 900 assessments (6x105=630 using the pair comparison method plus 18x15=270 using the magnitude estimation method).

3 Results and discussion

As mentioned before, we only assessed 6 of 18 sausage samples using the pair comparison method, while all samples were assessed using the magnitude estimation method. Figure 1 shows the relationship between average results found for the 6 samples assessed by both methods. We note that using the pair comparison method the range of answers is about 1-8 while using the magnitude estimation method it is about 3.5-8.5. This difference is not surprising because in the pair comparison method, sometimes the observer must choose between two very unpleasant illuminations, in such a way that using this method some 'points' are gathered even for most unpleasant illuminations. The parameters of regression models for results from pair comparison and magnitude estimation methods are statistically significant (average r=0.915, p<0.001).

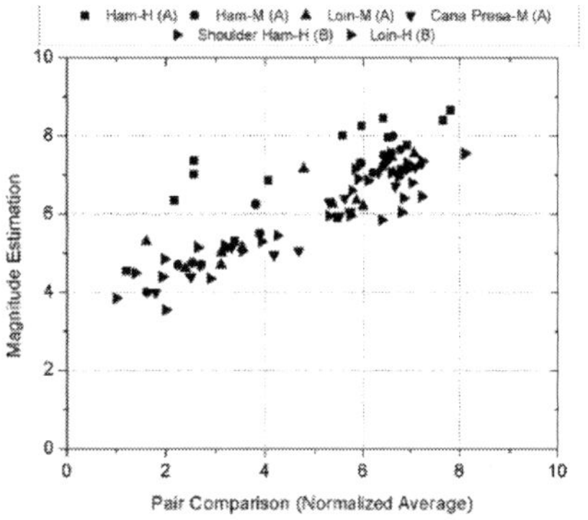

Fig. 1. Average visual preferences of 15 illuminations (Table 1) for 6 sausage samples assessed using the normalized pair comparison method and magnitude estimation method.

Based on the results from the magnitude estimation method, ANOVA tests revealed statistically significant differences amongst 15 illuminations tested. Subsequent Tukey tests identified 6 distinct homogeneous groups, each comprising 3 to 6 illuminations. Two-way ANOVA tests indicated no significant interaction between sample types and illuminations, allowing us to conclude that the best and worst illuminations are consistent across all types of sausage samples. However, we found some slight statistically significant differences for different samples of the same sausage in the cases of Ham-H (A), Ham-H (B) and *Caña de presa* (A).

Table 3 shows values of Pearson's linear correlation coefficients considering the four main parameters (CCT [6], GAI [7], R_a [8], and R_f [9]) used to characterize illuminants or light sources and average visual preferences of samples under these illuminations, for the pair comparison (6 samples) and magnitude estimation (18 samples) psychophysical methods. From Table 3, it is evident that the CCT and GAI indices provided the best correlations with the visual preferences of sliced Iberian sausages, although the correlation coefficients are not particularly high. This suggests that regrettably, none of these indices of withe light can accurately predict on their own the visual preferences for sliced Iberian sausage samples observed in our experiment. Interestingly, Table 3 also reveals that the CIE general color-rendering index R_a [8] offered very poor predictions of visual preferences in our experiment. While the predictions made by the CIE color fidelity index R_f [9] were somewhat better than those from R_a [8], they were still notably inadequate. The correlations for R_f [9] shown in Table 3 can be slightly improved by excluding the values of 100 corresponding to the three CIE standard illuminants (A, D50 and D65); however, the correlations using R_a [8] remain unimproved by such an exclusion.

Table 3. Pearson's correlation coefficients considering visual preference results from two psychophysical methods against four quality indices of white lights (CCT, GAI, R_a and R_f).

Parameters of 15 illuminants/sources	Pair comparison	Magnitude estimation
CCT (K) [6]	0.703	0.579
GAI [7]	0.662	0.766
R_a [8]	0.209	0.176
R_f [9]	0.328	0.310

Figures 2-5 show average visual preferences with their standard deviations (error bars) for the 18 sausage samples assessed by the magnitude estimation (ME) method, against values of GAI [7], CCT [6], R_a [8], and R_f [9], respectively, for the 15 illuminants/sources tested (Table 1). From Fig. 2 we can see that visual preferences changed with GAI values in a considerably wide range of 2.8/10 approximately, and lights with values below about 70 led to low visual preferences.

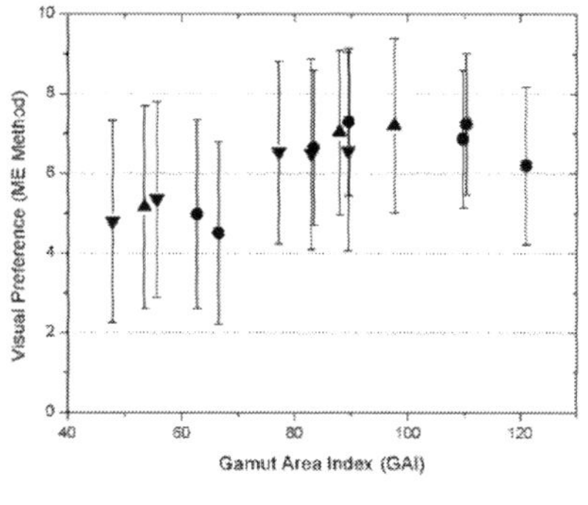

Fig. 2. Average visual preferences and standard deviations (error bars) of 18 sausage samples using the magnitude estimation (ME) method against values of GAI [7] for the 15 illuminations in Table 1: CIE standard illuminants A, D50 and D65 (up triangles); CIE illuminants LED-B1 to LED-B5 (down triangles); light sources S1-S7 (circles).

Figure 3 shows average visual preferences against correlated color temperatures of lights, indicating that CCT values below 3000 K may lead to low preferences of sausage samples. Figures 4 and 5 show that correlations between visual preferences and values of R_a [8] or R_f [9] are very low, as indicated in Table 3.

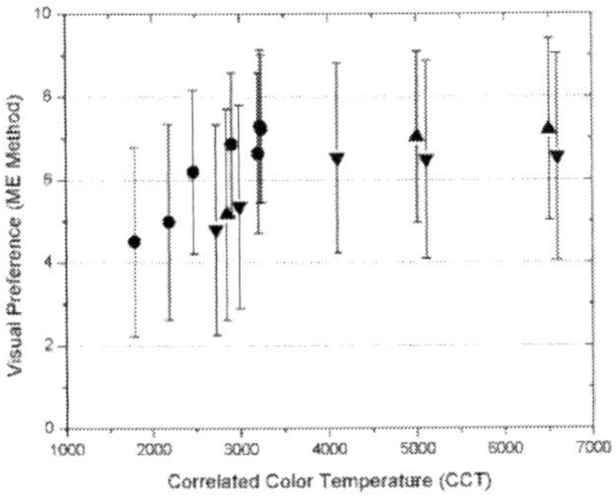

Fig. 3. Idem to Fig. 2 but considering visual preferences of sausage samples against CCT values (Table 1).

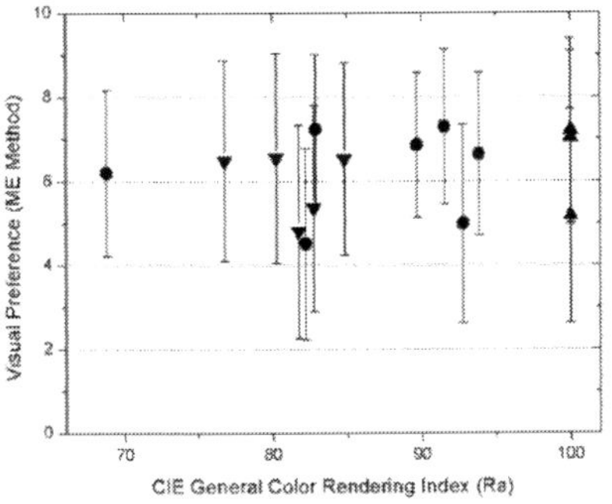

Fig. 4. Idem to Fig. 2 but considering visual preferences of sausage samples against R_a [8] values (Table 1).

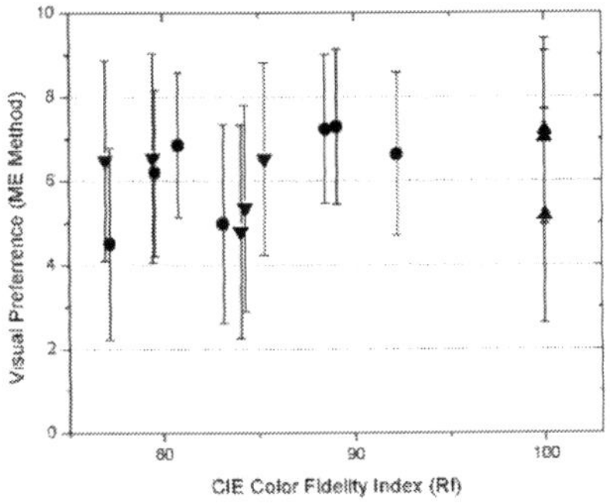

Fig. 5. Idem to Fig. 2 but considering visual preferences of sausage samples against R_f [9] values (Table 1).

4 Conclusions and future work

We have used a methodology allowing the generation of high-resolution realistic images of sliced Iberian sausages under any light with known spectral power distribution. We employed such images to perform an experiment concluding that preferences of sliced Iberian sausages may be significantly influenced by selected illumination. Unfortunately, traditional indices to measure the quality of white lights (e.g. CCT, GAI [7], R_a [8] and R_f [9]) are not very well correlated with visual preferences of sliced Iberian sausages. In the future, we will report detailed colorimetric analyses of current samples, additional quality indices of lights [10, 11], and results from an analogous experiment to this one using 20 Japanese adult observers.

Grant PID2022-138031NB-I00 funded by MICIU/AEI/10.13039/501100011033 and by ERDF/EU. JSPS KAKENHI Grant Number 22H03896.
Discussions of current results with María del Mar Gandolfo (Signify, Madrid, Spain) and Javier Oneca Torres (Saltoki, Pamplona, Spain) are acknowledged.

References

1. M. D. Fairchild, Color Appearance Models, Third Edition, Chapter 3 (Wiley, 2013).
2. J. L. Caivano and M. del P. Buera (Editors), Color in Food – Technological and Psychophysical Aspects. https://doi.org/10.1201/b11878. CRC Press (Boca Raton, 2012).
3. S. Barbut, Frankfurters – acceptability under different retail lights. Can. J. Anim. Sci. **84**, 391-396 (2004). https://doi.org/10.4141/A03-098.
4. F.B. Özyürek, N. Karatas, M. Tapan, G.B. Var, M. Cakir, C.O. Özer, The effects of light sources and packaging types on the storage stability of fresh lean and chuck beef meat during refrigerated storage. J. Food Process. Pres. **45**, Special Issue (2020). https://doi.org/10.1111/jfpp.14525.
5. O. Masuda, S.M.C. Nascimento, Best lighting for naturalness and preference. J. Vision **13**, 4 (2013). https://doi.org/10.1167/13.7.4.
6. CIE 015:2018. Colorimetry, 4th Edition. CIE Central Bureau (Vienna), 2018. ISBN: 978-3-902842-13-8. DOI: 10.25039/TR.015.2018.
7. M.S. Rea, J.P. Freyssinier-Nova, Color rendering: A tale of two metrics. Color Res. Appl. **33**, 192-202 (2008). https://doi.org/10.1002/col.20399.
8. CIE 13.3-1995. Method of measuring and specifying colour rendering properties of light sources. CIE Central Bureau, Vienna (1995). ISBN: 978 3 900734 57 2.
9. CIE 224:2017. CIE 2017 Colour Fidelity Index for accurate scientific use. CIE Central Bureau, Vienna (2017). ISBN 978-3-902842-61-9.
10. W. Davis, Y. Ohno, Color quality scale. Opt. Eng. **49**, 033602 (2010). https://doi.org/10.1117/1.3360335
11. K.A.K. Smet, W.R. Ryckaert, M.R. Pointer, G. Deconinck, P. Hanselaer, A memory colour quality metric for white light sources. Energ. Buildings **49**, 216-225 (2012). https://doi.org/10.1016/j.enbuild.2012.02.008

A LiDAR based obstacle detection framework for railway

Susana Dias[1*], *Pedro* Sousa[1, 2], *Francisco* Afonso[1], *Nuno* Viriato[1], *Paulo* Tavares[1], and *Pedro* Moreira[1]

[1]INEGI, Rua Dr. Roberto Frias N° 400, 4200-465 Porto, Portugal
[2]Faculty of Eng., University of Porto, Rua Dr. Roberto Frias, 4200-465 Porto, Portugal

Abstract. Obstacle detection on the railway, a crucial operational safety concern, is a complex task that encompasses a multitude of challenges. While Machine Learning (ML) algorithms are commonly employed in analogous applications such as autonomous car driving [1] [2], the railway field faces a significant barrier due to the scarcity of available data (particularly images), rendering conventional ML approaches impractical.
In response to this challenge, this study proposes and evaluates a framework which uses LiDAR (Light Detection and Ranging) data for obstacle detection on the railways. The framework aims to address the limitations posed by image data scarcity while enhancing operational safety in railway environments.
The developed methodology combines the use of a long-range LiDAR capable of detecting obstacles at distances of up to 500 meters, with the train's GPS (Global Positioning System) coordinates to accurately determine its position relative to detected obstacles. The LiDAR data is processed using a data fusion approach, where pre-existing knowledge regarding the track topography is combined with a clustering algorithm, specifically DBSCAN (Density-based spatial clustering of applications with noise), to identify and classify potential obstacles at a pre-defined distance.
Tests of the proposed framework were conducted within the confines of a moving locomotive, specifically the CP 2600-2620 series, along a designated section of the Contumil-Leixões line. These tests served to validate the effectiveness and feasibility of the approach under real-world operating conditions.
Overall, the utilization of LiDAR data coupled with advanced algorithms presents a promising avenue for enhancing obstacle detection capabilities in railway operations. By overcoming the challenges associated with data scarcity, this framework holds the potential to significantly improve operational safety and efficiency within railway networks. Further research and testing are warranted to validate the framework's performance across diverse railway environments and operating conditions.

1 Introduction

The railway is still a dominant mean of transportation around the world, with an estimated, of at least, 1 300 000 km worldwide [3], 211 430 km in Europe alone. With such a vast expanse of railway lines, one of the biggest safety concerns is railway intrusion (when a potentially harmful object is within the region of interest, or ROI, of the railway line). The timely detection of these objects, obstacles, on the railway line can prevent serious accidents with potential loss of people and property, an area that has been studied over the last few decades [4] [5].

Over the years, several methods have been used to detect intrusions in the railway, such as infrared, vibrations, laser, radar, video, to name a few [6], [7], [8], [9]. Each of these technologies has inherent limitations that can compromise their effectiveness in ensuring the safety and reliability of railway systems.

Camera-based systems rely on visual data to detect obstacles, but they are heavily dependent on lighting conditions. In low-light environments, such as during night-time or in tunnels, cameras may struggle to accurately identify obstacles. Additionally, cameras are susceptible to environmental factors such as rain, fog, dirt, or glare from the sun, which can obscure the field of view and lead to missed detections.

Radar technology offers better performance in adverse weather conditions and can detect objects at longer ranges than cameras. However, radar generally lacks the high spatial resolution needed to distinguish between small or closely spaced objects. This limitation can result in less precise obstacle detection, particularly in complex environments with multiple potential hazards.

Ultrasonic sensors are typically used for short-range detection and are valued for their simplicity and cost-effectiveness. However, they are prone to inaccuracies due to noise and can be less reliable in varying environmental conditions. The limited range and resolution of ultrasonic sensors make them insufficient for comprehensive obstacle detection on railways, where early and accurate detection over longer distances is critical.

LiDAR (Light Detection And Ranging) is one of the technologies that are widely used in autonomous vehicles for detecting obstacles, usually in conjunction with other technologies, such as cameras (sensor fusion) [10] [9].

LiDAR is an optical remote sensing technology that uses light, in the form of a pulsed laser, to measure distances. The laser is sent from a source (transmitter) and is reflected by

* Corresponding author: sdias@inegi.up.pt

https://doi.org/10.1051/epjconf/202430500027

objects present within the laser's range. The reflected light is detected by the system's receiver and the time it takes for the reflected light to return to it (time of flight - TOF) is used to develop a distance map of the objects. LiDAR has some advantages over cameras [11] [12]:

- Allows to create a 3D map of the surrounding environment;
- It is not affected by night and low light conditions;
- Measures distances to objects with great precision.

The framework proposed in this paper combines LiDAR's high-resolution 3D mapping capabilities with GPS's precise geolocation data. This approach addresses many of the limitations of existing technologies by providing accurate 3D information with high resolution and precise positioning information. LiDAR enhances detection in diverse environmental conditions, while GPS ensures precise localization of obstacles relative to the train's position.

In this paper we will describe the proposed framework for obstacle detection in the railway as well as the tests performed during the development of the framework.

2 Method

Although Machine Learning algorithms are currently widely used to detect obstacles, the lack of data as well as the difficulty in carrying out acquisitions with LiDAR in the railway environment, made this solution unfeasible.

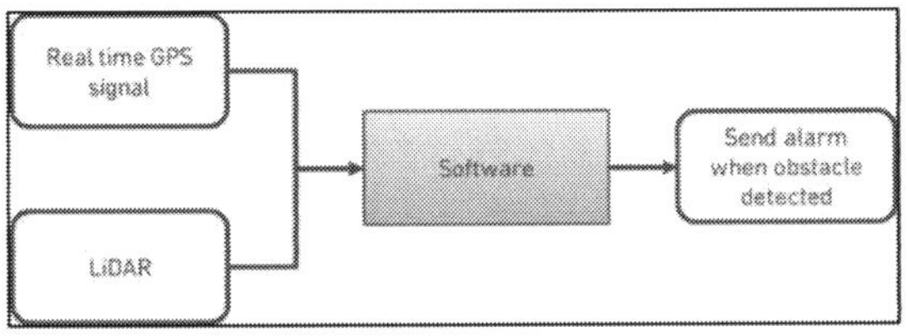

Fig. 1. Proposed Solution Overview.

The conceived solution, presented in Fig. 1, is based on GPS (Global Positioning System) coordinates and the train path. The GPS coordinates are used to determine the train's position and calculate, at a predefined distance, if there are obstacles on the track.

Obstacles were identified resorting to the use of the DBSCAN (Density-Based Spatial Clustering of Applications with Noise) clustering algorithm (Fig. 2). DBSCAN is a non-parametric density-based clustering algorithm proposed in 1996 by Martin Ester, Hans-Peter Kriegel, Jörg Sander, and Xiaowei Xu [13]. It is based on the idea that clusters are formed by dense regions of points, separated by less dense regions. In the presented solution, two DBSCAN parameters are used: the eps (the maximum distance between two points for them to be considered part of the same neighborhood) and the minimum samples (the minimum number of points required to form a dense region). Both parameters can be configured in order to be adjustable to different search distances. The regions identified by the DBSCAN algorithm will be considered obstacles if they have more than a configurable number of points within the same neighborhood. It is one of the clustering algorithms frequently used in Data Mining and Machine Learning [14, 15].

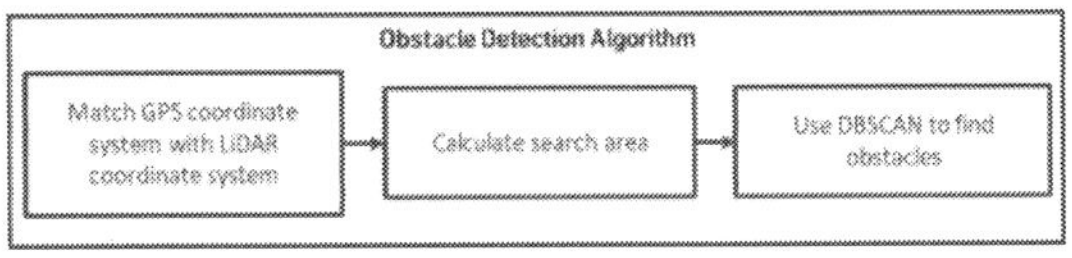

Fig. 2. Proposed Solution Overview.

As mentioned before, the devised solution relies on the GPS coordinate system and the fact that the travelled train path is always the same. Before the processing of the LiDAR data occurs, it is necessary to prepare a train path file. This file contains the GPS coordinates and altitude for the whole train path, with intervals of a configurable distance (for example, 5m distance between each GPS coordinate) and always in the middle of the railway. This path file is used to identify the location of the area where obstacles will be searched for, which is dependent on the current train location.

During the processing, for each LiDAR data acquisition the GPS train coordinates are also acquired. These coordinates are then searched for in the train path file, and the closest coordinates from the file are used as the train position. Using this position, the obstacle search area can be determined. The central point of the obstacle search area, the goal point, is determined by getting the point in the path (in the path file) in front of the train at a configurable distance (for example 100 m). The GPS goal point is converted into the LiDAR field of view coordinate system and obstacles are searched inside a predefined area around that point. The search area size, around the goal point, is defined in a configuration file (for example 4m in directions around the goal point, height, width, depth).

The decision to define a search area is related to the necessity of minimizing the processing time, considering the size of the point cloud acquired by LiDAR and the response time that is needed when obstacles are detected on the train track. Only obstacles that are directly on the path travelled by the train and within the field of view of the LiDAR will be searched for (Fig. 3 and Fig. 4). This method allows for an efficient search of obstacles even if the railway track follows a curved path.

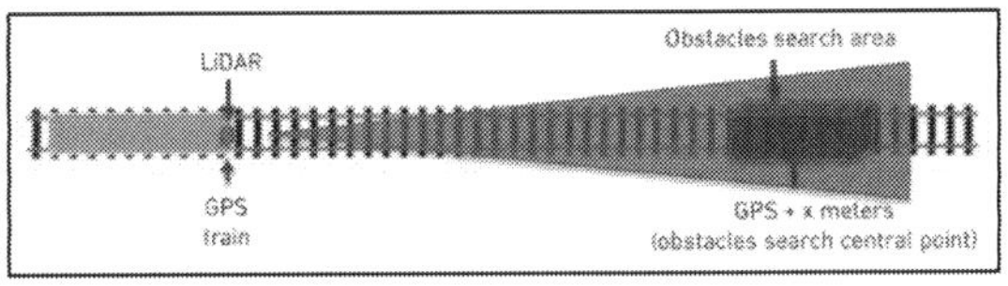

Fig. 3. Obstacle search inside LiDAR FOV.

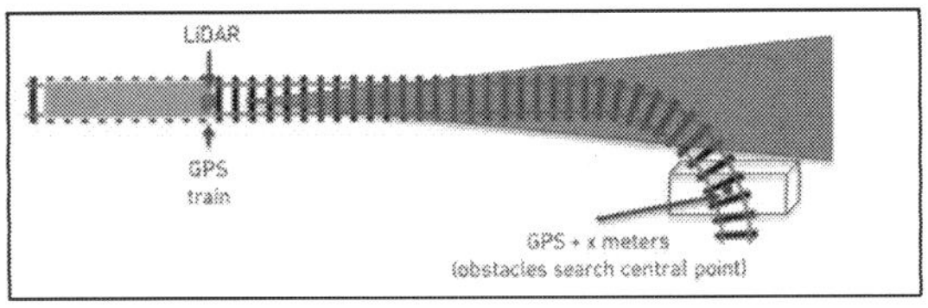

Fig. 4. Obstacle search outside LiDAR FOV.

3 Tests and Results

The tests and results presented in this paper were conducted using the LiDAR Livox Tele-15 [16], a long-distance detection LiDAR with a 500m range, and a GPS module from the u-blox, the ZED-F9P-02B [17], which provides multi-band GNSS to high-volume industrial applications.

In order to better understand the advantages and limitations of the LiDAR and how to integrate it into an obstacle detection system, initial acquisitions were performed. The first set of acquisitions was a controlled experience, at several distances, to collect information of what a person looks like in a point cloud. That information is then used to define, in the DBSCAN algorithm, what should constitute an obstacle, for a certain distance. This was a static acquisition, meaning, both the LiDAR and the person were stationary. Acquisitions were performed, with the person, at: 50m, 100m, 150m, and 200m (Fig. 5).

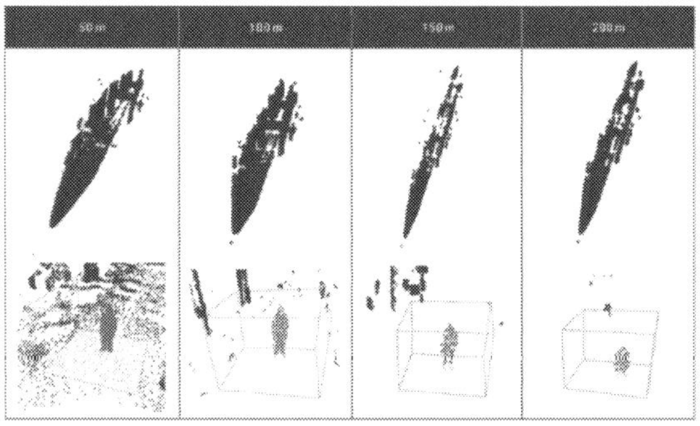

Fig. 5. Proposed solution – initial static tests.

The next step was the execution of static tests in a railway environment (Fig. 6). Tests with a person inside the rail track, near, and outside were performed and the obstacle successfully detected.

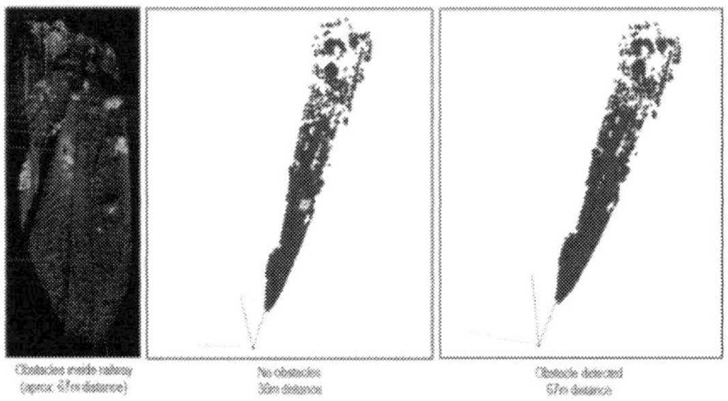

Fig. 6. Proposed solution – static tests in a railway environment.

The next step was the execution of tests in a moving environment, in this case inside a car (Fig. 7).

Fig. 7. Proposed solution – moving tests in a car.

The average speed of the car was 10-20 km/h. The obstacles were search at a distance of 100m in front of the car. In this scenario, the detection of obstacles was also successful, a person crossed the road during the test and was detected as an obstacle (Fig. 8).

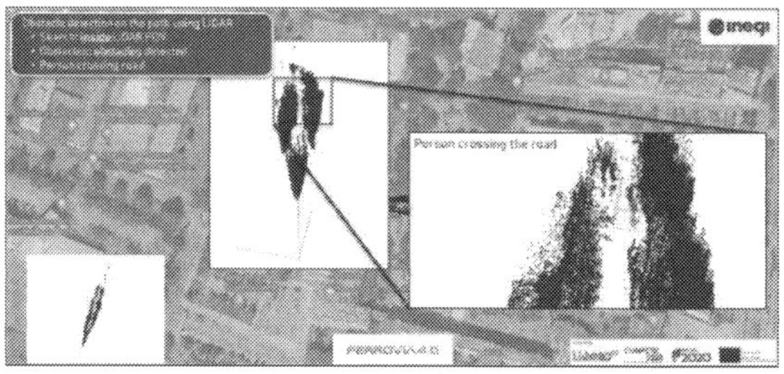

Fig. 8. Proposed solution – moving tests in a car (map from: www.google.com/maps).

The final tests of the proposed solution were conducted within the confines of a moving locomotive, specifically the CP 2600-2620 series, along a designated section of the Contumil-Leixões line. These tests served to validate the effectiveness and feasibility of the approach under real-world operating conditions.

The LiDAR and GPS were placed inside the train locomotive. For that purpose, a physical support was design and constructed, using aluminium 3.3315 (EN-AW 5005), as shown in Fig. 9.

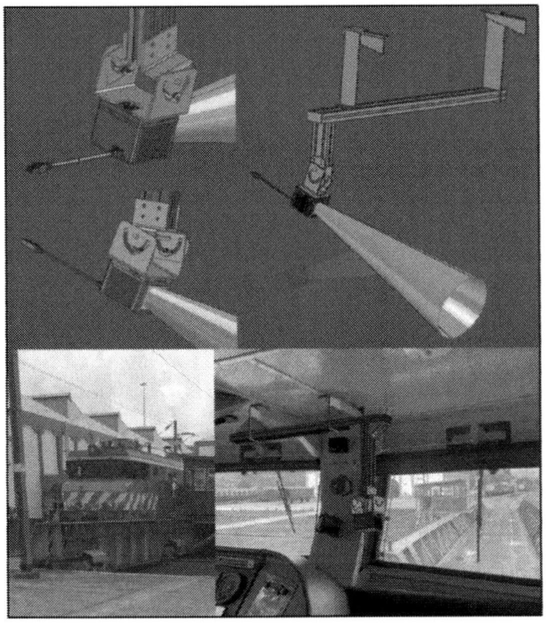

Fig. 9. LiDAR Physical Support.

As previously stated, the validation in relevant environment was conducted in the Contumil-Leixões line, Porto, Portugal (Fig. 10).

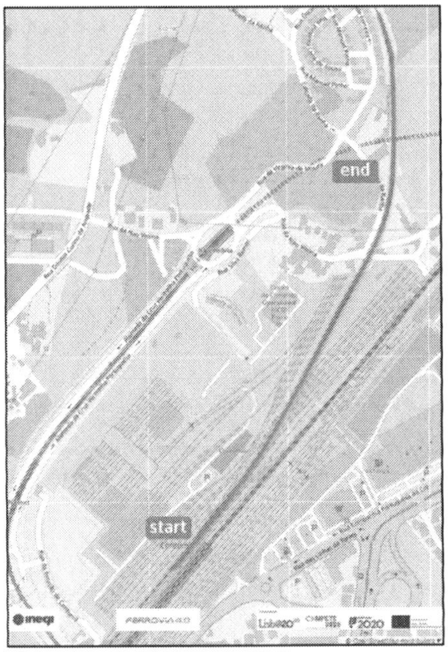

Fig. 10. Line track for validation in relevant environment (map from: www.openstreetmap.org).

During the test, the train reached a maximum speed of 80km/h. As in previous tests, the search for obstacles was executed at 100m distance from the LiDAR (in front of the locomotive). Fig. 11 and Fig. 12 illustrate results from the processing of the LiDAR acquisitions. Fig. 11 presents a search inside the LiDAR FOV (field of view). In this case no obstacles existed.

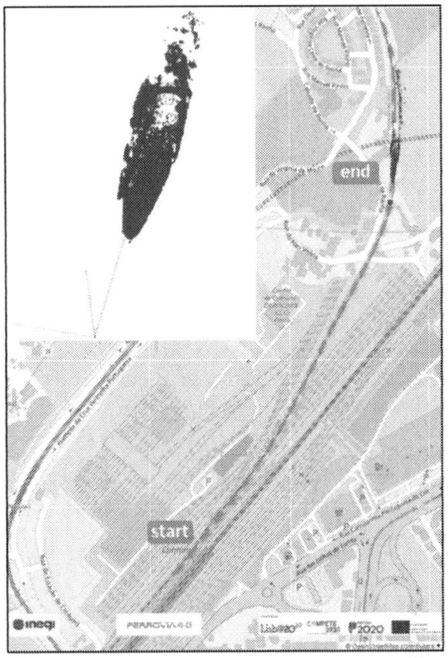

Fig. 11. Obstacle detection on the rail path using LiDAR - inside LiDAR FOV (map from: www.openstreetmap.org).

Fig. 12 represents the scenario were the search for obstacles is outside the LiDAR FOV. In this case, the search step is not executed, saving processing time, which is critical in a system like this one.

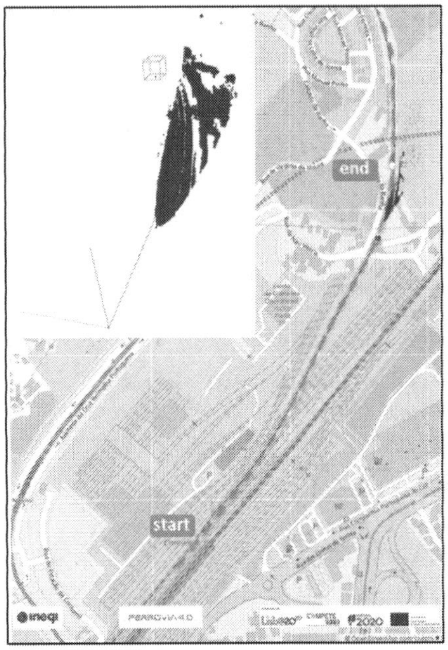

Fig. 12. Obstacle detection on the rail path using LiDAR - outside LiDAR FOV (map from: www.openstreetmap.org).

4 Conclusion

In this paper a framework for obstacle detection in the railway was described. The framework methodology combines the use of a long-range LiDAR capable of detecting obstacles at distances of up to 500 meters, with the train's coordinates to accurately determine its position relative to detected obstacles. The DBSCAN clustering algorithm was used to search for obstacles inside the determined area of interest. Several tests were conducted, including tests within the confines of a moving locomotive, along a designated section of the Contumil-Leixões line. These tests served to validate the effectiveness and feasibility of the approach under real-world operating conditions.

Overall, the utilization of LiDAR data coupled with advanced clustering algorithms presents a promising avenue for enhancing obstacle detection capabilities in railway operations. Further research and testing are warranted to validate the framework's performance across diverse railway environments and operating conditions

Acknowledgements: This work was developed in the scope of the project LAETA2023 - Refª UIDB/50022/2020, funded by "Programa Financiamento Plurianual de Unidades de I&D" of the "Fundação para a Ciência e a Tecnologia" and in the scope of the project FERROVIA 4.0, nº 46111 which has received funding from "ANI - Agência Nacional de Inovação, S.A" through the programme "Mobilizador COPROMOÇÃO_PT2020".

References

[1] L. T. A. A. I. P. G. Zamanakos, "A comprehensive survey of LIDAR-based 3D object detection methods with deep learning for autonomous driving," *Computers & Graphics,* p. 9577, 2021.

[2] J. E. B. S. Y. Alaba, "A survey on deep-learning-based lidar 3d object detection for autonomous driving," *Sensors,* pp. 153-181, 2022.

[3] "wikipedia," [Online]. Available: https://en.wikipedia.org/wiki/List_of_countries_by_rail_transport_network_size. [Accessed 07 2024].

[4] R. A. Kromer, D. J. Hutchinson, M. J. Lato, D. Gauthier and T. Edwards, "Identifying rock slope failure precursors using LiDAR for transportation corridor hazard management," *Engineering Geology,* vol. 195, pp. 93-103, 2015.

[5] H. Lan, C. D. Martin, C. Zhou and C. H. Lim, "Rockfall hazard analysis using LiDAR and spatial modeling," *Geomorphology,* vol. 118, pp. 213-223, 2010.

[6] Z. a. Q. Y. a. J. L. a. X. Z. a. G. Y. a. W. Y. a. L. P. a. Y. Z. Cao, "Railway Intrusion Detection Based on Machine Vision: A Survey, Challenges, and Perspectives," *IEEE Transactions on Intelligent Transportation Systems,* vol. 25, no. 7, pp. 6427-6448, 2024.

[7] M. P. a. W. B. da Silva, "State-of-the-art technologies for intrusion and obstacle detection for railroad operations," United States. Department of Transportation. Federal Railroad Administration, 2007.

[8] M. C. M. a. J. Y. Hasanujjaman, "Sensor fusion in autonomous vehicle with traffic surveillance camera system: detection, localization, and AI networking," *Sensors,* p. 3335, 2023.

[9] D. J. Yeong, G. Velasco-Hernandez, J. Barry and J. Walsh, "Sensor and sensor fusion technology in autonomous vehicles: A review," *Sensors,* vol. 21, no. 6, p. 2104, 2021.

[10] W. Zhangyu, Y. Guizhen, W. Xinkai, L. Haoran and L. Da, "A camera and LiDAR data fusion method for railway object detection," *IEEE Sensors Journal,* vol. 21, no. 12, pp. 13442-13454, 2021.

[11] P. Wang, "Research on comparison of lidar and camera in autonomous driving.," *Journal of Physics: Conference Series,* vol. 2093, no. 1, p. 012032, 2021.

[12] D. a. J. M. a. Y. W. a. K. A. Lee, "LiDAR odometry survey: recent advancements and remaining challenges," *Intelligent Service Robotics,* vol. 17, no. 2, pp. 95-118, 2024.

[13] M. Ester, H.-P. Kriegel, J. Sander and X. Xu, "A density-based algorithm for discovering clusters in large spatial databases with noise," in *Proceedings of 2nd International Conference on Knowledge Discovery and Data Mining*, Portland, EUA, 1996.

[14] A. C. A. B. a. I. B. Benabdellah, "A survey of clustering algorithms for an industrial context," in *Second International Conference on Intelligent Computing in Data Sciences (ICDS 2018)*, 2018.

[15] O. a. B. A. Kulkarni, "A Survey of Advancements in DBSCAN Clustering Algorithms for Big Data," in *3rd International conference on Power Electronics and IoT Applications in Renewable Energy and its Control (PARC)*, 2024.

[16] "Livox," [Online]. Available: https://www.livoxtech.com/tele-15.

[17] "u-blox," [Online]. Available: https://www.u-blox.com/en/product/zed-f9p-module.

 https://doi.org/10.1051/epjconf/202430500028

Design and fabrication of asymmetric Mach-Zehnder interferometers based on EpoClad and EpoCore strip waveguides

Tiago E.C. Magalhães[1,*], *Jérôme* Borme[1], *Temple* Douglas[1], *Christian* Maibohm[1], and *Jana B.* Nieder[1]

[1]INL - International Iberian Nanotechnology Laboratory, Av. Mestre José Veiga s/n, 4715-330, Braga, Portugal

Abstract. Integrated polymeric optical interferometers offer the possibility of developing sensors with low cost, scalability, and easy integration. Although they are not yet competitive with inorganic materials in terms of sensitivity, they have good biocompatibility in general, and suitable designs may provide sufficient sensitivity for biosensing. A new design for integrated Mach-Zehnder interferometers based on asymmetric arms with different widths has been proposed and experimentally demonstrated, avoiding the need for additional fabrication steps for an interaction window where biosamples are placed. The basic sensory principle is built upon the non-zero variation in the difference of the effective refractive index between the two arms due to their different dimensions, causing a phase variation in the output signal. In this work, we present a design optimization method and fabrication results by e-beam lithography for integrated asymmetric Mach-Zehnder interferometers based on strip waveguides made from EpoClad and EpoCore polymers. The operation wavelength was set to 650 nm. The optimization algorithm is based on open-source mode-solver simulations that return the optimal fabrication dimensions of the interferometer, avoiding high-order modes and enhancing single-mode confinement.

1 Introduction

Large-scale, cost-effective mass production, flexibility, and easy integration [1, 2] are some of the key advantages of polymer-based integrated optical platforms compared to other materials. Because of this, all-polymer optical sensors have been considered for the realization of disposable label-free detection of biomolecules [3]. Although polymer sensors are not yet competitive in terms of sensitivity when compared to sensors based on inorganic materials [3, 4], due to the low refractive index difference, they show good biocompatibility [2, 5] and suitable designs may provide sufficient sensitivity for biosensing applications. They have also been explored for new gas [6, 7] and temperature sensing devices [8].

Integrated Mach-Zehnder interferometers have been used and proposed for many applications, such as quantum computing [9], communications [10], digital signal processing [11] and sensing [12–15]. In the latter case, an interaction window is usually fabricated on one of the arms where the sample to be sensed is placed [16]. The basic principle is that the effective refractive index will change differently for both arms along the sensing region, causing a phase difference between both arms. To eliminate the need to fabricate an interaction window on only one arm, which typically requires additional alignment, lithography, and etching steps [17, 18], several types of asymmetric Mach-Zehnders interferometers (AMZIs) were proposed and developed. They generally consist of having arms with different lengths [19]. In 2016, Hofmann et al. [17] introduced a new design concept for an AMZI which consists of having waveguides of different widths in each arm, which yields different effective refractive indexes. Depending on the upper cladding refractive indices, the effective refractive index will change differently due to the different dimensions, causing a phase shift that alters the output intensity.

Here, we present a design optimization algorithm and the subsequential fabrication of AMZIs based on strip waveguides. The main requirements of the final design are the following: (i) maximization of the sensitivity, (ii) confinement of the fundamental mode, and (iii) maximization of higher-mode losses. For the cladding and core materials, we used EpoClad and EpoCore (Micro Resist Technology GmbH), which are epoxy-based polymers that have been explored in the fabrication of near-infrared waveguides [20, 21]. They present great mechanical properties [22] and chemical stability, showing a lower transmission degradation rate when compared to other polymers such as the SU-8 [23]. Regarding biocompatibility, some studies have also been performed [24], demonstrating that they are hemocompatible and thus attractive for biophotonic sensing systems for blood specimen analysis.

This work is structured as follows. First, in Section 2, we present a brief introduction to the theoretical background of the AMZI proposed by Hofmann et al. [17]. In Section 3, we will discuss the design optimization method used to obtain the fabrication dimensions. In Section 4, we present the simulation results and the final design. The

*e-mail: tiago.magalhaes@inl.int

fabrication process and morphology characterization are discussed in Section 5. Finally, in Section 6, we conclude and allude to the future work.

2 Theoretical background

In what follows, we will briefly introduce a simple theoretical model concerning the AMZI. We will start by introducing the structure of the interferometer and then present the calculations used for the sensitivity.

Consider the Mach-Zehnder interferometer where both arms have a length L_a (measured parallel to the z-axis), as illustrated in Fig. 1, and each arm has a different width (w_b and w_n, see top insert in Fig. 1). Since in this interferometer design, there is no distinction between the reference and sensing arm, we will denote the arm with a larger width as the broad arm and the other one as the narrow arm. Consider now the input intensity I_i on the interferometer with wavelength λ. Assuming that the integrated interferometer is a superposition of both arms, we can write the output intensity I_o as

$$I_o = a\, I_i\, e^{-\alpha_b L_a} + (1-a)\, I_i\, e^{-\alpha_n L_a} + 2\sqrt{a(1-a)\, I_i^2\, e^{-(\alpha_b+\alpha_n)L_a}}\, \cos\phi\,, \quad (1)$$

where a is the splitting ratio, α_b and α_n are the absorption coefficients of the broad and narrow waveguides, respectively, and ϕ is the phase difference between both arms at the output, given by

$$\phi = \frac{2\pi}{\lambda}\left[\int_{P_b} \beta_b(z)\mathrm{d}z - \int_{P_n} \beta_n(z)\mathrm{d}z\right], \quad (2)$$

where β_b and β_n are real numbers and represent the effective refractive index (for the fundamental mode) of the broad and narrow arms, respectively, integrated over the path lengths P_b and P_n. Assuming that β_b and β_n are constants (for a given waveguide design) and that the total length of each arm is L_a, Eq. (2) can be approximated as

$$\phi = \frac{2\pi}{\lambda}(\beta_b - \beta_n)\, L_a\,. \quad (3)$$

Note that, in reality, the bending and tapering in each arm will change the effective refractive index, but for the sake of simplicity, we will not consider here these effects.

We are interested in performing intensity measurements to extract the phase ϕ as a function of the change in the analyte refractive index n_a. Thus, the sensitivity can be written, in general, as

$$S = \frac{\mathrm{d}I_o}{\mathrm{d}n_a}\,. \quad (4)$$

All parameters in Eqs. (2) and (3) are constant, except for β_b and β_n, which depend on the upper cladding (or analyte). Thus, performing the derivative of Eq. (4) we obtain

$$S = C\, \sin\left[\frac{2\pi}{\lambda}(\beta_b - \beta_n)\, L_a\right] \frac{\mathrm{d}(\beta_b - \beta_n)}{\mathrm{d}n_a}\,, \quad (5)$$

where C is a constant defined as

$$C = \frac{4\pi L_a}{\lambda}\sqrt{a(1-a)\, I_i^2\, e^{(\alpha_b+\alpha_n)L_a}}\,. \quad (6)$$

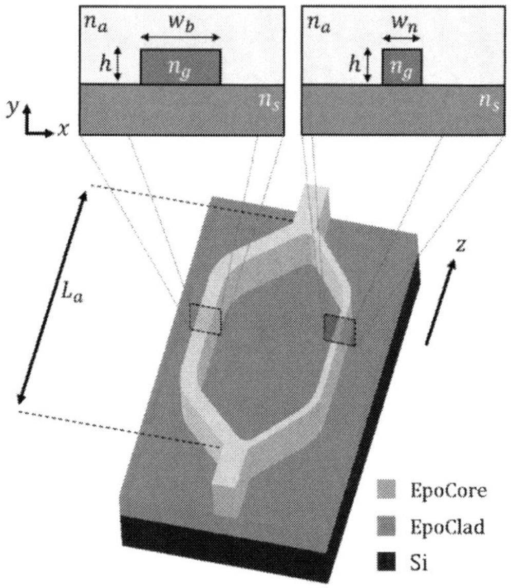

Figure 1. Illustration of an integrated asymmetric Mach-Zehnder interferometer (not to scale). Top left: cross-section of broad waveguide arm. Top right: cross-section of narrow waveguide arm. n_a, n_g, and n_s: refractive indexes of the upper cladding (analyte), core, and lower cladding, respectively. w_b and w_n: widths of the broad and narrow arms ($w_b > w_n$), respectively. h: height of the core. L_a: Length of each arm of the interferometer.

The first term in Eq. (5) is a constant and the second term is a periodic function that varies between 0 and 1, where the frequency is proportional to $(\beta_b - \beta_n)$. The third term is a derivative of $(\beta_b - \beta_n)$ with respect to n_a and, therefore, we can define a new sensitivity S_w of the sensing region as done in [17, 18]:

$$S_w = \frac{\mathrm{d}(\beta_b - \beta_n)}{\mathrm{d}n_a}\,. \quad (7)$$

Henceforth, we will use the term sensitivity when referring to Eq. (7). Assuming that $\Delta\beta$ varies slowly when changing n_a, we can approximate Eq. (7) as

$$S_w = \frac{\Delta\beta_b - \Delta\beta_n}{\Delta n_a}\,, \quad (8)$$

where Δn_a is the change in the analyte's refractive index and $\Delta\beta_{b,n}$ is the difference between the effective refractive index $\beta_{b,n}(n_a)$ when the analyte has refractive index n_a and the effective refractive index $\beta_{b,n}(n_a + \Delta n_a)$ when the analyte has effective refractive index $n_a + \Delta n_a$, i.e.,

$$\Delta\beta_{b,n} = \beta_{b,n}(n_a) - \beta_{b,n}(n_a + \Delta n_a)\,, \quad (9)$$

where the subscripts b, n denote the broader or narrow waveguide, respectively. We can rewrite Eq. (8) as

$$S_w = \frac{\Delta\beta}{\Delta n_a}\,, \quad (10)$$

where $\Delta\beta$ is defined as

$$\Delta\beta = \Delta\beta_b - \Delta\beta_n\,. \quad (11)$$

To increase the sensitivity in Eq. (10) for a given Δn_a, we need to increase $\Delta\beta$, which will depend on the dimensions of both arms of the interferometer. However, as the width and height of the waveguides increase, the number of modes that the waveguide supports also increases. We also note that the length L_a of both arms will play a role in the overall sensitivity through Eqs. (5) and (6), but since it is independent of the analyte, it can be treated separately after we have determined the optimal dimensions of the waveguides.

3 Sensitivity optimization

The ultimate goal of the optimization model is to return a set of values for the widths and heights of the broad and narrow arms of the AMZI, namely, w_b, w_n, and h. Note that we are imposing the same height for both arms due to limitations in the fabrication procedure, but a generalization of the model is straightforward. Nonetheless, we must also take into consideration that using different heights will also imply an additional bending in the yz plane and additional losses will have to be considered.

3.1 Mode-solving

The effective refractive index of each arm can be determined through mode-solver algorithms, which depend on the waveguide dimensions and refractive indices of the materials used. In this work, we will utilize *modesolverpy* [25], an open-source and Python-based waveguide mode-solver.

Throughout this paper, we will be considering quasi-TM modes, i.e., electric fields vertically polarized (parallel to the y-axis) with respect to the waveguide representation of Fig. 1. The same procedure and methods can be made for quasi-TE modes, i.e., assuming a horizontally polarized electric field.

With reference to Fig. 1 (top boxes), we start by defining the following computational parameters: spatial resolutions Δx and Δy of the meshing on both axes (x and y), wavelength λ, lower cladding height H_s and width W_s, upper cladding (analyte) height H_a, core refractive index n_g, lower cladding refractive index n_s, the upper cladding refractive index (analyte) n_a and the component of the input electric field (e.g., the y-component). Note that, these parameters correspond to input parameters, and depending on the mode-solver algorithm, different parameters may be needed. After setting up values for each parameter, we define four extra parameters, $w_{\min}$, $w_{\max}$, $h_{\min}$, and $h_{\max}$ which define the limits to the width and height of the waveguides such that $w_{\min} \leq w_{b,n} \leq w_{\max}$ and $h_{\min} \leq h \leq h_{\max}$. Moreover, we define the number of points N_w and N_h so we can create arrays $\mathbf{w}$ and $\mathbf{h}$, with size N_w and N_h, respectively, comprising the possible widths and heights for both arms. Finally, we define the analyte refractive index variation Δn_a such that the final analyte refractive index is $n_a + \Delta n_a$. The input parameters are summarized in Table 1.

After defining the input parameters, we set out to compute the modes and effective refractive indices for each

Table 1. Input parameters of the design simulations.

Quantity	Symbol	Value
Spatial resolution x-axis (μm)	Δx	0.06
Spatial resolution y-axis (μm)	Δy	0.06
Wavelength (nm)	λ	650
Core ref. index.	n_g	1.5794
Lower cladding ref. index	n_s	1.5926
Analyte ref. index	n_a	1.33
Analyte ref. index change	Δn_a	0.03
Minimum width (μm)	$w_{\min}$	1.5
Maximum width (μm)	$w_{\max}$	5.5
Minimum height (μm)	$h_{\min}$	1.5
Maximum height (μm)	$h_{\max}$	5.5
Num. of width values	N_w	41
Num. of height values	N_h	41
Lower cladding height (μm)	H_s	3
Lower cladding width (μm)	W_s	8
Upper cladding height (μm)	H_a	3
Confinement threshold for qTM$_0$	$\gamma^{(0)}_{\min}$	0.8
Energy difference threshold	$\Delta\gamma_{\min}$	0.06

pair (w_i, h_j) with $i, j = (1, \ldots, N_{w,h})$. We thus define two arrays containing both computed quantities. We first define the effective refractive index array $\mathbf{B}$ for a given analyte refractive index n_a:

$$\mathbf{B} = \begin{bmatrix} \begin{pmatrix} B^{(0)}_{11} \\ B^{(1)}_{11} \end{pmatrix} & \cdots & \begin{pmatrix} B^{(0)}_{N1} \\ B^{(1)}_{N1} \end{pmatrix} \\ \vdots & \ddots & \vdots \\ \begin{pmatrix} B^{(0)}_{1N} \\ B^{(1)}_{1N} \end{pmatrix} & \cdots & \begin{pmatrix} B^{(0)}_{NN} \\ B^{(1)}_{NN} \end{pmatrix} \end{bmatrix}, \quad (k = 0, 1), \quad (12)$$

where $B^{(k)}_{ij}$ represents the numerically obtained effective refractive index of the kth mode for a core width w_i and height h_j, respectively, for an analyte refractive index n_a. The dependence of $B^{(k)}_{ij}$ on n_a is not written explicitly in order to avoid excessive subscripts. We define the first (fundamental) mode ($k = 0$) as the mode with the highest effective refractive index. Only the effective refractive indices of the first two modes are stored. To compute the sensitivity, we need two effective refractive index arrays, $\mathbf{B}_1$ and $\mathbf{B}_2$, for two different analytes with refractive indices n_a and $n_a + \Delta n_a$, respectively. Then, using Eq. (10), we can define a three-dimensional sensitivity array $\mathbf{S} = [\mathbf{S}^{(1)}, ..., \mathbf{S}^{(j)}]$ $(j = 1, ..., N_h)$, where $\mathbf{S}^{(j)}$ is the two-dimensional sensitivity array for a given waveguide height

h_j, defined as

$$\mathbf{S}^{(j)} = \begin{bmatrix} S^{j}_{11} & \cdots & S^{j}_{N1} \\ \vdots & \ddots & \vdots \\ S^{j}_{1N} & \cdots & S^{j}_{NN} \end{bmatrix}, \tag{13}$$

where $S^{j}_{i_1 i_2}$ $(i_1, i_2 = 1, \ldots, N_w)$ is the sensitivity for widths w_{i_1} and w_{i_2} of each arm. The sensitivity values can thus be calculated through Eq. (10), i.e.,

$$S^{(j)}_{i_1,i_2} = \Delta n_a^{-1} \left\{ \left[B^{(0)}_{i_1,j}(n_a) - B^{(0)}_{i_2,j}(n_a) \right] - \left[B^{(0)}_{i_1,j}(n_a + \Delta n_a) - B^{(0)}_{i_2,j}(n_a + \Delta n_a) \right] \right\}. \tag{14}$$

3.2 Mode confinement

To later evaluate how well each mode is confined in the core, the mode of each electric field can also be stored in an array $\mathbf{E}$ defined as

$$\mathbf{E} = \begin{bmatrix} \begin{pmatrix} E^{(0)}_{11} \\ E^{(1)}_{11} \end{pmatrix} & \cdots & \begin{pmatrix} E^{(0)}_{N1} \\ E^{(1)}_{N1} \end{pmatrix} \\ \vdots & \ddots & \vdots \\ \begin{pmatrix} E^{(0)}_{1N} \\ E^{(1)}_{1N} \end{pmatrix} & \cdots & \begin{pmatrix} E^{(0)}_{NN} \\ E^{(1)}_{NN} \end{pmatrix} \end{bmatrix}, \quad (k = 0, 1), \tag{15}$$

where $E^{(k)}_{ij}$ represents the electric field of the kth mode for a core width and height of w_i and h_j, respectively. We can then compute the energy confinement factor $\Gamma^{(k)}_{ij}$ of a given mode k and for a given pair of dimensions (w_i, h_j) using the following definition [26, 27]:

$$\Gamma^{(k)}_{i,j} = \frac{\iint_\rho \left| E^{(k)}_{i,j}(x,y) \right|^2 \mathrm{d}x\,\mathrm{d}y}{\iint_{-\infty}^{+\infty} \left| E^{(k)}_{i,j}(x,y) \right|^2 \mathrm{d}x\,\mathrm{d}y}, \tag{16}$$

where the double integral in the nominator is performed over the core waveguide area ρ. Thus, we can build a $2 \times N_w \times N_h$ array $\mathbf{\Gamma}^{(k)}$ containing all the energy confinement factors.

3.3 Optimal dimensions

After computing the sensitivity array $\mathbf{S}$ and the energy confinement factor array $\mathbf{\Gamma}^{(k)}$, we can devise a method to search for the optimal sensitivity and, at the same time, avoid high-order modes $(k > 0)$ and reduce the losses of the fundamental mode. Next, we will present a simple straightforward method, but other approaches could be used.

We start by defining two thresholds: $\gamma^{(0)}_{\min}$ for the energy confinement of the fundamental mode; and $\Delta\gamma_{\min}$ for the difference between the fundamental and first high-order mode. The meanings are as follows. Firstly, the fundamental mode must have an energy confinement factor greater than $\gamma^{(0)}_{\min}$, i.e.,

$$\Gamma^{(0)}_{i,j} > \gamma^{(0)}_{\min}. \tag{17}$$

Secondly, the difference between the energy confinement factor of the fundamental mode and the first high-order mode must be greater than $\Delta\gamma_{\min}$, i.e.,

$$\Gamma^{(0)}_{i,j} - \Gamma^{(1)}_{i,j} > \Delta\gamma_{\min}. \tag{18}$$

After imposing these two constraints, we add two final ones: (i) the effective refractive index of the fundamental mode should be larger than that of the substrate and (ii) the effective refractive index of the high-order mode should be smaller than that of the lower cladding. Running a loop on each value of the sensitivity array $\mathbf{S}$ and imposing these 4 constraints, we will obtain two pairs of values for the width and height of both arms, i.e., (w_b, h) and (w_n, h), which maximize the sensitivity.

4 Design and simulation results

In what follows, we will present the optimization results for all-polymer AMZI using the method described previously in Section 3. The input parameters are summarized in Table 1, namely, in the 3rd column.

4.1 Optimization results

The values for the lower cladding and core refractive index were chosen according to the values given by the manufacturer for EpoClad 20 and EpoCore 2, respectively, for a wavelength of 650 nm (see Table 1). The difference in the analyte refractive index, Δn_a, was chosen to be 0.03, as done in reference [18]. The number of values for the possible widths and heights (N_w and N_h) was chosen to be 41 each so that the $2 \times N_w \times N_h$ mode-solver simulations do not take too long to run. Due to the fabrication recipe followed, we imposed a height of $h = 1.8\,\mu\text{m}$. With regards to the constraint parameters, we chose $\gamma^{(0)}_{\min} = 0.8$ and $\Delta\gamma_{\min} = 0.06$ for an analyte refraction index of $n_a = 1.33$. $\Delta\gamma_{\min}$ ensures that the first higher-order mode has always greater losses than the fundamental mode. Figure 2 illustrates the main results obtained, which we will next explain in more detail.

First, we started by computing $4 \times N_w \times N_h$ modes (fundamental and first higher-order mode) and respective effective refractive indices, corresponding to the combinations of possible pairs of widths and heights, i.e., (w_b, h) and (w_t, h), for the two analyte refractive indices, n_a and $n_a + \Delta n_a$. Figure 2(a) shows the effective refractive indexes for all pairs $(w_{b,n}, h)$ for the case of $n_a = 1.33$. Figure 2(b) shows a cross-section of (a) for $h = 1.8\,\mu\text{m}$. Then, using Eq. (16), we compute for each mode k and for each pair $(w_{b,n}, h)$ the energy confinement factor matrix $\mathbf{\Gamma}^{(k)}$. Figure 2(c) shows the energy confinement factor for the fundamental mode, while Fig. 2(d) shows the first high-order mode. In Fig. 2(e), the difference in energy confinement factors is shown, showing the region where the fundamental mode is more prevalent when compared to the first high-order mode. The sensitivity array $\mathbf{S}$ is shown in Fig. 2(f) and a particular case for $h = 1.8\,\mu\text{m}$ is shown on the right side. Notice that the sensitivity is higher at the edges, where the difference between the widths of the

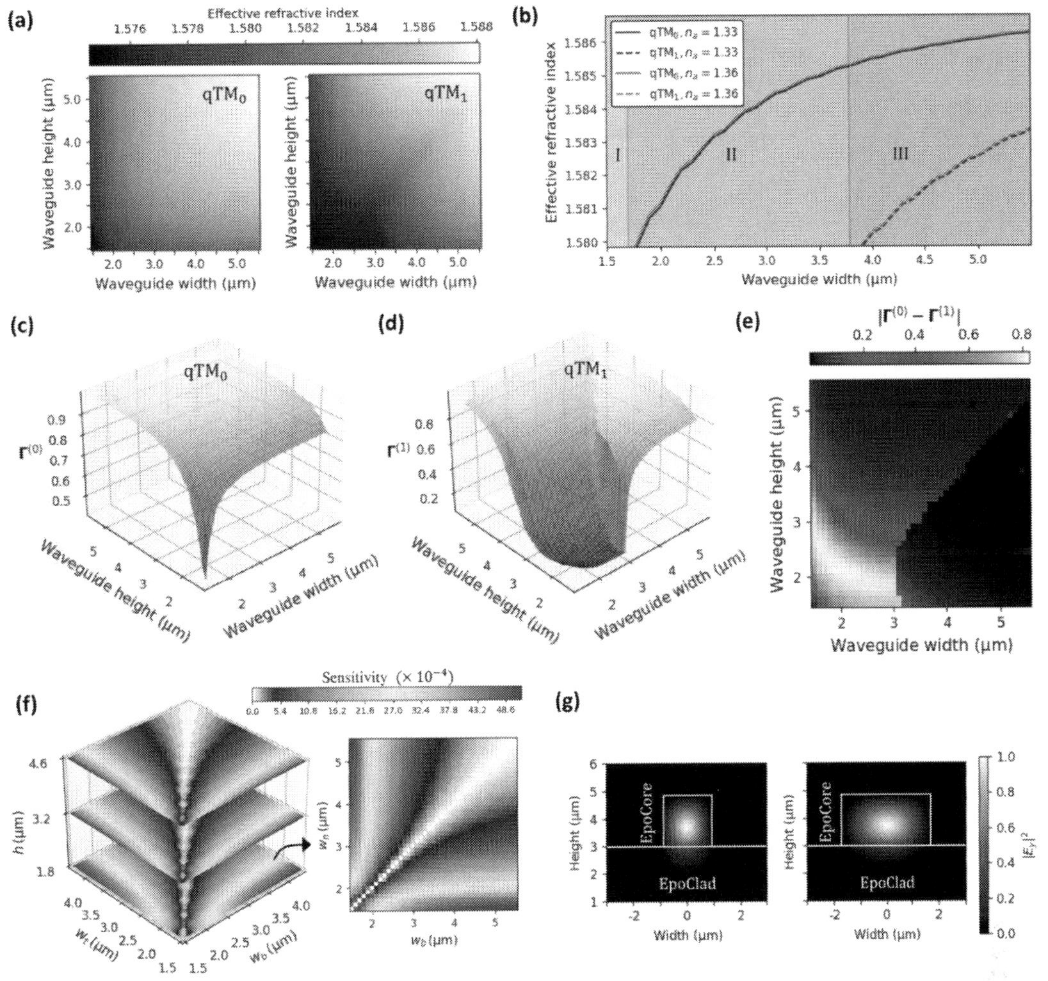

Figure 2. Design optimization results. (a) The effective refractive index of the fundamental (TM_0) and first high-order mode (TM_1) as a function of the waveguide width and height for $n_a = 1.33$. (b) Effective refractive index for a height of 1.8 μm. The three background colors represent 3 different regions, namely, I, II, and III. Region II starts when the effective refractive index of the fundamental mode is larger than that of the lower cladding. Region III starts when the effective refractive index of the second mode (TM_1) is larger than that of the lower cladding. By definition, region I does not propagate light and ends when region II starts. (c) Energy confinement factor $\mathbf{\Gamma}^{(0)}$ for the fundamental mode as a function of the waveguide width and height. (d) Energy confinement factor $\mathbf{\Gamma}^{(1)}$ for the high-order mode as a function of the waveguide width and height. (e) Difference between both energy confinement factors, $\mathbf{\Gamma}^{(0)}$ and $\mathbf{\Gamma}^{(1)}$. The larger values are optimal since high-order modes have more losses. (f) Sensitivity as a function of the height and widths of the waveguides (sensitivity array $\mathbf{S}$). On the right, we can observe the sensitivity array for a height $h = 1.8$ μm. (g) Fundamental modes of the narrow (left) and broad (right) waveguide for retrieved values: $w_b = 3.4$ μm and $w_n = 1.8$ μm.

broad and narrow waveguides is larger. As expected, the diagonal is zero since the widths and heights of the waveguides are the same. The fundamental modes for the retrieved dimensions are shown in Fig. 2(g). The optimal values returned were $w_b = 3.4$ μm and $w_n = 1.8$ μm, with a sensitivity of $S_w = 0.0025$ [see Eqs. (14) and (10)].

4.2 Final design

After selecting the proper dimensions of both arms, two additional steps are required: (i) choose the arm length L_a (which comprises the length of the bending curve), and (ii) choose the bending function. Although the length L_a of the arms of the interferometer does not affect the difference in the effective refractive index of both arms, it does affect the overall phase [see Eq. (3)]. Therefore, depending on the length L_a, the phase will vary differently and, consequently, the output intensity I_o. Moreover, since the term L_a is inside the cosine function, we will have to take into account the number of cycles we want for the difference in analyte refractive index Δn_a we are expecting. For more details on this, we refer the reader to reference [18].

The AMZI input waveguide starts with the broader dimensions, i.e., with width w_b and then the waveguide will

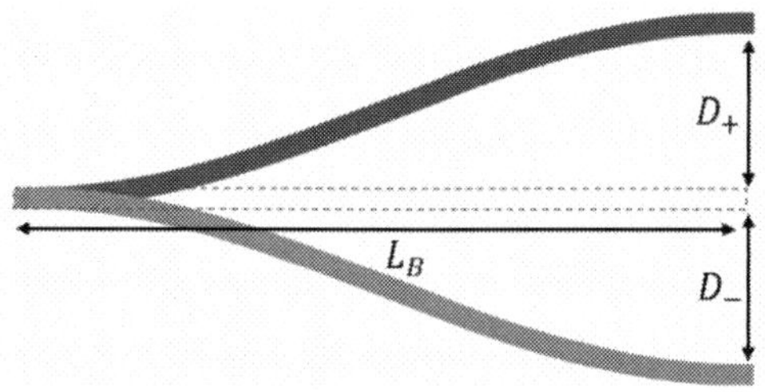

Figure 3. Notation used for the Y-coupler bending of the AMZI.

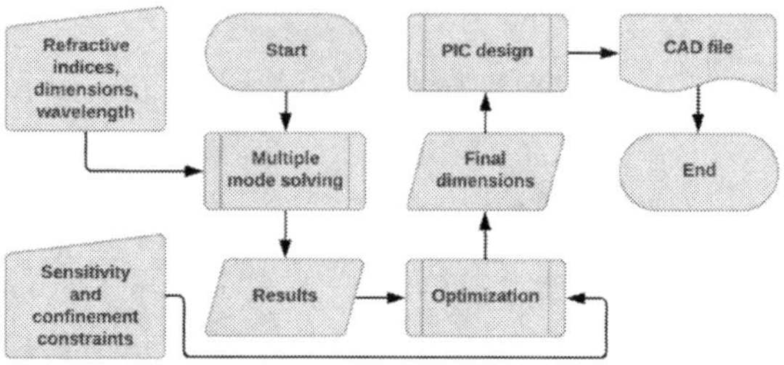

Figure 4. Flowchart showing the main steps for the design of the AMZI. PIC: Photonic integrated circuit.

split into two using a Y-coupler. Thus, we need to account for the bending function. This evaluation of the best bending design can be done using numerical analysis techniques, for example, beam propagation method (BPM) and Finite-difference time-domain (FDTD) [28]. Here we will use two cosine S-bends, as depicted in Fig. 3, due to their low bending losses [4, 29]. The expression for the cosine S-bend can be written as

$$x_\pm(z) = \pm\frac{D_\pm}{2}\left[1-\cos\left(\frac{\pi z}{L_B}\right)\right], \quad z \in [0, L_B], \tag{19}$$

where the plus and minus signs correspond to the top and bottom arms, respectively, and L_B is the bending length, which corresponds to the half-cycle of the cosine function. Since one arm of the AMZI has the same dimensions as that of the input and output waveguide, only the narrow arm will change its dimension. To minimize the losses, a linear taper function is applied in the narrow arm, as done in reference [18].

Finally, to generate the CAD file with the interferometer design, the open-source framework Nazca Design [30] is used to generate the GDS file. Note that, in this step, several interferometers can be placed within the same CAD file. For the bending length we chose $L_B = 4000\,\mu m$ and for the separation parameters we chose $D_+ = 30\,\mu m$ and $D_- = 10\,\mu m$ [see Eq. (19)]. In both cases, we took into consideration the simulations presented in references [18, 29] in order to reduce bending losses. The total length chosen for each arm (measured parallel to the z-axis) was 20 mm, which already includes the initial and final bending. The length of both the input and output waveguide of the interferometer was chosen to be 10 mm. Thus, the complete AMZI device measures 40 mm. A single Python script that comprises all of these tasks, including the creation of the optimization procedure and the creation of the CAD file, was developed and its flowchart is summarized in Fig. 4.

5 Fabrication and morphological characterization

The various fabrication steps are outlined in Fig. 5 and all of them were executed in a clean room environment at INL. First, a 200 mm silicon wafer was cleaned with acetone and treated with plasma ashing. Then, the wafer was spin-coated with EpoClad 20 at 2500 RPM for 40 seconds, and baked for 5 minutes at 50 °C and 10 minutes at

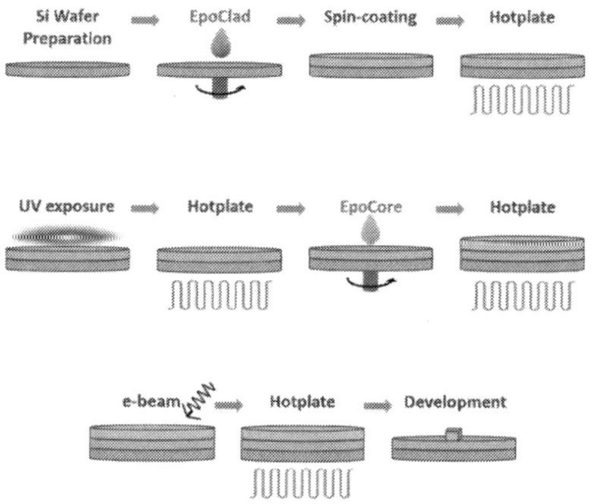

Figure 5. Brief summary of the fabrication procedure (order from left to right and top to bottom).

120 °C. Then, after resting for 20 minutes at room temperature, the wafer with EpoClad 20 was flood-exposed to UV light with a dose of 600 mJ/cm^2 and then transferred to a hotplate at 120 °C for 120 minutes. The wafer was then removed and rested for 5 minutes at room temperature. Then, Epocore 2 was placed on top of the wafer with the EpoClad 20 layer, spin-coated at 2500 RPM for 40 seconds, and then baked on a hot plate at 50 °C and 90 °C, 2 minutes each.

The designed patterns were then transferred to the EpoCore layer. Twelve patterns, containing several AMZI, S-bends, and straight waveguides, were written on the EpoCore 2 layer via e-beam lithography using different doses, ranging from 6 μC/cm^2 to 11.5 μC/cm^2, with steps of 0.5 μC/cm^2. The minimum dose, 6 μC/cm^2, has already been reported in the past [31] as an adequate dose for EpoCore 2. The reason for this variation of the dose was to see if there would be morphological differences in the AMZIs. The exposure was performed using a beam acceleration of 100 kV, with a beam step size of 20 nm and a beam current of 1 nA. To limit the possibility of stitching errors, the write field was limited to 220 μm^2. After the e-beam lithography writing process, the samples were baked on a hot plate at 50 °C for 2 minutes and 85 °C for 3 minutes. Then, the sample was immersed in the devel-

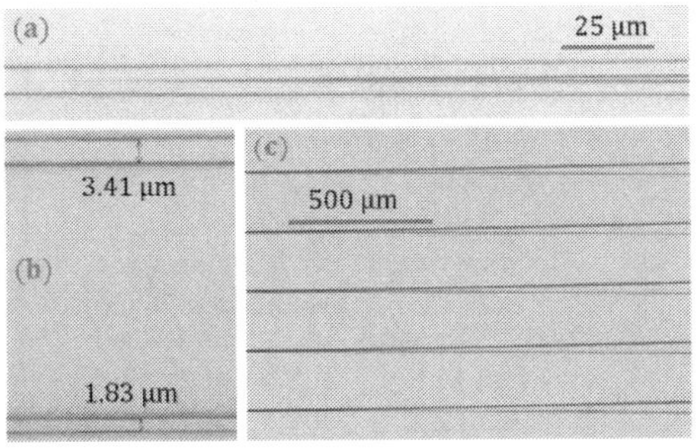

Figure 6. Microscope images of the fabricated AMZI. (a) Y-coupler section. (b) Broad and narrow arms of the AMZI. (c) Y-coupler section of several fabricated AMZI.

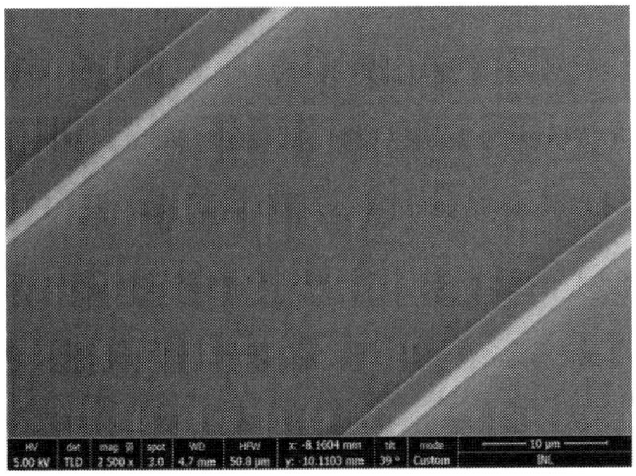

Figure 7. Tilted SEM image of both arms, namely, the broad arm (top) and narrow arm (bottom).

oper mr-Dev 600 (Micro Resist Technology GmbH) for 2 minutes, revealing the transferred pattern.

For the morphology characterization, we first used an optical microscope to inspect the fabricated MZI and evaluated the lateral dimensions (see Fig. 6). The widths of the AMZI coincide with those of the CAD file that was used in the e-beam process. Moreover, no significant difference was observed between the AMZI fabricated with different doses ($6-11.5\,\mu C/cm^2$). Then, by using a contact profilometer (KLA Tencor P-16+), we measured a height of 1.8 µm, as expected. For SEM imaging, one of the samples was coated with 10 nm of gold through sputtering (UHV PVD, Kenosistec). Figure 7 shows a tilted SEM image of both arms of the AMZI.

As a preliminary test, a 635 nm laser (HLS635 Thorlabs), coupled to a single-mode fiber, was used to couple light into a straight waveguide with the same dimensions as that of the broad arm ($w_b = 3.4\,\mu m$). Although the wavelength of the laser is slightly blueshifted by 15 nm when compared to the wavelength used in the optimization procedure, the dimensions of the waveguide are sufficient to allow confinement of the fundamental mode for this wavelength. Figure 8 shows the final portion of the

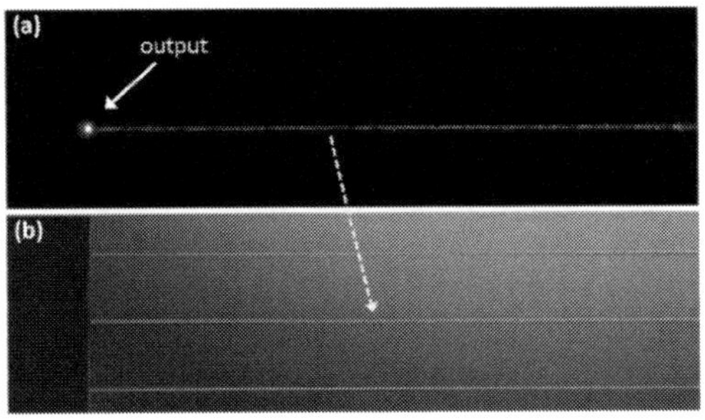

Figure 8. Photograph of the output of the waveguides with light from a laser (635 nm) coupled with a single-mode fiber. (a) Light propagation from the EpoCore strip is observed when the microscope's lamp is switched off. (b) When switched on, several fabricated waveguides can be observed. The yellow dashed arrow indicates the waveguide which was used for light coupling.

waveguide, where light propagation through the EpoCore strip can be observed.

6 Conclusion

In this work, we have first described a method based on open-source Python modules to retrieve optimal dimensions for the design of an AMZI in which the waveguides of both arms have different widths, avoiding the need to fabricate an interaction window on one of the arms. The algorithm maximizes the sensitivity, single-mode confinement, and high-order mode losses. Refractive indices of two commercially available polymers, EpoClad 20 and EpoCore 2, for an operation wavelength of 650 nm, were considered, along with other input parameters such as energy confinement constraints. A sensitivity of 0.0025 was retrieved along with the corresponding dimensions, which were then used to design an AMZI using S-bends. In the fabrication procedure, the core of the waveguides was written using e-beam lithography. Morphological characterization was performed and matched the design values. In future work, we aim to perform loss characterization and refractive index measurement to evaluate the performance of the device and compare it with the simulation results.

The authors acknowledge the financial support of the project HfPT – Health from Portugal, with the reference n.ºC644937233-00000047, co-funded by Component C5 – Capitalisation and Business Innovation under the Portuguese Resilience and Recovery Plan, through the NextGenerationEU Fund.

References

[1] M.A. Butt, Integrated optics: platforms and fabrication methods, Encyclopedia **3**, 824 (2023).
[2] B. Lin, Y. Yi, Y. Cao, J. Lv, Y. Yang, F. Wang, X. Sun, D. Zhang, A polymer asymmetric mach–zehnder interferometer sensor model based on electrode thermal writing waveguide technology, Micromachines **10**, 628 (2019).

[3] R. Bruck, E. Melnik, P. Muellner, R. Hainberger, M. Lämmerhofer, Integrated polymer-based mach-zehnder interferometer label-free streptavidin biosensor compatible with injection molding, Biosensors and Bioelectronics **26**, 3832 (2011).

[4] Y. Xiao, E. Pichler, M. Hofmann, K. Bethmann, M. Köhring, U. Willer, H. Zappe, Towards integrated resonant and interferometric sensors in polymer films, Procedia Technology **15**, 691 (2014).

[5] J. Han, X. Wu, X. Ge, Y. Xie, G. Song, L. Liu, Y. Yi, Highly sensitive liquid mz waveguide sensor based on polymer suspended slot waveguide structure, Polymers **14**, 3967 (2022).

[6] C. Lemieux-Leduc, R. Guertin, M.A. Bianki, Y.A. Peter, All-polymer whispering gallery mode resonators for gas sensing, Optics express **29**, 8685 (2021).

[7] L. Zheng, N. Keppler, H. Zhang, P. Behrens, B. Roth, Planar polymer optical waveguide with metal-organic framework coating for carbon dioxide sensing, Advanced Materials Technologies **7**, 2200395 (2022).

[8] D. Niu, L. Wang, Q. Xu, M. Jiang, X. Wang, X. Sun, F. Wang, D. Zhang, Ultra-sensitive polymeric waveguide temperature sensor based on asymmetric mach–zehnder interferometer, Applied Optics **58**, 1276 (2019).

[9] D. Pandey, L. Bellentani, M. Villani, G. Albareda, P. Bordone, A. Bertoni, X. Oriols, A proposal for evading the measurement uncertainty in classical and quantum computing: application to a resonant tunneling diode and a mach-zehnder interferometer, Applied Sciences **9**, 2300 (2019).

[10] Z. Zhao, Z. Li, J. Niu, G. Zhang, H. Chen, X. Fu, L. Yang, Eight-channel LAN WDM (de) multiplexer based on cascaded Mach–Zehnder interferometer on SOI for 400GbE, in *photonics* (MDPI, 2022), Vol. 9, p. 252

[11] S. Kumar, G. Singh, A. Bisht, S. Sharma, A. Amphawan, Proposed new approach to the design of universal logic gates using the electro-optic effect in mach–zehnder interferometers, Applied optics **54**, 8479 (2015).

[12] A. Psarouli, A. Salapatas, A. Botsialas, P. Petrou, I. Raptis, E. Makarona, G. Jobst, K. Tukkiniemi, M. Sopanen, R. Stoffer et al., Monolithically integrated broad-band mach-zehnder interferometers for highly sensitive label-free detection of biomolecules through dual polarization optics, Scientific reports **5**, 17600 (2015).

[13] M. Janik, M. Koba, A. Celebańska, W.J. Bock, M. Śmietana, Live e. coli bacteria label-free sensing using a microcavity in-line mach-zehnder interferometer, Scientific reports **8**, 17176 (2018).

[14] M. Angelopoulou, D. Kourti, K. Misiakos, A. Economou, P. Petrou, S. Kakabakos, Mach-zehnder interferometric immunosensor for detection of aflatoxin m1 in milk, chocolate milk, and yogurt, Biosensors **13**, 592 (2023).

[15] H. Wang, Z. Chen, T. Li, H. Xie, B. Yin, S.H.D. Wong, Y. Shi, A.P. Zhang, Optofluidic chip with directly printed polymer optical waveguide mach-zehnder interferometer sensors for label-free biodetection, Biomedical Optics Express **15**, 3240 (2024).

[16] A.L. Siarkowski, L.F. Hernandez, B.H.V. Borges, N.I. Morimoto, Sensing based on mach-zehnder interferometer and hydrophobic thin films used on volatile organic compounds detection, Optical Engineering **51**, 054401 (2012).

[17] M. Hofmann, Y. Xiao, S. Sherman, U. Gleissner, T. Schmidt, H. Zappe, Asymmetric mach–zehnder interferometers without an interaction window in polymer foils for refractive index sensing, Applied optics **55**, 1124 (2016).

[18] Y. Xiao, M. Hofmann, Z. Wang, S. Sherman, H. Zappe, Design of all-polymer asymmetric mach–zehnder interferometer sensors, Applied optics **55**, 3566 (2016).

[19] R. Bernini, G. Testa, L. Zeni, P. Sarro, in *Sensors and Microsystems: AISEM 2009 Proceedings* (Springer, 2009), pp. 373–376

[20] E. Fiedler, N. Haas, T. Stieglitz, Suitability of SU-8, EpoClad and EpoCore for flexible waveguides on implantable neural probes, in *2014 36th Annual International Conference of the IEEE Engineering in Medicine and Biology Society* (IEEE, 2014), pp. 438–441

[21] M. Čehovski, J. Becker, O. Charfi, H.H. Johannes, C. Müller, W. Kowalsky, Single-mode polymer ridge waveguide integration of organic thin-film laser, Applied Sciences **10**, 2805 (2020).

[22] P. Gijsenbergh, K. Wouters, K. Vanstreels, R. Puers, Determining the physical properties of epoclad negative photoresist for use in mems applications, Journal of Micromechanics and Microengineering **21**, 074001 (2011).

[23] T. Guan, F. Ceyssens, R. Puers, An epoclad/epocore-based platform for moems fabrication, Journal of Micromechanics and Microengineering **23**, 125005 (2013).

[24] S. Hessler, M. Rüth, C. Sauvant, H.D. Lemke, B. Schmauss, R. Hellmann, Hemocompatibility of epocore/epoclad photoresists on coc substrate for optofluidic integrated bragg sensors, Sensors and Actuators B: Chemical **239**, 916 (2017).

[25] J. Tambasco, Modesolverpy (2018), `https://github.com/jtambasco/modesolverpy`

[26] G. Wang, J. Zhu, D. Wei, F. Jiang, Y. Huang, Enhanced air microcavity of channel spp waveguide halby graphene material, Plasmonics **14**, 313 (2019).

[27] Y. Liu, F. Li, C. Xu, Z. He, J. Gao, Y. Zhou, L. Xu, The design and research of a new hybrid surface plasmonic waveguide nanolaser, Materials **14**, 2230 (2021).

[28] C. Xu, E. Heller, M. Bahl, R. Scarmozzino, K.N. Ku, Y. Zhou, T. Su, P.C. Chang, C.Y. Lin, S.C. Chen et al., Suitability of BPM simulation for silicon pho-

tonics, in *2021 IEEE 17th International Conference on Group IV Photonics (GFP)* (IEEE, 2021), pp. 1–2

[29] A. Kumar, S. Aditya, Performance of s-bends for integrated-optic waveguides, Microwave and Optical Technology Letters **19**, 289 (1998).

[30] R. Broeke, X. Leijtens, Nazca design (v0.5.13), `https://nazca-design.org/`

[31] F.J. Gordo, J. Tátá, J. Borme, M.A. Geday, M. Caño-García, J.B. Nieder, Design and manufacture of an all-polymeric integrated multimode interferometer for visible photonics, Optics Express **30**, 31147 (2022).

AUTHOR INDEX

AUTHOR INDEX